Forensic History: Crimes, Frauds, and Scandals

Elizabeth A. Murray, Ph.D., D-ABFA

PUBLISHED BY:

THE GREAT COURSES
Corporate Headquarters
4840 Westfields Boulevard, Suite 500
Chantilly, Virginia 20151-2299
Phone: 1-800-832-2412
Fax: 703-378-3819
www.thegreatcourses.com

Printed in the United States of America

Elizabeth A. Murray, Ph.D., D-ABFA

Forensic Anthropologist
and Professor of Biology
Mount St. Joseph University

Dr. Elizabeth A. Murray is a forensic anthropologist who is also a Professor of Biology at Mount St. Joseph University, where she teaches doctoral-level cadaver-based human gross anatomy and undergraduate-level forensic science, vertebrate anatomy and physiology, musculoskeletal anatomy, and medical terminology. She is also an adjunct affiliate of the University of Cincinnati College of Medicine. Dr. Murray received her bachelor's degree in Biology from the College of Mount St. Joseph, and she received her master's degree in Anthropology and her doctoral degree in Interdisciplinary Studies in Human Biology from the University of Cincinnati.

Dr. Murray is an award-winning teacher who has received the Sears-Roebuck Foundation Teaching Excellence and Campus Leadership Award and who has twice earned the Clifford Excellence in Teaching Award. She has also served as an instructor for numerous professional organizations, including the training academy of the U.S. Department of Justice/National Institute of Justice's National Missing and Unidentified Persons System (NamUs), the American Academy of Forensic Sciences (AAFS) Student Academy, the Armed Forces Institute of Pathology, the Wayne State University School of Medicine's Medicolegal Investigation of Death program, the Ohio State Coroners Association, and the International Association of Coroners & Medical Examiners.

Most of Dr. Murray's regular forensic casework has been in Ohio and Kentucky, where she has participated in hundreds of forensic investigations involving skeletal, decomposing, burned, buried, and dismembered human remains. A Fellow of the AAFS, she is one of fewer than 100 anthropologists who are certified as Diplomates by the American Board of Forensic Anthropology (ABFA). Dr. Murray is on the Board of Directors of the ABFA and is the Forensic Anthropology Consultant on the "Ask the Experts" Panel of the National Organization of Parents of Murdered Children. She is also on the Mass Disaster Team for the Cincinnati/Northern Kentucky International Airport and has served on the Board of Trustees for The Forensic Sciences Foundation.

In 1994, Dr. Murray was recruited by the Armed Forces Institute of Pathology to participate in morgue operations after the crash of American Eagle Flight 4184 in Roselawn, Indiana. She also served as a visiting scientist to the U.S. Army Central Identification Laboratory, for which she led a team of military personnel in the recovery of a Vietnam War–era plane crash site in the jungle of Laos. As a consultant and on-screen personality for the National Geographic Channel's *Skeleton Crew* (aired internationally as *Buried Secrets*), Dr. Murray was dispatched to observe and participate in fieldwork with the Guatemalan Forensic Anthropology Foundation. This four-part miniseries showcased the uses of forensic anthropology in analyzing historical mysteries and modern forensic contexts. In addition, Dr. Murray served as a regular cast member on the Discovery series *Skeleton Stories* and has appeared on such television shows as *America's Most Wanted*, *Forensic Files*, *The Decrypters*, and *The New Detectives: Case Studies in Forensic Science*.

Dr. Murray's publications include numerous entries in the *Proceedings*, published by the AAFS, as well as books and chapters covering both historical and forensic anthropological analyses. Dr. Murray's book *Death: Corpses, Cadavers, and Other Grave Matters* was named one of the 2011 top 10 summer reads for students by the American Association for the Advancement of Science. Her second book, *Forensic Identification: Putting a Name and Face on Death*, was selected by the National Science Teachers Association/Children's Book Council as one of the Outstanding Science Trade Books for Students K–12 for 2013. Her most recent book is *Overturning Wrongful Convictions: Science Serving Justice*. For The Great Courses, Dr. Murray also taught *Trails of Evidence: How Forensic Science Works*. ■

Table of Contents

Table of Contents

Table of Contents

Acknowledgements

First of all, many thanks to my sister, Kathy Isaacs, who served as my research assistant during the development of this course and loves a great story as much as I do. I'd also like to thank those at New Dominion Pictures for introducing me to the Colorado Cannibal and other amazing historic forensic cases through our filming of the *Buried Secrets* miniseries in 2004, including coverage of the Forensic Anthropology Foundation of Guatemala (FAFG). I wish to honor Fredy Peccerelli and those at FAFG for their humanitarian work and express my gratitude for the opportunity to learn from them. Appreciation is due to my forensic anthropology colleagues Dr. Norm Sauer and Brian Spatola for specific information about Nazi artifacts. I also owe a great debt of thanks to the professionals at the Hamilton County Coroner's Office, where I first began my adventure into forensic casework in 1986. Their support for Project Identify has been invaluable, in particular, the work of Andrea Hatten and Dr. Bill Ralston; thanks to all those at the office for allowing me to relate the stories of this project and others in this series. I am also grateful to Investigator Mike Ratliff of the Scioto County Coroner's Office, as well as the talented forensic artist Catyana Skory-Falsetti. My work with the unidentified has been enriched by the inception of the National Missing and Unidentified Persons System (NamUs), and I hope that the families of Arthur R. Flowers and Paula Beverly Davis, especially her sister, Stephanie Clack, may find peace. My students at Mount St. Joseph University also deserve recognition for discovering some amazing historic case stories in our forensic science courses that became the bases for several of these lectures. Last but certainly not least, I am extremely grateful for the keen eye of Susan Dyer, who did a superb job of editing all my lecture scripts; Nancy Eskridge for her ongoing support; and all the other wonderful people at The Great Courses with whom I worked on this project. ■

Forensic History: Crimes, Frauds, and Scandals

Scope:

This course is a journey through time and place from the perspective of forensic science and criminal history. It examines technological advancements and political and legal issues and demonstrates how science and society relate to each other, especially with regard to criminal investigations. The lectures use a case-based approach—including some of Dr. Murray's own forensic casework—to focus on historic forensic issues and show how new evidence or more advanced technology can sometimes be used to develop alternative conclusions or finally solve cold cases. At times, we consider how historic crimes would have different outcomes if they occurred today.

The first lecture looks at the Jack the Ripper murders of 1888 in London's East End. A surprising amount of forensic science was in use at the time, yet the identity of the killer is still unknown. In Lecture 2, we examine the case of the woman known as the Black Dahlia, found in Los Angeles in 1947, comparing her death to those in the Jack the Ripper series. We then turn to three mysterious Hollywood deaths—those of George Reeves, Bob Crane, and Bruce Lee—and review a brief history of the use of autopsy to resolve suspicious deaths. Lecture 4 highlights the fact that initial appearances can sometimes be deceiving and illustrates how the decomposition process can affect autopsy results. We next explore two infamous "family feuds" that took place nearly 100 years apart: the cases of Lizzie Borden and the Menendez brothers. Lecture 6 examines the 1982 Tylenol murders, in which seven people died after taking cyanide-laced acetaminophen capsules, and provides a rich backstory on the prime suspect.

Lectures 7 and 8 cover a series of copycat crimes, hoaxes, and frauds, including copycat product-tampering cases, the famous Piltdown fossil hoax, the attempt of a British politician to fake his own death, and art and other forgeries. Lecture 9 presents sports scandals, including the use of performance-enhancing drugs by Lance Armstrong and others, as well as Olympic figure skater Tonya Harding's sabotage of a competitor and the infamous Chicago "Black Sox" scandal. Old and new political sex scandals are examined in Lecture 10, with a discussion of how DNA testing and other forensic methods are now used to document them. In Lecture 11, we move to the Wild West to see how criminals can become cult figures. Cases include the gunfight at the O.K. Corral, the James Gang, and the Colorado Cannibal. In Lecture 12, we follow the investigation of two amazing bank heists using a variety of forensic methods.

Moving in a new direction, Lectures 13 through 16 demonstrate how the legal system sometimes fails badly, charging or even convicting the innocent of crimes they did not commit. The pitfalls in such cases include faulty eyewitness testimony, mistaken or corrupt law enforcement agents, planted evidence, legal malpractice, and improper or misused forensic science. In these four lectures, we look at the causes behind false confessions and see how the criminal justice system can and has learned from its mistakes and reformed some police and legal procedures. Recent cases of exoneration in the United States, particularly through the use of DNA technology, illustrate how terrible wrongs can be addressed, even though the years spent in prison by the innocent can never be recovered.

Political assassinations are a global forensic phenomenon, and in Lecture 17, we look at the unusual murder of Bulgarian dissident Georgi Markov, the killing of Olaf Palme of Sweden, and the questioned death of Palestinian leader Yasser Arafat. Lecture 18 is devoted to the mysteries surrounding the murders of the Romanov family of Russia, including how forensic anthropology and DNA were used to identify the victims almost 100 years after their deaths. In Lecture 19, we examine acts of genocide, with special attention to ongoing work uncovering mass graves in Guatemala. In Lecture 20, we visit the Buchenwald work camp of the Nazis and analyze the allegations against Ilse Koch, also known as the Witch of Buchenwald. This set of politically related investigations concludes with coverage of famous spies, among them, Mata Hari and FBI agent Robert Hanssen.

With Lecture 22, we turn to kidnappings, using motive as part of the backdrop of our study; examples here include the massacre at the 1972 Munich Olympics, the kidnapping of J. Paul Getty III, infant abductions, and a case that introduces the use of the National Missing and Unidentified

Persons System (NamUs). Lecture 23 highlights six recent John and Jane Doe cases, showing how new evidence and information can be gathered to help identify the unknown dead. The final lecture focuses on three issues that have dramatically altered case resolution or the commission of crimes throughout history: advances in fingerprint technology, DNA profiling, and computerization.

Through this exploration of fascinating cases from both the history books and today's headlines, not only will you discover how science is applied to criminal investigations, but you'll also sharpen your analytical skills and learn to evaluate evidence critically—skills you can bring to bear when the next crime or forensic mystery hits the news. ■

The Infamous Jack the Ripper

Lecture 1

Outside of personal experience as either a victim or a perpetrator, how do we know what we know about crimes? As long as there has been written language, historians have recorded landmark events, including the lives—and deaths—of famous people and unusual events involving ordinary people. Today, much of our knowledge of crime victims comes from the media, which is also a source for history. Of course, the landmark case for media interest in crime is that of Jack the Ripper. This case has spawned hundreds of theories and countless publications and has been reexamined by some of forensic science's greatest minds. But despite all these efforts, we still do not know the identity of the perpetrator.

Basic Forensic Analysis

- Although the exact number of victims is not known, most authorities agree that five women killed in the Whitechapel area of east London between August 31 and November 9 of 1888 are the work of the same serial killer.

- The reason most agree that these five female victims are definitively linked is based on what forensic scientists call *modus operandi* (MO).
    - All of the definitive victims were prostitutes who lived and worked in the slums of the East End of London. Apparently, part of the Ripper's MO was to prey on women of the night.

    - In addition, no semen was discovered in any of these cases, which not only further links them but also suggests that rape was not part of the killer's MO.

    - The murders are also connected by increasing brutality, showing an escalation as time went on, a phenomenon seen in the careers of many criminals.

- Investigators look at these same factors today when they suspect that crimes may be linked—victim choice, weapon used, mutilation, and escalation. In a way, examining these factors—conducting forensic analysis—is similar to what our brains do every day in countless situations: compare data and recognize patterns.

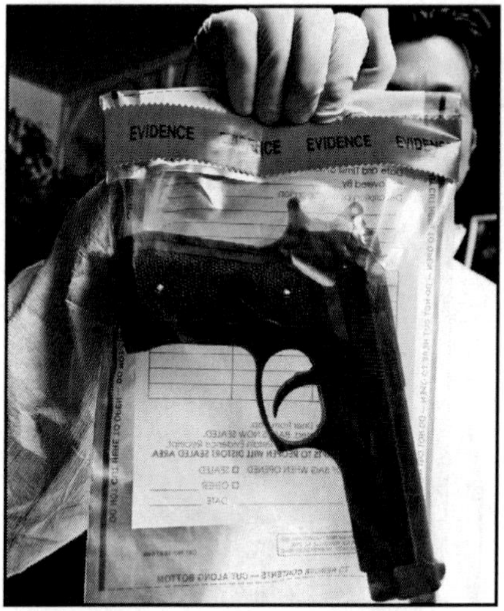

© Digital Vision/Thinkstock.

Investigators today look at many of the same factors that were examined at the time of the Ripper murders, including a killer's choice of victims and weapon.

Mary Ann Nichols and Annie Chapman

- The definitive Ripper victims (the "canonical five") were all females whose poverty and thirst for alcohol apparently led them to prostitution. Their ages varied somewhat, but all were young to middle aged, and all were Caucasian.

- The first victim, Mary Ann Nichols, was discovered on August 31. She was found on her back with her skirts pulled up; her throat had been slashed deeply twice, and her low abdomen had been cut several times. A physician who examined her body at the scene at around 4:00 a.m. concluded that she had probably been dead for about 30 minutes.

- The body of the second victim, Annie Chapman, was found around 6:00 a.m. on September 8. She also had two extremely deep cuts to her throat, which a physician later said had been made from left to right. Chapman's skirts had been raised, her belly had been cut open, her intestines had been pulled outside the body, and her entire uterus had been removed and taken.

Elizabeth Stride and Catherine Eddowes

- Almost a month later, on September 30, the body of Elizabeth Stride was found at around 1:00 a.m. But this case was different; the victim had only one cut to her neck, her skirts were not raised, and her belly wasn't mutilated.

- Several witnesses claimed to have seen Stride in the area between 11:00 p.m. and 12:45 a.m. in the company of a man on the same night. As is often the case, few of the witnesses could agree on the appearance of the man they claimed to have seen.

- One eyewitness, however, said that he saw Stride at 12:45—in the same location where her body was found—in an altercation with a man. According to this witness, the man had dark hair and a thin moustache; was about 5 feet, 5 inches tall and broad-shouldered; and was around 30 years old.
    - The onlooker thought he was watching a domestic argument and didn't want to get involved. The discovery of Elizabeth's body at 1:00 a.m., just 15 minutes later, suggested that this witness—a Hungarian Jew—may have stumbled on the Ripper in the process of killing one of his victims.

    - The authorities took the account of this Hungarian quite seriously because he had come forward despite anti-Semitic tensions in the area. He saw what he described as an altercation, including the woman's low-pitched cry, before the attacker noticed the bystander, causing the Hungarian to rush off.

    - Investigators already believed that the Ripper's MO was to strangle or first slit his victim's throats, which would prevent them from screaming during the rest of the attack. Police theorized that the Hungarian may have stumbled on the attacker as he was just beginning to execute his typical MO on

Stride. Once the Ripper realized someone was watching, he ran off after only a single cut to the victim's neck.

- Within moments of the homicide's discovery, a crowd of nearly 30 people had gathered around Stride's body. An officer told them to stand back, not because they might contaminate the crime scene, but because if they got blood on their clothes, they might become suspects. In fact, officers examined the hands and clothes of onlookers for bloodstains, took down names and addresses, and even checked people's pockets before allowing anyone to leave.

- About 45 minutes after the discovery of Stride's remains, the body of Catherine Eddowes was found not too far away. Eddowes had been released from jail for being drunk and disorderly at about 1:00 a.m., and it seems that the Ripper chose her to complete his interrupted MO.
    - Eddowes had her throat slit twice, her skirts lifted, her belly cut open, and portions of her intestines removed and put on her shoulder. Most of her uterus had been cut out, and her left kidney was removed and taken as a trophy. Her face was also cut up, including one ear.

    - Investigators later found a piece of cloth smeared with blood and feces not too far from the crime scene. The cloth turned out to be part of Eddowes's apron that the killer had cut from her clothing and apparently used to clean his knife and, possibly, his hands.

- Another unusual aspect of the Eddowes case was a message written in chalk on the wall in the busy marketplace next to where the piece of apron had been discarded. It said, "The Juwes [sic] are the men that will not be blamed for nothing." The meaning of this message is still open to debate, but it echoed the anti-Semitism that divided the community with regard to the murders. Instead of photographing the message, the police decided to copy it down, then obliterate it.

- Because of the anatomical knowledge some ascribed to the mutilations and trophy removal and the allegations that the Ripper might be a Jew, some began to suspect that the killer was a Jewish slaughterhouse worker. Perhaps to satisfy the public, the police rounded up the knives of these workers to see if any matched the suspected weapon. A doctor concluded that none of them did.

- Nearly two weeks after the night Stride and Eddowes were killed, the chairman of the Whitechapel Vigilance Committee received a package containing half a kidney, preserved in wine and assumed to be human, along with a note explaining that the other half had been eaten by the killer. Using microscopic examination, two doctors independently concluded that the kidney was human, but they were unable to say whether it belonged to Eddowes.

Mary Jane Kelly

- The next Ripper victim, Mary Jane Kelly, was not killed until November 9, almost six weeks after the murders of Stride and Eddowes. Her body was found in her home around 10:45 a.m. by a man sent to collect overdue rent. He knocked, then peered through the window and was stunned to see Kelly's mutilated body on the bed.

- This time, the crime scene investigation seemed much more thought out, perhaps owing to criticism of previous police actions. A telegram was sent to Scotland Yard to bring bloodhounds, the area was cordoned off to the public, and a doctor was called to the scene. For some reason, the decision to use the dogs was reversed, and at 1:20 p.m., police broke down the locked door and began to examine the murder scene.

- Crime scene photos show Kelly's body was on the bed, nude, with her legs splayed open. Her face had been mutilated beyond recognition; her throat was cut to her spine; her abdomen was completely eviscerated; and her heart had been cut out and was never recovered.

- At the autopsy, the attending physicians estimated the time of Kelly's death at somewhere between 2:00 and 8:00 a.m. They believed it might have taken the Ripper perhaps two hours to do that much damage to the body.

- The murder of Mary Kelly apparently ended the killer's rampage—at least in the London Whitechapel area. No one knows to this day who he was or why the murders ended. Despite thousands of books written on the subject and countless articles espousing different theories and possible suspects over 125 years of analysis, the killer's identity is unknown.

The Role of the Media

- The reason this series of brutal murders became the best-known criminal matter of its time was the concurrent boom in newspaper circulation in the second half of the 19th century. Advances in printing and tax reform in England allowed unprecedented low-cost production and distribution. London was arguably the most prominent of capital cities in the world at the time and had dozens of newspapers, as well as a true crime magazine. This was the era when journalism was born.

- Thus, news of London's East End crime wave spread quickly after the discovery of Nichols's mutilated body. In fact, the infamous name of this still-unknown killer was actually delivered to the newspapers in the form a letter written in red ink to London's Central News Agency on September 27. The letter writer took credit for the prostitute killings and signed the message "Jack the Ripper."

- Then, as now, many who have delved into the Ripper murders believe that the letter and a later postcard were part of a hoax. Most attribute them to someone in the media—a journalist who had inside knowledge of the events and wanted to sell newspapers. And sell papers they did. In fact, Mary Jane Kelly's boyfriend told police that she regularly asked him to read her news of the killer from the local papers. Neither could know that she would soon be front-page news as the Ripper's final victim.

Suggested Reading

Begg, *Jack the Ripper*.

Douglas and Olshaker, *The Anatomy of Motive.*

Federal Bureau of Investigation, *Serial Murder*.

Rumbelow, *The Complete Jack the Ripper*.

Sugden, *The Complete History of Jack the Ripper*.

Questions to Consider

1. How do you think the investigation of the Jack the Ripper case would be different (either positively or negatively) if those crimes happened today?

2. Do you think Jack the Ripper could have been a woman? Why or why not?

The Infamous Jack the Ripper

Lecture 1—Transcript

Outside of any personal experience, how do we know what we know about crime and the methods employed to solve it? As long as there's been written language, historians have recorded landmark events, especially surrounding the lives and deaths of famous people. In some cases, other people became famous because something unusual—either tragic or extraordinary—happened in their lives and was written down for posterity in our history books. This includes those who broke the laws of society, the unfortunate victims of crime, as well as those whose work helps seek justice.

I'm Beth Murray, and I'm here to take you on a journey through history by focusing on some compelling forensic cases, both famous and not so famous, including a few of my own. I've been a practicing forensic scientist—specifically a forensic anthropologist—for nearly 30 years. I like to say I was forensic before it was cool! But I'm also quite a history buff, and love a great story, thanks to my father. So this is a wonderful opportunity for me to blend those interests and focus on what I hope you'll agree are some fascinating cases.

Together, we'll re-examine some of the world's most notorious crimes, and at times, consider how they might be investigated had they happened today. We'll also see how modern forensic techniques have uncovered fresh or unexplored evidence and used it to re-analyze old cases, shedding new light on what we thought we knew. And in some lectures, we'll look at more recent forensic matters to see how far we've come fighting crime. Along the way you'll learn a little bit about the history of forensic science too.

Today, much of our knowledge of crime, criminals, their victims, and forensics comes from the media, which of course, is a form of recorded history. So do you have any idea what's considered by many to be the landmark case for crime reporting? The forensic evidence and investigation in this case created the first known major media crime sensation, which started locally, then spread across much of the world. It's a forensic analysis that began in the late 1880s and is still ongoing today. This case has spawned hundreds of theories, countless publications, and has had some of forensic science's greatest minds re-examine it. But, despite all of this effort, we still don't know the identity of the perpetrator. So he goes by the legendary name of Jack the Ripper.

Although the exact number of victims is not really known, most authorities agree that five women killed between August 31 and November 9 of 1888 in the East London area known as Whitechapel are confidently the work of one serial killer. Many sources also attribute a sixth victim from earlier that August to Jack the Ripper. Some theories raise the number of victims to 11 women, and still other sources go higher, perhaps including a 7-year-old boy, encompassing a larger group of attacks that took place in the same general vicinity between February 25 and December 29 of 1888.

But the earlier assaults during this broader 10-month period seem more haphazard, and the victims were not killed. As a result, most scholars believe those are not connected, but rather, just reflect the impoverished, dangerous, and crime-ridden East End of London at the time. It's also possible, though, that at least some of those earlier attacks were the more clumsy beginnings of a killing spree that Jack the Ripper ultimately mastered.

In addition to time, what other evidence has been used to connect some of these crimes? The reason most agree five female victims are definitively linked, and why some authorities insist others are decidedly not the work of the same killer, is what forensic scientists call modus operandi, or simply the perpetrator's MO. All of these five women were prostitutes who lived and worked in the slums of the East End of London. So apparently, part of the Ripper's MO was to prey upon women of the night, which is when all of these killings occurred. But no semen was discovered in any of these cases, which not only further links them, but also suggests that rape was not part of the MO. The murders are further connected by increasing brutality, showing an escalation of sorts as time went on, a phenomenon seen in the careers of many criminals.

Investigators look at these same facets today when they suspect crimes might be linked—victim choice, weapon used, level of brutality, escalation. Connecting similarities like these is, really, in a way, just common sense. Much of forensic science analysis, at least within the mind of the investigator, is similar to what our brains do every day in countless situations, compare data and recognize patterns. As an example, let's look at the evidence used to connect the Whitechapel casualties in 1888.

The definitive Ripper victims, known as the "canonical five," are the cases agreed on by all accounts; all five were females whose poverty and love of alcohol apparently led them to prostitution. Their ages varied somewhat, but all were young to middle aged, a range we'd assume typical for prostitutes. All were Caucasian, but that could be said of the entire neighborhood. The East End of London was rapidly becoming crowded with immigrants from Ireland, as well as Russian Jews, which resulted in economic downturn and social tension, as we'll see. But now, let's get to the crime scenes.

The first of the canonical five, Mary Ann Nichols, also known as Polly, was discovered on August 31. She was lying on her back with her skirts pulled up; her throat was slashed deeply twice, all the way down to her vertebrae, and her low abdomen had been cut several times. Some of those belly wounds were so deep they literally ripped open her body cavity, exposing her intestines. When a physician examined her remains at the scene, at around 4:00 in the morning, he considered body temperature, as we would today. Since her body and legs were still somewhat warm compared to her cold hands, he concluded Nichols was probably dead for about a half hour.

The body of the second victim, Annie Chapman, was found around 6:00 on the morning on September 8. She also had two extremely deep cuts to her throat, which a physician later said at the inquest had been made from left to right. I assume they thought this would perhaps reveal whether the perpetrator was behind the victim or in front of her. But that would also depend on whether he was left or right handed. Chapman's skirts had been raised, and this time, not only was her belly cut open, but her intestines were pulled outside her body and her uterus had been removed and was missing. This is what forensic investigators call "taking a trophy."

Chapman's autopsy also showed bruises that the doctor knew must have come from a previous altercation. Bruising doesn't happen at or after death, since once the heart stops beating, blood pressure no longer drives the circulation that pushes blood out of broken vessels and into tissues. Dead skin lacks what's known as a vital reaction.

Later, witnesses described seeing Annie in a bar fight earlier in the week, substantiating the ante mortem bruises. Anyway, the MO made it far more logical to link this killing to the same vicious perpetrator who had murdered Mary Ann Nichols just a week before, not suspect that it was the fallout from a barroom brawl.

Almost a month later, on September 30, the body of Elizabeth Stride was found at around 1:00 in the morning. But this case was different. The victim had only one cut to her neck, her skirts weren't raised, and her belly wasn't mutilated. So why did police connect her to these other murders? Several witnesses claimed to have seen Stride in the area between around 11:00 p.m. and 12:45 on the same night she was murdered, and in the company of a man. One witness told police the couple was kissing in the recessed doorway of a building and that he heard the man say to Stride, "You would say anything, but your prayers." Another witness who claimed to have seen Stride about 12:30 a.m. was a young police constable who said the man with her carried a package about 18 inches long, wrapped in newspaper.

So, eyewitnesses are, without fail, a big lead for investigators, right? Well, not always. In this case, there were numerous people willing to say they saw a couple in the area that evening, or at least saw Elizabeth Stride with a man, but few of them could agree on the appearance of the guy they claim to have seen or the clothing he was wearing. And eyewitness testimony can be notoriously unreliable. Plus, there's the well-known phenomenon of people falsely reporting events to get attention or to jump on the media-frenzy bandwagon around incidents like this.

But one eyewitness said he saw Stride at 12:45 a.m. in the same location where her body was found in an altercation with a man. The witness said the man had dark hair, a thin moustache, was about 5′ 5″ tall, broad-shouldered, and around 30 years old. This was fairly consistent with the description of the man made by the police constable. The onlooker thought he was watching a domestic argument, and he didn't want to get involved. Especially after the attacker took notice of the witness, so the bystander decided to just keep moving. We've all heard of cases in the news in which people don't want to get involved, or maybe they only realize the gravity of what they saw after they hear later that they witnessed a crime. The discovery of Elizabeth's body, just 15 minutes after this man saw what he thought was a domestic dispute, suggests that the witness, who was a Hungarian Jew and had to be interviewed through an interpreter, may not have only seen Jack the Ripper, but actually stumbled upon him in the process of killing one of his victims.

Authorities took the account of this Hungarian quite seriously for two main reasons. First, the witness came forward despite significant anti-Semitic tensions in the area. Second, he saw what he described as an altercation, including the woman's very low-pitched cry, before the attacker noticed the bystander, causing the Hungarian to rush off. Investigators already believed that the Ripper's MO was to strangle or slit his victim's throats, cutting them twice, so deeply that their airway was cut, disabling them from screaming during the rest of the attack. Police theorized that the Hungarian may have stumbled upon the attacker as he was just beginning to execute his typical MO on Stride. But once the Ripper realized someone was watching, he ran off after only a single cut to the victim's neck, abandoning the rest of his standard mutilation.

An examination of Stride's body in the courtyard where she lay showed the cut in her neck was so deep that there was a spray of blood extending some distance from her body. Today's blood spatter analysis wasn't standard protocol at that time, but my read of that jet of blood sounds like a major artery was cut, causing the force of the heart to send the blood quite a distance away. An officer on the scene said the woman's face was still warm, corroborating her recent death.

Within moments of the discovery of Stride's body, as people began shouting in the streets that yet another murder had taken place, a crowd of nearly 30 people had gathered around. This prompted an officer to tell them to stand back. He didn't seem as concerned about them contaminating the crime scene with their own trace evidence, as we would today, but rather, he told them if they got blood on their clothes, they might be suspect. In fact, officers examined the hands and clothes of onlookers for bloodstains, just as might be done today. Investigators took names and addresses, and even checked people's pockets before allowing anyone to leave.

Stride's clothing was in place from head to toe with one exception; the bow that had tied her scarf was askew, as though the perpetrator had used it to grab her from behind. A doctor at the scene estimated the time of death at around 20 to 30 minutes prior. Based on the depth of the cut to the victim's neck, he said she would have bled to death in a minute and a half or so. Pretty sophisticated for 1888, isn't it? Keep in mind, though, that most common people were familiar with slaughtering animals, and a standard method was cutting the major vessels of the neck to bleed them to death.

Now, why else connect Stride with the Ripper? What might a serial killer do if he's built up a frenzy, but then been thwarted? Well, about 45 minutes after the discovery of Stride's remains, the body of Catherine Eddowes was found not too far away. In a stroke of unbelievably bad fortune, Eddowes had been in jail that night for being drunk and disorderly, and she was released about 1:00 that the morning, around the same time Stride was being murdered. If the Ripper couldn't complete his full MO on Stride, he quickly did so on Eddowes to satisfy himself. She had her throat slit twice, her skirts lifted, her belly cut open from nearly her navel to her rectum, and portions of her intestines removed and placed up on her shoulder. This time, in addition to most of her uterus being cut out, her left kidney was also removed and taken as a trophy. Her face was cut up, too, including one ear.

Investigators later found a piece of cloth smeared with both blood and feces, not too far from the crime scene. The cloth turned out to be part of Eddowes's apron that the killer had cut from her clothing and then apparently used to clean his knife and possibly his hands. Had that happened today, perhaps DNA from somebody other than Eddowes might have been found on that fabric, especially if the perpetrator had used it to wipe off his hands. But when you consider how infrequently people likely washed their clothing in those days, well, the analysis would certainly have been complicated. Besides, without a suspect, whose DNA would be gathered for a comparison sample? Minimally, though, the victims could have been definitively linked if the same DNA profile was found in each case, even if the owner of that DNA was not known.

Another unusual aspect of the Eddowes case was a message, neatly written in chalk on the wall in the busy marketplace, right next to where the piece of apron had been discarded. It said, “The Juwes [sic] are the men that will not be blamed for nothing.” The meaning of this message is still open to debate and interpretation, but it echoed the anti-Semitism that was dividing the community with regard to the murders. Terror in society often stirs up wild emotions, including the desire to blame and demonize some group or some person, and as we’ve seen throughout history, we most often point the finger at those who are not like us.

How would evidence like a chalk-written message on a building entryway be captured? It should have been saved then, as it would be today, using forensic photography. Photography was well established by then, and there are many images related to the Ripper murders available, including morgue photos of victims. But unbelievably, because of the darkness of the early-morning hour, the wait time for the arrival of a photographer, and the fact that the shop owners were beginning to open up their market stalls, the police decided instead to just copy down the writing and then obliterated the message at 5:30 in the morning. They feared if crowds saw the message, there would be a riot against the Jewish community, and more lives might be lost.

While I can certainly understand their anxiety, destroying potential handwritten evidence like that leaves me shaking my head. Many Ripper scholars believe, though, that the graffiti may have already been there, especially since some observers said it was blurry, and it may have been a complete coincidence that the piece of Eddowes’ apron was dropped in the same location. Either way, there was heavy criticism of the police decision to eradicate the message.

Another interesting forensic aspect related to this writing was the early use of forensic linguistics; that’s the study of language. Based on the spelling and grammatical errors, the police commissioner felt strongly the graffiti did not come from a native English speaker. He wrote in his report, “The idiom does not appear to be English, French, or German, but it might possibly be that of an Irishman speaking a foreign language. It seems to be the idiom of Spain or Italy.” Given the odd phrasing, this makes complete sense, but it's interesting to note the use of language to help profile the suspect. The nonsensical nature of the wording led authorities to two possible conclusions, if the killer actually wrote it: Jack the Ripper was Jewish and bragging of it, or the killer was trying to thwart the investigation by blaming a Jew.

Now, what about other police profiling of the likely killer? Due to the anatomical knowledge some ascribed to the mutilations and the trophy removal, and the allegations that the Ripper may be a Jew, some began to suspect the killer was a Jewish slaughterhouse worker. So, perhaps to help satisfy the public, the police rounded up the knives of the local Jewish slaughtermen to see if any matched the suspected weapon. A doctor concluded that none of them seemed to be the murder weapon, so we see evidence of early attempts to match a sharp tool with a wound, something common in forensics today, including in my own work, although typically far more reliable when analyzing trauma to bone, rather than soft tissue.

Nearly two weeks after the night Stride and Eddowes were killed, the chairman of the Whitechapel Vigilance Committee got a package containing half a kidney. It was preserved in wine and assumed to be human. Along with this was a note full of spelling errors that explained the other half had been eaten by the killer. We'll talk about notes attributed to the Ripper in a few minutes. But with regard to the kidney, how could forensic science establish this organ as belonging to a human?

Based on my many years of teaching anatomy, using both human and non-human organs for dissection and microscopy, I can tell you that a pig kidney, or that of any similar-sized animal, is very like that of a human. Today's DNA testing, as well as much faster antibody testing, could allow investigators to fairly quickly determine the animal source, although I have no idea how pickling an organ in wine might affect those tests. But the best they could do in 1888 was a microscopic examination, something commonly done in autopsies today to look for pathology, but not to try to figure out the species. Two doctors independently verified the kidney was human, but based on the portion they had, the doctors were not specifically able to match it to Eddowes' body at her truncated renal artery. So they couldn't say for certain it was her missing kidney, but they also couldn't prove it wasn't.

Now, despite the rapid succession of four victims in the month between August 31 and September 30, the last of the Ripper five, Mary Jane Kelly, was not killed until November 9, almost six weeks later. Kelly is also the only one of the canonical five murdered indoors. She lived in a multiple-unit building, and two neighbors claimed to have heard a woman cry out "murder" once, at around 4:00, but said the building was always fairly noisy. A man, sent by the landlord to collect some overdue rent that morning, found Kelly's body around 10:45. After knocking and then peering through the window, he was stunned to see Kelly's mutilated body on the bed and pieces of flesh lying on a table near the window; he ran to get his boss. Then the landlord sent his man on to the police station.

This time, the scene investigation seemed much more thought out, perhaps due to criticism of the previous actions. Police sent a telegram to Scotland Yard to bring in the bloodhounds, since the crime scene hadn't been disturbed. The area was cordoned off to the public; no one was allowed to either enter or leave; and a doctor was called in. Because the door was locked, which is curious to me, but I don't know how the door lock worked, the doctor looked through the window, and he immediately realized there was nothing he could do for the victim. So they waited for the dogs. But for some reason, the decision to use the dogs was reversed, and at 1:20 in the afternoon, police broke down the door and began to examine the murder scene.

This time, crime scene photos were taken; they're available and really awful. Kelly's body was on the bed, nude, with her legs splayed open. Her face had been mutilated beyond recognition, with parts of her nose, cheeks, eyebrows, and ears removed. Kelly's throat was cut to her spine, her abdomen completely eviscerated of its organs, which were found in various places on the bed. Her heart had been cut out and was never recovered. Pieces of skin and flesh had been removed from her thighs down to the bone, as well as from her abdomen, and her breasts were excised. As I mentioned before, some of her tissue was piled on a table near the window. This shows the escalation of the Ripper's actions, or maybe he just had more time and privacy since he was inside.

Investigators noticed the fireplace was still warm and found that a large quantity of women's clothes that had been burned, perhaps to give the killer light by which to work, since there was only one candle in the room. They searched for evidence of anything unusual and found a man's pipe, but it ended up belonging to Kelly's boyfriend. Police padlocked the door to her room to protect the scene, in case they later wanted to re-examine things. And although they lacked the technology we have today, investigators did consider some of the same things we do now. What's present and what's absent at the scene? What assumptions might we derive from that evidence?

Later that day, Kelly's body was autopsied. There were so many cuts to the victim's neck, that investigators couldn't establish the direction of the weapon's path. Because rigor mortis was setting in, the attending physicians estimated the time of Kelly's death at somewhere between 2:00 and 8:00 that morning. They believed it might have taken the Ripper perhaps two hours to do that much damage to her body. But unlike earlier reports, their conclusion was that the killer did not have any anatomical knowledge based on the haphazard way the organs had been removed. They estimated the size of the blade used at an inch wide and about six inches long—though today it's known that estimates of blade size from soft tissue wounds are notoriously difficult and unreliable.

The murder of Mary Kelly apparently ended the killer's rampage, at least in the London Whitechapel area. No one knows to this day who he was or why the murders ended. Did he die? Was he incarcerated for something else? Did he wind up in a mental institution? Did he move away? Despite thousands of books written on the subject and countless articles espousing different theories and possible suspects—over 125 years of analysis—his identity is not known.

Why was the Jack the Ripper case such a worldwide media sensation, even back in the 1880s—I mean, in addition to how horrific it was? Any idea why these murders are among the first to prompt what investigators call copycat killings? Why this case ultimately included a series of letters claiming to be from the murderer? This brings us to another major theme I hope you recognize throughout this course, and that's the continuous interplay between science and society. For example, how forensic advances, in both crimes and the technology used to solve them, are fairly directly correlated with other changes in society.

The simple reason this series of brutal murders became the best-known criminal matter of its time was the concurrent boom in newspaper circulation in the second half of the 19th century. Advances in printing and tax reform in England allowed unprecedented low-cost newspaper production and distribution. And since London was arguably the most prominent of capital cities in the world at that time, it had dozens of newspapers—*The Lancet*, *Times*, *Daily Telegraph*, *Daily News*, *Evening News*, *Standard*, *Star*, *Echo*, *Lloyd's Weekly*, *East London Observer*, *East London Advertiser*, *Jewish Chronicle*, *Jewish Standard*, a whole host of others. Even one publication called the *Illustrated Police News*, which was an early type of true-crime magazine.

In addition to the newspaper industry, a big reason word of the Ripper's killings went worldwide so quickly was that this was also the time of advances in the telegraph technology, including intercontinental communications. The 19th century also witnessed many advances in photography. So in essence, the Ripper murders coincided with the era when modern journalism was born.

So, after the discovery of Polly Nichols' mutilated body, news of London's East End crime wave spread quickly. Newspapers across the globe picked up that story and the related ones that came with each killing. In fact, the infamous name of this still-unknown killer was actually delivered to the newspapers. It came in the form of a letter written in red ink to London's Central News Agency organization on September 27. That was between the second and third murders among the canonical five. The letter's author took credit for the prostitute killings, saying he would cut off the ears of the next victim, and was signed, Jack the Ripper. Later, on October 1, a postcard in the same red ink and handwriting was delivered to the news agency about the "double event," alluding to the deaths of Elizabeth Stride and Kate Eddowes in the same night, even saying that "number one squealed a bit, couldn't finish straight off. Had not the time to get ears for police." And if you remember, although Eddowes' ears were not removed, one was definitely cut.

But then, as now, many who have delved into the Ripper murders believe the letter and the postcard were part of a hoax. Most attribute them to someone in the media, a journalist who had inside knowledge of the events and wanted to sell newspapers. And sell papers they did. In fact, Mary Jane Kelly's boyfriend told police she regularly asked him to read her the news of the killer from the local papers. Little could either one know that she would soon be front-page news as the Ripper's final victim.

Analyzing the Black Dahlia Murder

Lecture 2

On January 15, 1947, in the Leimert Park area of Los Angeles, a mother out walking on an errand with her daughter saw what she thought were two pieces of a store mannequin laying in a vacant lot. She was shocked when she realized it was actually the nude body of a woman who had been cut in two at the waist. The victim had black hair, and her skin was porcelain white, even more so from a near complete lack of blood. Some said her hair color was the reason this woman came to be called the Black Dahlia. In this lecture, we'll compare her murder and the ensuing investigation to the earlier case of Jack the Ripper.

Lust Murder

- Although they're rare, murder-mutilation cases are considered by many forensic behavioral analysts to be among those crimes that most lend themselves to psychological profiling. A special crime category known as *lust murder* involves fantasizing about, and deriving erotic pleasure from, killing in an intimate way, sometimes during sex and not uncommonly followed by genital mutilation or evisceration of the victim.

- Other common features of these crimes are posing the body after death, consuming the blood or tissue of the victim, or keeping body parts as trophies. Frequently, the killer's actions escalate over time in a string of murders. Torturing the victims before death is typically part of the serial killer's sick fantasy; some also engage in necrophilia.

- Many psychologists believe that this type of *paraphilia*, or sexual deviance, relates to childhood trauma: The killings represent a way to retaliate against the person who harmed the perpetrator as a child. When the murders don't satisfy the killer's needs, they often lead to a vicious cycle of increasing violence.

- The vast majority of lust murderers are men, many of whom kill only women. Well-known cases include those of Ted Bundy, who murdered more than 30 women in the 1970s, and Gary Ridgway, the Green River Killer, who is estimated to have killed more than 90 women and girls in the 1980s and 1990s. Other serial lust killers prey on couples or males. Infrequently, couples commit lust murders together.

- Although there have been serial killers who are women, they typically don't fit the lust murder profile and are more likely to kill lovers or their own children or to become delusional "angels of mercy," killing feeble or disabled people. A notable exception is Aileen Wuornos, a woman who killed seven men in Florida between 1989 and 1990.

Examining the Crime Scene

- In both the Black Dahlia and most of the Ripper murders, onlookers, the press, and even the police themselves severely compromised crime scene evidence and its recovery. Such contamination used to be all too common at crime scene locations in populated areas or outdoors. Today, we have come a long way in our understanding of crime scene contamination and trace evidence.

- *Trace evidence* refers to tiny bits of material that might be found at a crime scene. This evidence directly translates to pieces of information—but only with the recognition that it's present, even if difficult or impossible to see with the naked eye. Valuable trace evidence can be unwittingly destroyed or removed from the scene, even by investigators.

- Another problem with overaggressive interest in a scene is the introduction of unrelated material, such as the hairs and clothing fibers that fall off observers. Such material complicates analysis and wastes time because it focuses attention on evidence that is unrelated to the victim or perpetrator.

- In the Black Dahlia case, the lead detective was a senior LAPD officer who had literally written the book on protecting crime scenes and preserving evidence, but significant damage had already been done. Journalists were walking around the area, snapping photos of the body and throwing cigarette butts and used flashbulbs on the ground.
    - The fact that there was no blood on the ground around the remains told the police that Dahlia's body had been brought to the site, rather than killed where it was discovered, as the Ripper's victims had been.

    - Also unlike the Ripper victims, the Black Dahlia had rope marks on her wrists, ankles, and neck, causing police to suspect she had been tortured before she was killed. The body appeared to be literally empty of blood, and the detectives wondered if ropes had been used to hang the victim like a meat carcass to drain her blood.

    - As in the Ripper killings, police thought the perpetrator might have some type of anatomical knowledge—perhaps from experience as a butcher, physician, or medical student.

    - By comparing the amount of dew in the grass around the victim versus the amount beneath her, investigators surmised that the body parts were deposited around 2:00 a.m. on the day the body was discovered.

    - Detectives believed that the perpetrator had carried the two body parts to the lot one at a time, suggesting a single killer. Debris on the upper half of the body indicated that it was originally face down but had been turned face up for display.

    - The lower half of the body had been carried on an empty cement bag, which was found nearby with traces of watered-down blood on it. Police also saw a heel print, but it had been damaged by tire tracks.

- Before the body parts were removed from the scene, detectives collected some trace evidence from them: tiny bits of a bristle-like material, which they thought could be broom straw or fibers from a car floor mat.

The Black Dahlia's Autopsy

- The autopsy revealed that the Dahlia's body had been drained of blood and washed clean. Like more than half of the Ripper victims, the Dahlia had been eviscerated; her intestines were found tucked underneath her body. Bruising of the face and right side of the head showed that she had been bludgeoned while still alive. Investigators suspected that she may have been tortured for days before her death. The cause of death was ruled to be blunt force trauma to the head and subsequent loss of blood.

- Although the body was further examined for trace evidence, none was found; there was no trace of semen or evidence of vaginal intercourse. However, the victim had been sodomized, and the medical examiner believed the penetration to have been postmortem.

- By measuring the lengths of both body segments, the woman's height was recorded as 5 feet, 5 inches tall, and together, the two body parts weighed 115 pounds. Her age was estimated to be between 15 and 30 years.

- Morgue photos were given to a police artist to generate a likeness of the victim for release to the media in the hopes of a quick identification. When the images appeared in the newspaper, calls started pouring in about possible missing persons, but none of them checked out.

Identifying the Victim

- The victim had been fingerprinted as soon as she was brought into the morgue at 2:45 p.m. so that the prints could be sent by special-delivery airmail to the FBI in Washington, DC. But snowstorms

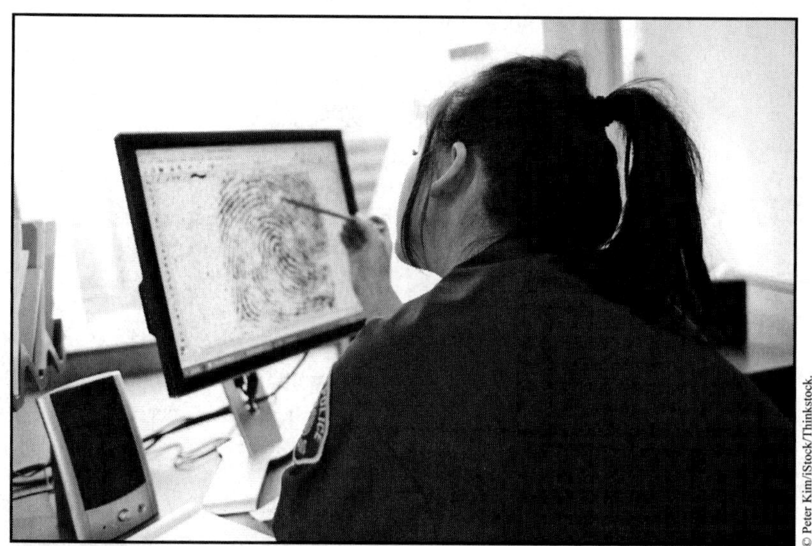

© Peter Kim/iStock/Thinkstock.

The Integrated Automated Fingerprint Identification System (IAFIS) allows digitized prints to be compared within a matter of seconds using computer technology.

were delaying plane flights, and estimates were that it could take as long as a week before the prints were actually delivered.

- Faced with this potential delay, the editor of the *Los Angeles Herald-Examiner* got the idea to send the prints using a Soundphoto machine, an early means of sending images over telegraph wires. The images arrived in Washington the next morning, but they were blurry and impossible to classify into the searchable fingerprint codes that were in common use in comparisons at the time.

- The photographers at the *Herald-Examiner* then enlarged the fingerprint images and resent them. The results were good enough to be confidently classified by FBI agents. Within minutes, the Black Dahlia was known to be 22-year-old Elizabeth Short, originally born in Massachusetts. Her prints were on file because she had worked about four years earlier as a civilian employee at Camp Cooke army base in California.

- Short had moved to California when she was 19 years old, with hopes of becoming an actress or a model. She later worked at the army base and as a waitress. She moved around frequently and was always short of cash. A party girl, Short went out on dates almost every night, although there was never any evidence of prostitution.

Suspects in Dahlia's Case

- As they had in the Ripper killings, police immediately began to canvass the neighborhood to look for potential witnesses or possible suspects. Investigators interviewed Short's father, personnel at the army base, and a number of former boyfriends.

- The last person claiming to see Short alive was a 25-year-old married salesman named Robert Manley, who had met Short about a month before her death. Police held Manley as a suspect, but he passed a polygraph test and had a solid alibi for the night Short's body was dumped.

- Eight days after Short's body was found, a man phoned the editor of the *Herald-Examiner*, claiming to be the killer and saying that he would send the newspaper some of Short's belongings. The next day, a package arrived with several items, including photos, Short's birth certificate, and an address book with the name Mark Hansen embossed on it.

- Hansen was a nightclub owner who had befriended Short. He admitted the book was his, but clearly Elizabeth had been using it. He and all the others listed in the address book were questioned, with no results. In 1951, during a follow-up inquest, Hansen was still among the main suspects, but no charges were ever filed, and Hansen died in 1964.

Conclusions about Dahlia

- Strangely, Short's brutal murder appears to have been an isolated incident. No other similar cases in the Los Angeles area have been definitively linked to Short's death. Given the typical pattern of multiple killings, why or how could a lust murderer limit himself to a single killing?

- It seems unlikely that Short's killer continued to prey on women but became more careful about body disposal. Such a change doesn't fit with the killer's MO of dumping the body in a public place and posing the victim. The killer obviously wanted Short's body to be found and to shock the community.

- With Jack the Ripper, investigators theorized that the killings abruptly stopped because the murderer died, was incarcerated for another crime, or moved on. In the Dahlia case, it's possible that a serial offender operating somewhere else came to Los Angeles and committed this single horrible crime. Investigators looked at similar crimes elsewhere in the United States but found no links that seem likely today. As with the Ripper killings, we may never know whose lust and rage caused the terrible end of Elizabeth Short.

Suggested Reading

Douglas and Olshaker, *The Anatomy of Motive.*

Ellroy, *The Black Dahlia.*

Federal Bureau of Investigation, *Serial Murder.*

Gilmore, *Severed.*

Pietras, *Unanswered Evidence.*

Questions to Consider

1. Had you heard of the Black Dahlia killing before this course? If so, from what source?

2. How likely do you think it is that the Black Dahlia's death was the perpetrator's sole murderous act?

3. What types of evidence and other forensic leads do you think investigators use to link multiple crimes to each other?

Analyzing the Black Dahlia Murder

Lecture 2—Transcript

On January 15 of 1947, in the Leimert Park area of Los Angeles, California, a young mother was out walking an early morning errand with her three-year-old daughter. She saw what she thought were two pieces of a store mannequin lying in a grassy vacant lot. She was shocked to realize it was actually a nude woman's body, cut in two at the waist.

Both body parts were face up, and the woman's arms were stretched out over her head at about a 45-degree angle. The lower half of the body was situated about a foot away from the rest of the torso, and the victim's legs were spread open. In a hideous and ironic display of mutilation, the woman's mouth had been slashed in both corners, forming what resembled a grotesque smile. She had dark black hair, and her skin was almost porcelain white, even more so from near complete lack of blood. The woman came to be known as the Black Dahlia. Some say her hair color was the reason for the name; others said it was a play on title *The Blue Dahlia*, a 1946 movie in which a man killed his wife. Regardless, the media has a penchant for giving nicknames to killers, and in some cases their victims.

We all know that cases of brutality against women, as well as men, cut across time and place, as do horrible postmortem mutilations. Although they're rare, murder-mutilation cases are considered by many behavioral analysts to be among those crimes that most lend themselves to psychological profiling, and they typically have sexual underpinnings. A special forensic category known as *lust murder* involves fantasizing about and then deriving erotic pleasure from killing in an intimate way, sometimes during sex and not uncommonly followed by genital mutilation or evisceration of the victim.

Another common feature of these types of horrible crimes is posing the body after death, often to shock those who discover the corpse, as well as consuming the blood or tissue of the victim, or keeping body parts for trophies. Frequently the killer's actions escalate over time in a string of murders and mutilations. Torturing the victims prior to death is typically part of the serial killer's sick fantasy. Some also engage in necrophilia after the murder. Many psychologists believe this type of so-called *paraphilia*, or sexual deviance, relates to childhood trauma, and the killings represent a way to try to get back at the person who harmed the perpetrator as a child. And when the murders don't satisfy the killer's needs or make him whole again, they often lead to a vicious cycle of increasing violence.

Now, I say "he" because the vast majority of these kinds of perpetrators are men, many of whom kill only women, such as Jerry Brudos, the "Shoe Fetish Slayer" of Oregon, who killed and mutilated 4 women in the late 1960s; Ted Bundy who murdered over 30 women in a reign of terror across the United States in the 1970s; and Gary Ridgway, "the Green River Killer," who is estimated to have killed over 90 women and girls in Washington State in the 1980s and '90s, often returning to their dead bodies for further sexual gratification.

Other lust murderers killed couples, but usually only sexually abused or mutilated the women of the pair, such as Richard Ramirez, "The Night Stalker," who killed at least 13 people in California in the 1980s. In the U.S.S.R., "The Butcher of Rostov," Andrei Chikatilo, sadistically took the lives of over 50 people in the 1980s. His victims included women, but also children of both sexes. Others seemed less discriminate, like Peter Kurten, known as the "vampire of Dusseldorf," who killed an unknown number of men, women, and children in 1929 and 1930 period, perhaps as many as 60 people.

Some serial lust killers preferred males. That was the case with John Wayne Gacy, the "Killer Clown," who murdered at least 33 boys in the 1970s in Chicago. Jeffrey Dahmer killed 17 males in the Midwestern U.S. from the late 1970s to early '90s, and he cannibalized some of them. Moscow's Sergey Golovkin, called "The Fisher," butchered 11 boys in the 1980s and '90s. Infrequently, couples commit lust murders together, like Fred and Rosemary West, who often, in their own home, tortured, raped, and murdered around a dozen young women and girls in England, mostly during the 1970s. Ian Brady and Myra Hindley committed what are called the Moors Murders of five children in England during the mid-1960s.

Although there have been serial killers who are women, they typically don't fit the lust-murder profile and are more likely to kill lovers or their own children, not strangers. Other women become these delusional "angels of mercy," who kill feeble or disabled people. A notable exception is Aileen Wuornos, a lesbian who killed seven men in Florida between 1989 and 1990. Wuornos was alleged to have suffered childhood trauma, like most lust murderers, but she shot her victims, which is not an intimate killing, and she didn't mutilate or cannibalize them.

Interestingly, a book released in 2012 claims that the reason London's Jack the Ripper has never been identified is because the killer was a woman, but I don't buy it. Despite many suspects and much re-examination, no one will likely ever know the identity of that most notorious lust murderer. That's one of the commonalities between the Ripper cases of 1888 and the 1947 Black Dahlia killing that opened this lecture. Although authorities know who all of the victims were, they have never been able to figure out who brutally killed and mutilated of any of those women. So, now let's return to the scene where the Black Dahlia's body was discovered and look for other ways in which this case is and is not like those of the legendary Jack the Ripper. First off, in both the Black Dahlia and most of the Ripper murders, onlookers, the press, even the police themselves, quickly and severely compromised scene evidence and its recovery. This is, or used to be, all too common at locations where crimes occur or bodies are disposed of in populated areas, especially outdoors. Since those days, we have come a long way in recognizing how easy it is to contaminate a scene; that's because we now understand so much more about and can do a lot more with what's called *trace evidence*.

When found at a crime scene, tiny bits of material directly translate to pieces of information, but, only with the recognition that it's there, even if difficult or impossible to see with the naked eye. Valuable trace evidence can be trampled on, even by investigators; it can stick to people's shoes or clothing and be unwittingly removed from the scene, or larger pieces of evidence even taken as ghoulish souvenirs by curious onlookers. Stranger things have happened.

Another problem with overaggressive interest in a scene is the introduction of lots of other nonessential data, like hairs and clothing fibers that fall off of observers. When those samples are discovered, it complicates analysis and wastes time, since it focuses attention on evidence that has really nothing to do with the victim or the perpetrator. But really, when you consider the discovery of a body in that vacant lot, where people routinely dumped trash, the odds of finding trace evidence related to the crime, other than maybe on the body itself, would be pretty remote.

Still, the entire area and all things in it had to be carefully examined. As soon as they reached the scene, detectives immediately knew they were dealing with a highly unusual and high-profile case. The lead detective was a senior LAPD officer who had literally written the book on protecting crime scenes and preserving evidence. But significant damage had already been done, since people, especially journalists, were already walking around the area snapping photos of the body. They were throwing cigarette butts and used flashbulbs all around. Even though police tried to move onlookers away, some parked nearby and literally stood on the roofs of their cars to see what was going on.

So let's look at what police discovered when examining the scene. Well, for one thing, there was no blood on the ground around the remains. Now, what does that tell us? It tells us Dahlia's body had to be brought to the site, rather than killed where it was discovered, as Jack the Ripper's victims had been. Also, unlike the Ripper victims, the Black Dahlia had rope marks on her wrists, ankles, and her neck, causing police to suspect she had been tortured before she was killed. And the body appeared to be literally empty of blood. The detectives even wondered if ropes had been used to hang the victim like a meat carcass to drain her of her blood. Now, as was suspected in the Ripper killings, police thought the perpetrator might have some type of anatomical knowledge, was maybe a butcher, or perhaps even a physician, or maybe a student at the local university medical school.

By comparing the amount of dew in the grass around the victim, versus what was beneath her, scene investigators surmised the body parts were deposited around 2:00 in the morning on the day the body was discovered. Detectives figured the perpetrator had parked in a nearby driveway that led into the lot. They believed the two body parts were carried in one at a time, perhaps suggesting a single perpetrator. There were bits of debris on the upper half of the body that indicated to them it was originally face down, but had been turned face up for display. The lower half of the body had been dragged in on an empty cement bag, which was found nearby with traces of watered-down blood on it. They also saw a heel print, but it had already been run over by tire tracks. Before the body parts were taken from the scene, detectives did collect some trace evidence off the body. What they found were tiny bits of a bristle-like material, which they thought could be from maybe broom straw or possibly from the floor mat of the car.

What about Dahlia's autopsy reports? Not only had the body been drained of blood, it appeared to have been washed clean. Like more than half the Ripper victims, the Dahlia had been eviscerated; her intestines were found tucked underneath her body before they transported her to the morgue. She had bruising of the face on the right side of the head that showed she had been bludgeoned while still alive, since bruising doesn't happen after death. Investigators suspected she may have been tortured for even days before dying. The cause of death was ruled to be blunt force trauma to the head and resulting subsequent loss of blood. It was obvious the manner of death was homicide.

Although the body was examined for trace evidence, like foreign hairs or other fibers, they didn't find anything other than those tiny bristles I mentioned, and that's perhaps because the body had been washed; even the victim's hair appeared to have been cleaned. And just like in the Ripper murders, there were no traces of semen found in or on the body and no evidence for vaginal intercourse. In the Dahlia case, the victim had instead been sodomized and the medical examiner believed the penetration to have been postmortem, in other words, it was done after she was dead. And there's a myth that often circulates that the Dahlia was unable to have vaginal intercourse due to undeveloped sexual organs, but the autopsy report documents her genitalia as normal.

Now, by measuring the lengths of both body segments, the woman's height was recorded as 5′ 5″ tall, and together, the two body parts weighed about 115 pounds. They estimated her age to be between 15 and 30 years, but I can tell you, as an anthropologist, that some decent X-rays or an examination of the skeleton could have narrowed that range down quite a bit. But when trying to find the identity of an unknown person, it's always better to be broad than narrow, just to make sure you capture the normal amount of human variation.

The Dahlia had a slight build, a small nose, a round chin, and gray-green eyes. Her teeth were said to be in a poor state of dental care, but her toenails were brightly polished red. Because of the cut marks and bruises to her face, morgue photos were given to a police artist to generate a more presentable likeness to release to the media in the hopes of a quick identification. We still do that today, only now often using computer technology. The artist's images were released in that afternoon's paper, and calls started pouring in about possible missing persons, but none of them checked out.

So how did they ultimately discover the identity of the woman known as the Black Dahlia? This is a pretty cool part of the story. The victim had been fingerprinted as soon as she was brought into the Los Angeles County Morgue at 2:45 in the afternoon, even before the autopsy. They did that so the prints could be sent that night by special-delivery airmail to the FBI. But this was January, and snowstorms throughout the U.S. were delaying plane flights. They estimated that it could be as long as a week before the prints would be actually delivered to the FBI, and in kind of a horrendous case, that was a delay the investigation just couldn't afford.

Today we have what's known as the Integrated Automated Fingerprint Identification System, better known as IAFIS, which allows digitized prints to be compared within a matter of seconds to fingerprints anywhere in the world using computer technology. They had nothing like that back in 1947, but the latest technology did play a role in the Black Dahlia's ultimate identification. Faced with the potential delay in airmailing the fingerprints, the editor of the *L.A. Herald-Examiner* got the bright idea to send the prints using some cutting-edge newspaper technology. Now, I can tell you from my many years in forensics, at times the media can be your worst enemy in an investigation, but in other situations, it can be your best ally, and that was certainly the case here for the LAPD.

Have you ever heard of a thing called a Soundphoto machine? It was an early means of sending images over telegraph wires. By the time the journalists and investigators considered this option, the news wires were already closed for the night. But the Los Angeles International News Photowire would open at 4:00 the next morning, which would be 7:00 in the morning in Washington, DC. So at 4:02 a.m. the victim's prints were sent over the wires to an FBI agent waiting at the Hearst News Bureau at the other end.

This was the first time anything like this type of evidence transfer had ever been tried. But there was a problem; the printed results in DC were blurry, and they were really hard to read, let alone classify into the types searchable fingerprint codes that were in common use in comparisons, and the comparisons were still done by hand at that time. So the photo guys at the *L.A. Herald-Examiner* put their heads together. They decided maybe they could use the fingerprints like negatives, then enlarge the images, and perhaps get better results.

So they created a single 8 × 10 image of each individual fingerprint and then wired each of the 10 prints separately. The resulting images were good enough to be confidently classified by the FBI agents on the other end, and within minutes the Black Dahlia was known to be 22-year-old Elizabeth Short, originally born in Massachusetts, but who had worked for about four years before as a civilian employee at Camp Cooke Army Base in California, and that's why her fingerprints were on file.

Short was one of five daughters born to a Massachusetts couple. During the depression, her father lost his business and ultimately parked his car on a bridge to fake a suicide and abandon his family. He headed off to California and a while later contacted his wife, but she refused to take him back. And when Elizabeth Short was 19 years old, she decided to join her father in California. Short was very pretty and had hopes of making it big in the movies or maybe as a model, but when her dad threw her out due to her laziness and late nights, she went to work at the nearby army base; that's where she earned the nickname Camp Cutie, because she was such a flirt. In 1943, Short was arrested in Santa Barbara for underage drinking, and her mug shot from that arrest has become iconic in the case.

Now, Short sometimes worked as waitress and lived in either cheap boarding houses or apartments with other single girls. She was said to have lots of acquaintances, but really no close friends. Short moved around a lot and was always low on cash. People said whatever money she had she'd spend on clothes, rather than buy food, and she often dressed all in black.

Short went out almost every night and allegedly preferred the company of strangers. She ate when men took her to clubs or to restaurants; she accepted their gifts; and because she was very pretty, Short always had a lot of dates. There was never any evidence of prostitution, though; she simply appeared to take advantage of the kindness in others. In short, and pardon the bad pun, she was known as a "beautiful freeloader."

So how does Short's victim profile compare with Jack the Ripper's targets? All of them were females living in poverty and relying on the generosity of others; all were known to drink and cavort with men, often strangers. And if I've learned one thing about victim profiles in my many years in forensics, it's that people who regularly engage in high-risk behaviors are much more likely to wind up in the morgue than those who don't.

Another thing is that Los Angeles, especially the Hollywood area, was a beacon for women who wanted to use their female attributes to strike it big, sort of the way the East End of Ripper's London was flooded with its own immigrants seeking a better life in the big city. The results are not often good, and the stories too frequently do not have happy endings.

Now, what about suspects? Or, who may have been seen or known anything about Elizabeth Short's death or body disposal? As in the Ripper killings, police immediately began to canvass the neighborhoods to look for potential witnesses or possible suspects. In the Black Dahlia case, they quickly went to interview Short's father, who, despite living about three blocks away, hadn't talked to his daughter in almost three years. Investigators talked to people at the Army base and lots men who were former boyfriends. But curiously, and unlike the Ripper murders, dozens of people came forward to confess to Elizabeth Short's death, including both men and women. Now clearly, these were attention seekers, something Hollywood is known for, and none of these leads checked out.

The last person claiming to see Short alive was a 25-year-old married salesman named Robert Manley, who went by the nickname "Red." He had met Short about a month before her death. One week before her body was found, Manley picked Short up from a friend's home, and the two went out partying and then spent the night in a hotel together. Manley insisted they did not have sex, and the next day, he drove Short to the very glamorous Biltmore Hotel in L.A., where she said she was meeting her sister. Manley said when he last saw Short, she was on the phone in the hotel lobby, and he went home. Police did hold Manley as a suspect, but he not only passed a polygraph lie-detector test, but he had a solid alibi for the night the Dahlia's body was dumped.

Eight days after Short's body was found, a man who said he was the killer phoned the editor of the *L.A. Herald-Examiner*, similar to the way the London newspapers received notes in 1888 allegedly from Jack the Ripper. Now, this caller said he was disappointed that the news around the Black Dahlia case was dying down, and so he was going to send them some of Short's belongings. The next day a package arrived with several items, including photos, Short's birth certificate, and an address book with the name Mark Hansen embossed on it; now, we'll come back to that in a minute. Police tried to find fingerprints on a note that came with the package, but it had apparently been washed with gasoline, ruining any type of potential evidence.

So the next day, they found a black purse and one black shoe that was observed on top of a trash can several miles from where Short's body had been discarded. Red Manley was ultimately the one who identified those items as hers, saying that the purse smelled of the perfume she wore, and the shoe was one that he himself had paid to have resoled for her. After years of depression, Manley's wife had him committed in 1954, and, at that point, they gave him sodium pentothal, which is better known as truth serum, but he passed a polygraph, so most still think he was not the killer.

So who was this Mark Hansen, the guy whose address book was sent to the newspaper? Well, he was a nightclub owner that befriended Short. In fact, between May and October of 1946, Short had actually lived at his home, sharing a room with Hansen's girlfriend. Hansen admitted the book was his, but clearly, Elizabeth had been using it. Police questioned him, and they also questioned all the others listed in the address book, but with no results. Now, during a follow-up inquest in 1951, Hansen was still among the main suspects, but no charges were ever filed, and he died in 1964. All types of true-crime websites and many books about the case detail many other possible suspects, but it appears we may never know who actually killed Elizabeth Short.

Personally, what I find most interesting and unusual about the Black Dahlia case, as contrasted with all the other lust murders I briefly reviewed at the start of the lecture, is that her brutal death and violent mutilation appears to have been a single, isolated incident. Strangely, no other similar cases in the Los Angeles area have been definitively linked to Short's death. And given the typical pattern of multiple killings, often starting in many classic cases with animal cruelty during childhood or teenage years, why would, or even how could, a lust murderer limit himself to just a single killing? Why did the brutality start or stop with Short?

Well, we can't know for certain it did. The Hollywood area is notorious for having somewhat of a transient population, and I'm sure there were countless young women listed as missing persons around that time. So, is it possible the killer continued to prey on women but just got more careful about body disposal? Well, in my mind, that doesn't fit with the killer's MO of dumping the body in a public place and posing the victim. I mean, he wanted Short's body to be found, to show off his perversion, and to shock the sensibilities of the community. So, if he murdered others, it wouldn't make sense to go underground with his work all of a sudden.

The Jack the Ripper killings also ended abruptly, and either the thinking is the murderer either died, or was incarcerated for another crime, or maybe moved elsewhere. But in the Dahlia case, maybe the reverse is also possible; maybe a serial killer operating elsewhere came to Los Angeles, perhaps for a vacation or other visit, just to commit this single, horrible crime. This line of reasoning led investigators to start looking at other similar incidents throughout the United States at the time.

Now many suspected the Black Dahlia case could be linked to what are known as the Cleveland Torso Murders; that's a string of at least a dozen unsolved killings and mutilations in Ohio in the 1930s. But is the Dahlia murder of 1947 a good fit? I don't see it. One thing, the Cleveland Torso Murderer chose both male and female victims who were mostly just poor drifters. Some of them were cut across the waist, as with the Black Dahlia case, but in every single one of the Cleveland Torso Murders, the victim was decapitated.

Police also tried to link the Dahlia case to three murders in Chicago that happened in the two years prior to Dahlia's death, in 1945 and 46. Those murders were two women and then a six-year-old girl. Together those three cases have been called The Lipstick Murders due to a message written in lipstick on the wall at the scene of the apartment of the second woman's death. Well, these crimes do all precede the murder of Elizabeth Short, and the cases of the two women do show the escalation typical of lust murders. The first victim was repeatedly stabbed, and the second was also brutally stabbed, but she was also shot, and the note on her wall in lipstick read, "For heaven's sake, catch me before I kill more, I cannot control myself." But, neither of these victims was dismembered.

The last of those three Chicago victims, though, was a six-year-old girl who was abducted from her bedroom. Following an anonymous tip, the girl's body parts were found dispersed in various street sewers near her home. They also found a dismemberment site discovered in the laundry room of a local apartment building. But I have a real problem with trying to link this case to the other two Chicago women, let alone Dahlia. The little girl's family received a ransom note, not a lipstick message, and then there were associated phone calls. Now, a local teenager, who actually had nothing to do with the case, actually made the fake ransom calls, but the source of the note is still unknown. But the girl's kidnapping shows a completely different motive than the typical lust murder.

The three Lipstick Murders were eventually tied to a 17-year-old Chicago boy named William Heirens, who ultimately confessed to the crimes in exchange for a plea bargain. Heirens was a known and convicted serial burglar, but in none of the Lipstick Murders was anything taken, except the six-year-old girl. Nevertheless, Heirens went to jail for life, but nearly immediately recanted the confession, saying it was coerced by extreme police brutality. It certainly seems like Heirens' claims could be true, I mean, the community was begging for a perpetrator, and police delivered.

Given the scant and contradictory evidence, I find it difficult to link the three Lipstick Murders together at all, let alone to Heirens, and I strongly believe that had those killings happened today, the child abduction would not have been linked to the other two. To me, that case sounds like a kidnapping gone bad, after which the perpetrator had to dispose of the little girl's body parts.

In my experience, dismemberment usually simply reflects the desperation of someone to hide a killing; it's not the aftereffect of a lust murder. And, as an MO, does hiding body parts in sewers as in the six-year-old Chicago girl's case compare to leaving them out for display, like the Dahlia and Ripper cases?

Even if by some stretch Elizabeth Short's killing in Los Angeles is linked to the Lipstick Murders, Chicago's Heirens is definitely not the Dahlia's killer. He had been in custody for over six months when her body was found. Though Heirens died in prison, he was likely the wrong guy, even for the murders of which he was convicted. If so, perhaps the true Lipstick Murderer did kill the Black Dahlia as well, but I doubt it. Like the notorious Ripper killings, I think the world will never know whose lust and rage caused these terrible murders.

Dissecting Hollywood Deaths

Lecture 3

Hollywood is a place filled with movies and stars, stories of success and failure, and unfortunately, in some cases, untimely and suspicious deaths. In this lecture, we'll examine several historical cases from Hollywood and see how scene investigation and autopsy are crucial in analyzing mysterious deaths. We'll first explore the case of George Reeves, one of three actors who played Superman—all of whom died under unusual circumstances. Then, we'll look at the cases of Hollywood bad boy Bob Crane and martial arts expert Bruce Lee. In all these Hollywood cases, we'll see that autopsy results were essential in trying to establish both cause and manner of death.

George Reeves

- George Reeves, the actor who played Superman beginning in 1952, was born George Brewer in a small Iowa town in January of 1914. As a teenager, he began boxing and competed in the 1932 Olympics. In 1939, renamed George Reeves by a studio, he played one of Scarlet O'Hara's admirers in *Gone with the Wind*.

- After serving in World War II, Reeves eventually landed in New York, the hub of the early television industry. When a show about Superman was planned, Reeves auditioned and won the leading role. Filming began in the summer of 1951; the first episode aired in September 1952, and the show was an instant success, running for six years.

- In the early 1950s, Reeves began a longstanding affair with Toni Mannix, a former Ziegfeld Follies girl and the wife of a successful MGM producer. Reeves and Mannix were often seen as a couple in public, and Mannix was said to be very possessive of Reeves.

- In 1958, Reeves met Leonore Lemmon, a well-known New York party girl. When Reeves took up with Lemmon, Mannix began to harass him with daily phone calls. Although Reeves and Lemmon had a rocky relationship, they planned to marry in the summer of 1959.

- On the evening of June 16, 1959, three days before the wedding, a group of friends and neighbors gathered with Reeves and Lemmon at Reeves's home.
  - According to the guests, Reeves went upstairs to bed in a bad mood around midnight, only to come down later to complain about the noise the remaining party guests were making. Reeves had another drink before going back to bed.
  - About 2:00 a.m., the guests heard a single gunshot; one of the neighbors ran upstairs to find Reeves lying on his back, naked, with a 9-millimeter Luger pistol between his feet and a fatal gunshot wound to his right temple.
  - The houseguests didn't call police until 30 or 40 minutes after the shooting. Officers reported that the witnesses were all drunk and gave no excuse for the delay in calling.
- A single bullet casing was found beneath Reeves's body, which suggested that he was probably sitting on the edge of the bed when shot, then fell back onto the casing. The bullet that struck Reeves was recovered from the ceiling.
- Initially, Reeves received only a cursory autopsy at the funeral home before being embalmed. Later, a more thorough autopsy was done at the L.A. County morgue. Blood samples taken before the embalming revealed a blood-alcohol level of 0.27. Toxicology tests showed painkillers in Reeves's system that had been prescribed following a recent car accident.
- The fracture patterns in the skull were said to be consistent with a close shot. There was reportedly no gunpowder on Reeves's temple, but that's not uncommon in contact wounds. Because the gun had

been recently oiled, no clear fingerprints were found on it. The L.A. County coroner ruled Reeves's death a suicide.

- Helen Bessolo, Reeves's mother, wasn't satisfied with the ruling; she refused to have her son's body cremated and hired her own detective to investigate his death. In January of 1960, Reeves's body was sent to Cincinnati for a third autopsy, which again supported a finding of suicide. Reeves's mother finally agreed to cremate her son's body three years after his death and shortly before her own.

- During the investigation, rumors surfaced that Mannix had hired a hit man to kill Reeves. Another account, attributed to a neighbor present the night Reeves died, alleged that Lemmon instructed the other witnesses to tell police she was downstairs during the shooting, but she wasn't. Though suspicions may still exist, the official ruling of suicide in Reeves's death still stands.

Bob Crane

- The actor Bob Crane was best known for his starring role as Colonel Hogan on the sitcom *Hogan's Heroes*. The show ran from 1965 to 1971, after which Crane's career went downhill.

- Crane had a reputation as a bad boy, stemming from his insatiable sexual appetite. He constantly picked up women and often enjoyed group sex. Crane and a friend, John Carpenter, recorded Crane's sexual exploits using home video equipment, often without the knowledge of the other participants. Crane also had an extensive photo album of his many partners.

- The jealous boyfriends or husbands of the women involved with Crane sometimes caused him major problems. Crane's wife, Patricia Olson Crane, wasn't amused either. She and her husband were in the process of a divorce in 1978, when Crane met his death in Scottsdale, Arizona.

- On the night he was last seen alive, Crane appeared in a play at a Scottsdale dinner theater. Later that night, he went to a bar with two women and John Carpenter. At about 2:00 am, the four went into a coffee shop; Carpenter left about a half-hour later, before the other three.

- Around 2:00 the next afternoon, Crane's body was discovered in his apartment by his co-star in the play, Victoria Berry. Crane was lying in the fetal position on his side with an electric cord wrapped around his neck. Blood was spattered on the adjacent wall.

- Investigators determined that the scene showed no signs of Crane struggling during his death. Smears on the sheets looked like someone had wiped blood on them—possibly while cleaning off the murder weapon. The only thing known to be missing from the apartment was the photo album of Crane's sexual conquests.

- At around 3:15, while Berry was giving her statement to the police in Crane's apartment, John Carpenter called. The police lieutenant who spoke with Carpenter told him that police were investigating an "incident" in Crane's apartment. Carpenter never asked the lieutenant where Crane was or what the investigation was about.

- An autopsy concluded that Crane was likely sleeping on his right side when the killer delivered two blows to the left side of his head with a heavy object. The time of death was estimated between 3:00 and 3:30 a.m. The electric cord had been wrapped around Crane's neck after his death. The medical examiner speculated that the killer was probably a strong man, based on the blood spatter pattern analysis.

- Carpenter came under suspicion, and police brought the rental car he had used during his visit to Scottsdale in for analysis. On the inside of the passenger door, a small amount of blood was found; it was determined to be type B, the same as Crane's, but the match wasn't enough evidence for an arrest.

- However, in 1992, Carpenter was arrested for Crane's death based on a reexamination of photos of the rental vehicle that showed a fragment of what could have been human tissue. The defense argued that something that small could not be adequately identified from a photo, and Carpenter was acquitted. Still, he lived under a cloud of suspicion until his death in 1998.

Bruce Lee

- Bruce Lee was born Li Jun Fan in San Francisco's Chinatown but grew up in Hong Kong. At age 13, he started formal martial arts training. He later began a career as a martial arts hero in Asian films. By 1971, he finally earned recognition in Hollywood.

- Lee was young, apparently healthy, and in top physical shape, but while working on the film *Enter the Dragon* in May of 1973, he collapsed into a seizure. He was taken to the hospital and diagnosed with cerebral edema—swelling of the brain due to excess accumulation of fluid. Lee told his doctor that he had used hashish from Nepal the day before and wondered if there was a connection. Lee's doctor warned him against using the drug.

- Lee recovered, and *Enter the Dragon* was successfully completed. Plans immediately began for his next film, *Game of Death*. Lee was to co-star with actress Betty Ting Pei, with whom some alleged he was having an affair.

- On the afternoon of July 20, 1973, Lee and director Raymond Chow went to Pei's home to review the new script. At some point after the director left, Lee told Pei he had a headache, and she gave him a prescription painkiller called Equagesic. Later that evening, when Lee couldn't be awakened from a nap, an ambulance was called, and he was taken to the hospital, where he was pronounced dead.

- An autopsy showed no signs of injury and no blocked or broken blood vessels, but Lee's brain had swelled some 13 percent. Toxicology showed cannabinoids in Lee's blood, in addition to the Equagesic Pei had given him.

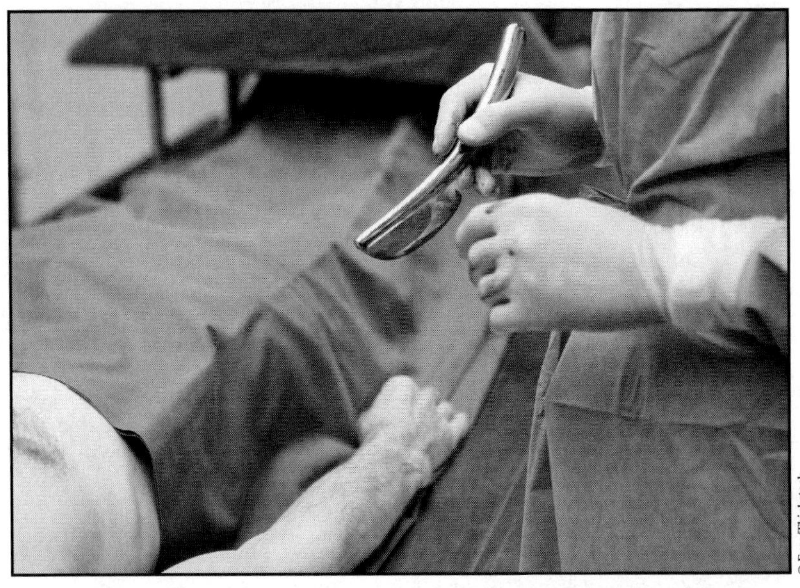

Scene investigation and autopsy are crucial in analyzing mysterious deaths.

- A dispute immediately arose among the physicians involved about whether the Equagesic or cannabis had caused Lee's death. In the end, authorities called Lee's case "death by misadventure," which is neither a recognized cause nor manner of death.

- In 2006, a Chicago pathologist asserted a different cause of death: *sudden unexpected death in epilepsy* (SUDEP). Statistics indicate that SUDEP deaths are more common in young adult males, and although Lee had not been diagnosed with epilepsy, he did have a previous seizure, observed at the time of his first episode of cerebral edema.

History of Death Investigation

- In all these Hollywood cases, autopsy results were essential in trying to establish both cause and manner of death. Assessing suicidal, homicidal, accidental, and natural deaths requires a thorough examination of tissues, body fluids, potential weapons, scene clues, and witness testimony, when possible.

- But these methods are nothing new in forensics, especially in death investigations involving famous people. For instance, in 44 B.C.E., Julius Caesar's physician examined the 23 stab wounds on the emperor's body and deduced that only one of them—a specific knife wound to his chest—was actually fatal.

- Early pioneers in the field of forensics couldn't even imagine today's computerized chemical analyzers or scanning electron microscopes. And even familiar evidence, such as blood, carries much more power now than it did even just a few decades ago. We don't know where the future of forensic science will lead us, but it's easy to see how current technology would have aided justice in the past and avoided the mystery surrounding some well-publicized deaths.

Suggested Reading

DiMaio and Dana, *Handbook of Forensic Pathology*.

Di Mambro, *True Hollywood Noir*.

Graysmith, *The Murder of Bob Crane*.

Henderson, *Speeding Bullet*.

Lee, *Life and Tragic Death of Bruce Lee*.

Murray, *Death*.

Parish, *The Hollywood Book of Death*.

Pietras, *Unanswered Evidence*.

Schroeder and Fogg, *Beverly Hills Confidential*.

Spitz and Spitz, *Spitz and Fisher's Medicolegal Investigation of Death*.

Questions to Consider

1. Do you believe George Reeves's death was a suicide? Why or why not?

2. Who do you think killed Bob Crane? Why?

3. Can you name other suspicious Hollywood deaths that made headline news?

Dissecting Hollywood Deaths

Lecture 3—Transcript

When I think of Hollywood I don't just think of movies and stars. I think of a place that's filled with stories of success and failure, where untimely and suspicious deaths are unfortunately just waiting to happen. In this lecture, we'll examine several historical cases from Tinseltown and see how scene investigation and autopsy are crucial in analyzing mysterious deaths.

If you're around my age or older, you likely remember the opening of the Superman television show. It began, "Faster than a speeding bullet, more powerful than a locomotive, able to leap tall buildings in a single bound." Well, unfortunately, it turned out poor Superman, as played by George Reeves beginning in 1952, was not faster than a speeding bullet.

In fact, George Reeves was one of three Superman actors who all died by unusual circumstances. Actor Christopher Reeve, that's without an "S," played Superman in four movies released between 1978 and 1987, but fell from a horse in 1995, became paralyzed, and eventually died from his injuries in 2004. And remember the infant Kal-el, who Marlon Brando sent to Earth in a space pod in the first Christopher Reeve movie? That baby Superman was played by Lee Quigley, who died in 1991 at age 14 while huffing solvents. You know, that's when kids inhale things like glue or gasoline. Because of these and other coincidences, some have suggested the role of Superman is cursed. I don't know about that, but the original TV Superman, George Reeves, certainly died a mysterious death that continues to haunt forensic history.

George Reeves was born George Brewer in a small Iowa town in January of 1914, although much of his life, he thought his birthday was the following April. That's because his mother fudged his birth date since George was born only five months after she married his father, who she quickly divorced. After moving to California, George's mother married a man named Bessolo, claimed her new husband was her son's father, and gave the boy Bessolo's last name. When she decided to divorce her second husband, the man George thought was his father, rather than tell her son the truth, his mother told George that Bessolo committed suicide.

As a teenager, George began boxing, including in the 1932 Olympics. His mother made him quit, though, so he wouldn't ruin his handsome face, and in 1939, that face made its first major on-screen appearance as one of Scarlet O'Hara's admirers in *Gone with the Wind*; now, by that point Warner Brothers Studios had renamed him George Reeves. In 1943, George was drafted into the World War II Army Air Corps, but rather than a combat assignment, he acted in Broadway Army plays in New York and made training films.

After the war ended, Reeves returned to Hollywood, but after he couldn't land any good roles, he moved back to New York, which I didn't realize was the hub of the early television industry. When Superman, which was already a popular radio series, was proposed as a TV show, Reeves hesitated to audition, since television was a brand-new medium—and who knew it would catch on? But Reeves tried out, and his self-assurance and muscular build won the role. Filming began in the summer of 1951; the first episode aired in September of 1952 and was an instant success. It ran for six years, eventually in 21 countries.

The role of Superman typecast Reeves, and he couldn't seem to get any other meaty roles. So instead, he capitalized on the marketing that followed superhero status. I'm sure some of you out there had Superman lunch boxes, Halloween costumes, or maybe action figures. Reeves seemed to really love the personal appearances, and he made more money from them than his actual salary. In public, he tried to embody the Superman persona; he never smoked in front of kids, and he was involved in several charitable organizations. He kept his personal life as a big partier and drinker private.

Many willing women threw themselves at him, and in the early 1950s Reeves began a longstanding affair with Toni Mannix, who was eight years older than he was, an ex-Ziegfield Follies girl, and the wife of a successful MGM producer. Apparently, Mannix's husband didn't interfere with the relationship because he had his own girlfriend. Reeves and Mannix were involved in charities together and often seen as a couple in public. Mannix was said to be very possessive of Reeves and bought him a house and a car, but the man of steel still had a mind of his own.

In 1958, Reeves met Lenore Lemmon at a New York nightclub. She was a well-known party girl from the city's in-crowd and allegedly the only woman ever thrown out of the Stork Club because of a fistfight. When Reeves took up with Lemmon, Toni Mannix began to harass him with daily phone calls. Although Reeves and Lemmon had a rocky relationship, they planned to marry in the summer of 1959. But that speeding bullet I mentioned earlier took George Reeves' life in the early morning hours of June 16, three days before the wedding, when he was just 45 years old.

The evening of his death, George and Lemmon went out to eat with writer Richard Condon and were overheard arguing during dinner. Then they all went back to Reeve's house, the one Toni Mannix bought him. Condon was working on a book about prizefighter Archie Moore, who Reeves was scheduled to box the very next day. Later, two neighbors showed up at the party. According to the houseguests, Reeves went upstairs to bed in a bad mood around midnight, only to come down later to complain about the noise the rest of the group was making. Reeves had another drink before going back to bed. And later on, Lemmon was quoted as saying, "Oh he'll probably go shoot himself now." About 2:00 in the morning they heard a single gunshot, and one of the neighbors at the party, William Bliss, ran upstairs and found George lying on his back, naked, part-way on the bed, with his feet on the floor, and a nine-millimeter Luger pistol between his feet. There was a single, fatal gunshot wound to his right temple.

What evidence do you think investigators would seek to help establish whether the shooting was a suicide, homicide, or accident? Well, they would definitely start with a thorough examination of the bedroom, collect any evidence there, including the gun; interview anybody in the house at the time of the shooting; and ultimately autopsy Reeves' body. So let's see what was discovered.

For one thing, the houseguests didn't call police until 30 or 40 minutes after Reeves' shooting. Officers reported the witnesses were all really drunk, difficult to interview, and gave no excuse for the delay in calling. Were they buying time, maybe, while they straightened out their stories? A single bullet casing was found beneath Reeves' body, which suggested he was probably sitting on the edge of the bed when shot and then fell backward onto the casing. The bullet that struck Reeves was recovered from the ceiling, but there were two additional bullet holes in the bedroom floor, which Lenore Lemmon claimed were from a few days before when she was, quote, playing with the gun. Interestingly, after telling police Reeves had been depressed, Lemmon left California the very next day; she didn't even attend Reeves' funeral.

Initially, since the witnesses all agreed the death was suicide, Reeves only received what was called a "cursory autopsy" at the funeral home before being embalmed. But Reeves' mother raised suspicions, and because Reeves had reported incessant phone calls to authorities a few months prior, which he attributed to Mannix, but could really only be traced to local pay phones, well, they conducted a more thorough autopsy at the L.A. County Coroner's morgue. Luckily, blood samples had been taken before the embalming, and they revealed a blood alcohol level of 0.27; that's over three times today's U.S. legal driving limit. Toxicology tests also showed painkillers in Reeves's system that had been prescribed following a recent car accident.

The fracture patterns in the skull were said to be consistent with a close shot. And while that type of wound can leave traces of gunpowder around the bullet entrance, there was reportedly no powder on Reeves' temple. In contact wounds to bare skin, though, most of the gunpowder enters the body, leaving little skin residue.

Now, if this case happened today, they'd do residue testing on Reeves' hands, as well as anyone else at the party, to try and verify who was holding the gun when it was fired. But the scanning electron microscope method that we currently use for such tests wasn't available in those days, neither were the types of elemental analyses that could detect lead, antimony, and barium from gunshot residue. What about the gun itself? Well, it was recently oiled, so no clear fingerprints were found anywhere on it. Based on everything investigators discovered, the L.A. County Coroner ruled Reeves' death a suicide.

Helen Bessolo, Reeve's mother, wasn't satisfied; she refused to have her son's body cremated and buried, and hired her own detective to investigate his death. In January of 1960, six months after the shooting, Reeves' body was sent to my hometown of Cincinnati for yet a third autopsy. Although his remains were deteriorating, this autopsy, again, supported a finding of suicide. Reeves's mother finally agreed to cremate his body three years after his death, and shortly before her own.

During the death investigation, rumors surfaced that Mannix had hired a hit man to kill Reeves for abandoning her. Lenore Lemmon was furious when she learned that $50,000 of Reeves's estate was to go to Toni Mannix. A third-hand account, attributed to a neighbor present the night Reeves died, William Bliss, alleges Lemmon instructed the other witnesses to tell police she was downstairs during the shooting, but that she really hadn't been. All those people are now dead, and the facts about what actually happened died with them. As a superhero, George Reeves fought a never-ending battle for truth, justice, and the American way. Though suspicions may still exist, it looks like the official ruling of suicide will stand, whether it's true or not. If his death was not at his own hand, America's Superman didn't get the justice fought for by his character.

Now, if George Reeves tried to be a public angel in the 1950s, another Hollywood icon who made it big in the '60s, was known as a real bad boy. Bob Crane began his career as the "King of the L.A. Airwaves." He was the first DJ to earn over $100,000. But Crane was best known for his starring role as Colonel Hogan on the sitcom *Hogan's Heroes*, ironically set in a World War II German prisoner of war camp. The show ran from 1965 to its cancellation in 1971. From there, Crane's career went downhill. In 1973, he bought, acted in, and toured with the play *Beginner's Luck*, which performed mostly at dinner theaters around the U.S.

Crane's bad-boy status stemmed from his insatiable sexual appetite. He was constantly picking up women, and often enjoyed group sex. Crane frequently played drums in topless bars and was an avid fan of pornography. He would brag about his sexual addiction and exploits to anybody who would listen, even when it made people uncomfortable. Crane's sex-capades and too-public private life may be why he didn't get many roles after *Hogan's Heroes*.

But during the sitcom's heyday, Bob Crane met and became friends with a guy named John Carpenter. Carpenter worked in the video department of Sony Electronics during the time when home-video equipment was first popular. He and Crane began recording Crane's sexual exploits, often without the knowledge of others involved. Crane also had an extensive photo album of his many partners. The jealous boyfriends or husbands of the women involved sometimes caused major problems for Crane. Crane's wife at the time, Patricia Olson Crane, who went by the stage name Sigrid Valdis, wasn't amused by any of this, either. She had played the *Hogan's Heroes* POW camp secretary, Hilda, which is how she met and married Crane. They were in the process of a divorce in 1978, when Crane met his untimely death at age 49 in Scottsdale, Arizona.

On the night he was last seen alive, Crane appeared in his play *Beginner's Luck* at a Scottsdale dinner theater. While signing autographs after the show, he was overheard to remind his co-star, Victoria Berry, about a meeting they had at noon the next day. Later that night, his apartment neighbors heard Crane having a heated phone argument with his wife, who was in Washington at the time. Following that, Crane went to a bar with two women and John Carpenter, who was visiting Scottsdale. At about two in the morning, the four were seen going into a coffee shop. Carpenter left a half hour later and before the other three, since he had to fly back to L.A. later that morning.

Around 2:00 the next afternoon, Victoria Berry came to Crane's apartment to check on him, since he had failed to show up for their noon meeting. The door was closed but not locked, unusual for Crane, and when Berry went into the darkened bedroom, she thought there was a woman with long, dark hair in Crane's bed. Turned out what Berry thought was dark hair was actually streams of blood. Crane was lying in the fetal position on his side with an electric cord wrapped around his neck. Blood was spattered on the adjacent wall.

Investigators determined the scene showed no signs that Crane had struggled during his death. Smears on the bed sheets looked like someone had wiped blood on them, possibly while cleaning off the murder weapon, before then pulling the sheets up over Crane's head. Blood was also found on the inside of the apartment's door. Crane's wallet was there with money inside, so robbery was ruled out as a motive. The only thing known missing was the large photo album of Crane's sexual conquests. A huge collection of pornography was found at the scene, and the bathroom had been turned into a makeshift darkroom, full of photo-developing equipment.

At around 3:15, while Victoria Berry was giving her statement to the police in Crane's apartment, the phone rang, and she was instructed by the cops to answer it. The caller was John Carpenter. The police lieutenant took the phone, identified himself, and told Carpenter police were investigating "an incident that had happened in Crane's apartment." Carpenter said he had been with Crane until 1:00 a.m. that morning before returning to his hotel to leave on his morning flight. Carpenter later revised the time he last saw Crane to 2:30, but never even asked the lieutenant where Crane was or what the investigation was about.

The Maricopa County Medical Examiner's autopsy concluded that Crane was likely sleeping on his right side when the killer delivered two blows to the left side of his head with a heavy, blunt object. The time of death was estimated between 3:00 and 3:30 a.m. The lack of a reaction in the soft tissues showed that the electric cord, which had been cut from a videocassette recorder, had been tightly wrapped around Crane's neck after his death. When asked if the perpetrator was likely a man or a woman, the medical examiner said the killer was probably a strong man, due to the blood spatter pattern analysis.

In these types of assaults, once blood gets applied to a weapon and the blunt object is raised for another blow, the blood gets cast off onto other surfaces, like the walls, or even the ceiling above the victim. The crime scene showed little height to the spatter, suggesting to the medical examiner that the killer was strong enough to deliver a second bone-crushing blow with something heavy without needing to really swing it back and get momentum behind it. In fact, they speculated a camera tripod could have been the weapon, so obviously, things were pointing to John Carpenter, who by the way, some said was bisexual and wanted more than a friendship with Crane. People said Crane was trying to get rid of Carpenter; there were reports of them arguing at a local restaurant just days before Crane's death.

Carpenter's prints would be expected in his friend's apartment, but investigators had another place to look for clues. Carpenter had rented a car during his visit to Scottsdale, so police brought the vehicle in for investigation. On the inside of the passenger door, they found a small amount of blood. Now, 1978 was before DNA profiling was available. The best analysts could do was determine the blood was type B, and while that was Crane's blood type, a little over 15 percent of the world's population is also type B.

Had this crime happened today, DNA testing might have made this an open-and-shut case, but at the time, blood typing just wasn't enough evidence for police to arrest Carpenter. When DNA technology became available in the mid-1980s, it prompted an international reopening of so-called cold cases. And in 1989, investigators tried to get a DNA profile from the blood sample preserved from the rental car, but by then it was too degraded.

Nevertheless, in 1992, Carpenter was arrested for Crane's death, but the Maricopa County Prosecutor just couldn't make the charges stick. This was prompted by a re-examination of photos from the vehicle that revealed a tiny fragment of red matter on the electric window button on the front passenger side. Several forensic experts testified it could be human tissue, in fact, might have been brain matter, but the defense rightly argued that something that small could not be adequately classified from just a photograph. Carpenter was acquitted, but he lived under a cloud of suspicion from Crane's death until his own in 1998.

Okay, we've talked about a superhero and a Hogan's Hero—now, how about a martial-arts hero? And no, I don't mean the 2009 death of David Carradine of the hit 1970s TV show *Kung Fu*, although that was a forensic issue ultimately determined to be accidental autoerotic asphyxia. I want to discuss, instead, the 1973 death of actor and kung-fu legend Bruce Lee, whose first well-known Hollywood role coincidentally intersected the world of superheroes. Lee played Kato, the Green Hornet's sidekick—kung-fu pun intended there—on both the *Green Hornet* and *Batman* TV series of the '60s. But those weren't his first roles. Lee's father was a Cantonese opera and film star whose fame helped placed Lee in 20 films spanning the first 20 years of his life.

Bruce Lee was born Li Jun fan in San Francisco's Chinatown, but grew up in Hong Kong. As Lee's neighborhood went downhill, street gangs sprung up, and young Lee's fighting days began. At age 13, Lee's father started him in formal martial-arts training, but as the gang violence increased, his parents began to fear for Lee's life, and so in 1959, they sent him back to San Francisco to live with an older sister. Lee later gained the attention of a Hollywood producer at a martial arts competition.

In addition to playing Kato in the mid-'60s, Lee choreographed Hollywood fight scenes, but his heavy foreign accent held him back from leading roles. So Lee returned to Hong Kong figuring starring as a martial arts hero in Asian films might ultimately lead to a Hollywood career. He was right. By 1971, Lee had his breakout roles in Hong Kong. The next year, he was given complete control as writer, director, choreographer, and star of the hit *The Way of the Dragon* which includes, arguably, the most famous martial arts fight sequence of all, in which Lee fights his friend and competitor, Chuck Norris. That film got Lee the Hollywood attention he had been seeking.

Little did anyone suspect Lee's next film, *Enter the Dragon*, co-released by Warner Brothers and a Hong Kong company in 1973, would be his last. Bruce Lee was young, apparently healthy, and in top physical shape. He worked out rigorously, was a health-food fanatic, and incorporated traditional Chinese supplements into his diet. Lee pursued a serious mind-body-spirit-type connection. He wrote poetry and philosophical sayings. From all appearances, the man was healthy in both body and mind.

Now, working at the studio on *Enter the Dragon* in Hong Kong in May of 1973, Lee collapsed into a seizure. He was taken to the hospital and diagnosed with cerebral edema, that's swelling of the brain due to excess accumulation of fluid. He was treated with mannitol, which is a sugar alcohol that acts as a diuretic and increases fluid loss through urine.

Lee told his doctor he had used hashish from Nepal the day before and wondered if there was some kind of connection. Hashish is a concentrated resin from the buds of the cannabis plants that marijuana comes from, but hash contains more THC, which is the active ingredient in cannabinoids. Lee's doctor warned him that the purity of the Nepalese hash made it very potent. Lee's wife said the actor didn't use hash often, nor did he show extreme effects when he did.

Bruce Lee recovered, movie production continued, and *Enter the Dragon* was completed successfully. Plans, then, immediately began for his next film, ironically to be titled *Game of Death*. Lee was to co-star with actress Betty Ting Pei, with whom some alleged he was having an affair. On the afternoon of July 20, 1973, Bruce Lee and director Raymond Chow went to Betty Pei's home, where the three reviewed the new script. The director left, and Lee and Pei remained at her place. At some point, Lee told Pei he had a headache, and she gave him a painkiller that she had been prescribed, called Equagesic. At around 7:30 p.m., Bruce Lee allegedly lay down for a nap. It's been speculated that Pei was also in bed with him. When Lee missed a dinner appointment with him, film director Chow returned to Pei's home. Pei told Chow she had tried to wake Lee but couldn't. Lee's physician was called, and when he couldn't wake Lee either, they called an ambulance to take Lee to the hospital, where he was pronounced dead.

An autopsy showed no signs of injury and no blocked or broken blood vessels, but Lee's brain had swelled some 13 percent. Toxicology showed cannabinoids in Lee's blood, in addition to the Equagesic that Pei had given him. Equagesic was a combination of aspirin and the synthetic muscle relaxer and anti-anxiety medication known as meprobamate, which was America's first real blockbuster drug. Developed in the mid-1950s, meprobamate was a tranquilizer particularly popular with the Hollywood set. Some forms of Equagesic also contained opioids, derived from opium poppy plants unrelated to cannabis.

A dispute immediately arose among the physicians involved, including Lee's personal doctor, about whether the Equagesic or cannabis caused his death, especially considering the earlier cerebral edema incident that happened after Lee's hashish use. Famous London pathologist, Dr. Donald Teare, supported the Equagesic theory, stating that, although people can die from THC-related accidents, cannabis never directly causes death; it just doesn't have the same direct toxic effects as many other drugs. The authorities ended up calling Bruce Lee's case "death by misadventure," which is neither a recognized cause nor a manner of death.

Over 30 years later, in 2006, at an American Academy of Forensic Sciences meeting I attended in Seattle, a Chicago pathologist renewed an interest in Lee's death. He presented his belief that a condition first named in 1995 as sudden unexpected death in epilepsy, or SUDEP, may have been the cause of Bruce Lee's untimely demise. His re-examination of the reported autopsy results conclude that the levels of cannabinoids and meprobamate were pretty low in Lee's blood, and that allergic-type reactions to medications typically manifest as swelling of the airways and maybe the face and eyes, not swelling of the brain. Statistics indicate SUDEP deaths are more common in young, adult males, and although Lee had not been diagnosed with epilepsy, he did have that previous seizure that was observed on the set of *Enter the Dragon* at the time of his first episode of cerebral edema.

Sadly reminiscent of the Superman curse, in 1993, 20 years after Bruce Lee's death, his son, fellow martial artist and actor Brandon Lee, was killed at age 28 while filming the movie *The Crow*. A gun that had been used as a prop was loaded with blanks, but unknowingly, a real bullet had been lodged in the barrel and accidentally struck Brandon Lee Despite immediate surgery, he died later that day.

In all these Hollywood cases, autopsy results were essential in trying to establish both the cause and manner of death. Assessing suicide, homicide, accidental and natural deaths requires a thorough examination of tissues, body fluids, potential weapons, scene clues, and witness testimony, when that's possible. But these methods are nothing new in forensics, especially in death investigations involving famous people.

For instance, in 44 B.C.E., Julius Caesar's physician examined the 23 stab wounds on the emperor's body and deduced only one of them, a specific knife wound to his chest, was actually fatal. And by around the 5th century C.E., medical practitioners were fairly commonly used in Europe to help establish cause of death, expanding the practice beyond just the rich and famous.

The first known book specifically about forensic medicine was written in China in the mid-13th century; its title is often translated as *Washing Away of Wrongs*. This 53-chapter, case-based approach includes the often-cited story of a murder committed with a large blade. The investigator tested different kinds of blades on dead animals to try and determine which implement the killer used. Once he settled on a sickle as the most likely weapon, an official had all the local people bring in their sickles. When flies were attracted to just one man's sickle, the investigator chose him for questioning, and the killer confessed.

Opening and examining a body specifically for legal purposes was first reported in the early 14th century in Bologna, Italy, following the murder of a nobleman, and as the Middle Ages closed, using autopsy to analyze suspicious deaths became far more standard in Europe. Unlike today's forensic pathology specialists, tough, often times a general physician who performed the autopsy. Toxicology studies were added to the standard death investigation later as by the mid-1700s, the effects of poisons were being intensively studied in Europe—in particular arsenic, which by this time had such a history of being used for murder it was called inheritance powder.

There have been countless high-tech refinements related to autopsy and toxicology since then, but death investigation definitely gets my vote as the oldest field in forensics. Early autopsy pioneers couldn't even imagine today's computerized chemical analyzers or scanning electron microscopes. And evidence familiar to them, like blood, carries so much more power now than it did even just a handful of decades ago when our Hollywood deaths took place. I have no idea where the future of forensic science might lead us, but it's easy to recognize how current technology would have aided justice in the past and maybe avoided some of the mystery surrounding some historical deaths.

Decomposition and Confusing Interpretations
Lecture 4

In July 1933, 7-year-old Dalbert Aposhian and his 9-year-old friend Jackie Confar were playing near downtown San Diego, where both of their parents worked. Later that day, when something terrible happened, Jackie panicked and ran home alone, afraid to tell his parents what he'd seen. Dalbert's body was found six days later, floating in San Diego Bay. The coroner determined that Dalbert had been murdered, but no killer was ever found, and a later examination disputed the original findings. Fortunately, most records and images related to the case were maintained, enabling insights from new research and technology to be applied more than 70 years later.

The Disappearance of Dalbert Aposhian

- Dalbert Aposhian's parents—a Protestant minister and his wife—were worried when their son did not return home one evening in July 1933. They reported the boy's disappearance to authorities. Police questioned Dalbert's friend Jackie Confar, who nervously reported that the pair had been playing together early in the day but that he had later left Dalbert at a local store.

- Six days later, four sailors found Dalbert's body floating in the waters of San Diego Bay. Dr. Frank Toomey of the San Diego Coroner's Office released the results of the boy's autopsy: Dalbert had been sodomized, murdered, and mutilated. Toomey said that no water was found in the lungs, which meant that the boy could not have drowned but was dumped in the bay. He also reported traces of semen in the boy's distended rectum.

- The press ran with the story, and the community was up in arms, insisting that police find the vicious killer before he struck again. Reporters—not the police—identified Dalbert as the sixth victim in a string of crimes they were trying to link to a suspected sexual predator believed to be operating in the San Diego area. There was at least one false confession.

- In the midst of all the publicity, police questioned Jackie again, but now, the boy changed his story. He told them that the boys had sneaked onto the San Diego pier to try to fish, but Dalbert slipped and fell into the water. Jackie said he had been too afraid to tell anyone what really happened because the boys had gone beyond their designated play area. Some of the investigators believed this version of the story, but Toomey stuck by his autopsy findings.

- Two years later, when the crime was still unsolved, a Los Angeles pathologist performed a second autopsy on Dalbert's body. This pathologist disputed the original findings and believed the cause of death to be drowning.

- More than 70 years later, the San Diego Police Department received a grant to open cold cases using DNA technology. Although there was no DNA left from the Aposhian case, the San Diego medical examiner, Dr. Jonathan Lucas, read the conflicting autopsy reports and examined numerous photos.
    - Lucas determined the death to be the result of drowning and the mutilation of the boy's body to have been caused by marine life. He also noted that it was highly unlikely that sperm could survive in open water for six days; no one knows why the original autopsy report concluded that there was semen in the rectum.

    - As a result of this reexamination, Dalbert Aposhian's case was officially closed as an accidental death in December 2005.

Studying Decomposition

- The San Diego Medical Examiner's Office claims that it never discards anything from open homicide cases, which means that new research and technology can be applied to old cases, sometimes resulting in fresh conclusions.

- For instance, through modern studies, we now know much more about the decomposition process and the effects of animal activity than was understood or documented at the time of Dalbert's death. That knowledge is what allowed Dr. Lucas to reach a new conclusion.

- Surprisingly, even in the 1930s, it wasn't completely unheard of to use experimentation to assess what might have happened in a forensic case. At the time of the second autopsy, two years after Dalbert's death, investigators conducted an experiment using rabbits to attempt to show that sperm could not survive in water, as Toomey had first reported. Their results documented the unlikelihood of finding semen after six days, but apparently, no one paid attention to this conclusion.

- Since that time, there has been much groundbreaking—and sometimes controversial—research into the science known as *taphonomy*, that is, the study of what happens to living organisms once they die and are deposited in the environment. This field studies everything from the effects of marine life to those of climate variables on bodies.
    - The Anthropology Research Facility at the University of Tennessee (known as the Body Farm), which started in the 1970s, was the prototype for formal taphonomic studies on humans, but today, several research facilities devoted to human decomposition now exist around the United States.
    - Donated bodies are subjected to a number of

Conducting experiments and documenting the results are the cornerstones of scientific understanding, and today's forensic community takes its role in that process seriously.

settings in different parts of the country to enable new insights into the science of decay. In addition, many experiments use pig models to study decomposition because pigs have relatively hairless skin, like humans.

- Taphonomic studies often seem disturbing to members of the general public, but this kind of research can help demonstrate that initial appearances may be deceiving. For example, studies have shown that insect activity can have effects not only on carcasses but also on associated artifacts, such as clothing.

Case Study: A Modern-Day Mummification

- Two women—a mother, Susan, in her 80s and a daughter, Ann, in her 60s—lived together in an apartment in a rundown area near downtown Cincinnati. The landlord spoke with Ann when she brought the monthly rent check, but he never saw Susan. Over time, the landlord became suspicious and finally told Ann that he would accept her next rent check only if her mother delivered it in person.

- The next month, Ann pretended to be her mother and walked into the landlord's office with the rent check. The man accepted the payment, but as soon as Ann left, he called the police and told them his suspicions.

- Officers went to the apartment where the pair lived to question Ann and ask if they could conduct a search. The apartment was horrendously filthy, suggesting a hoarding situation, but there were no signs of foul play. Ann told the officers that her mother was visiting relatives in Detroit.

- Back at the police station, officers contacted family members in Detroit, who said they hadn't seen Susan for four years. Investigators returned to the apartment that same afternoon and noted that everything was exactly as it had been that morning with one exception: A large, freestanding metal cabinet that had been in the kitchen was now thrown over a ravine in the backyard. The officers transported the cabinet and the cloth-covered bundle inside it to the coroner's office.

- The officers asked Ann if the cabinet contained the remains of her mother, and she admitted that it did.
    - Ann said that Susan had fallen down the stairs and was badly hurt. She claimed that she had helped her mother into bed and attempted to nurse her back to health but with no success. Eventually, Ann bathed her mother's body with bleach, wrapped the body in bedclothes, and stored it in the cabinet.

    - When police asked why Ann hadn't reported the death or buried her mother, the woman said that they were descendants of the Blackfoot tribe, and burial was not their custom.

- The first step in the examination at the morgue—even before unwrapping the remains—was to X-ray the bundle. The X-rays revealed a number of small, round metallic objects scattered through the chest and neck area, with some embedded in the victim's skull—possibly buckshot or birdshot.

- The next step was to unwrap the remains and perform an examination, with two main objectives in mind: The pathologist's job was to try to establish the cause or manner of death, while the role of the anthropologist was to look at any anatomical features that might help scientifically establish the person's identity.

- Removing the bedclothes revealed a mummified corpse. All the soft tissues were completely desiccated. The wrapping, combined with the relatively constant environment inside the cabinet—and, perhaps, something about the bleach—had caused a modern-day case of mummification. Such atypical decomposition is generally a product of circumstances in which the temperature is relatively constant and insects and other organisms have limited access to the remains.

- Both the skeletal features and external anatomy suggested that the body was that of a woman. The jaws lacked teeth, a finding consistent with the identity of Susan. There seemed to be no entrance or exit wounds on the surface of the body. The metallic pellets embedded in the skull were found to have an unusual consistency, almost like putty and unlike birdshot made of lead or other compounds.

- When the police asked Ann whether her mother had been shot, she replied that Susan had—back in 1926 in Bullitt County, Kentucky, in connection with a feud among neighbors over the illegal distilling of alcohol. Investigators confirmed Ann's story through a 1926 newspaper account of a shooting involving Susan's family and their neighbors.

- The authorities decided not to charge Ann with desecration of a corpse because she didn't seem to be of sound mind in many ways. But she was charged with Social Security fraud because she had been cashing her mother's government checks for the four years she had hidden the body.

Case Study: Deceiving Appearances

- A couple came home from vacation to find a hole of about a foot in diameter in the middle of the deck in their backyard. The homeowners called the local authorities, originally thinking that perhaps a meteorite had crashed through the deck.

- Investigators found that the object that had made the hole appeared not to be a solid mass but a plastic bag. As a result of the impact, the bag had broken open, and around it, a powdery substance was slightly scattered under the deck. Clearly, the bag had to have fallen from a great height to cause such damage to the deck.

- A physical anthropologist later determined that the bag was filled with human cremated remains and had been dropped from a small airplane. The remains were supposed to be scattered to the wind during flight, but the bag had been accidentally let go before it was opened.

- The police eventually tracked down the family involved, who recovered the ashes of their loved one and had a second chance to scatter them to the wind.

Suggested Reading

Bass and Jefferson, *Death's Acre.*

Murray, *Death.*

———, *Forensic Identification.*

Rathbun and Buikstra, *Human Identification.*

Spitz and Spitz, *Spitz and Fisher's Medicolegal Investigation of Death.*

Steadman, *Hard Evidence.*

Questions to Consider

1. What types of evidence are used to estimate time since death?

2. What are some of the causes and variables of the decomposition process?

3. What is your opinion of the use of human cadavers in decomposition studies throughout the United States and elsewhere?

Decomposition and Confusing Interpretations
Lecture 4—Transcript

Seven-year-old Dalbert Aposhian and his nine-year-old friend, Jackie Confar, were playing together in July of 1933 near the downtown San Diego area, where both of their parents worked. Later that day, when something terrible happened, Jackie panicked and ran home alone, afraid to tell his parents or anybody what he'd seen.

That evening, Dalbert's parents—a Protestant minister and his wife, who both supplemented the family income by working at a San Diego dry cleaner—well, they were surprised when their son didn't come home. They reported their boy's disappearance to the authorities. Police questioned Dalbert's buddy, Jackie, who nervously told them the pair were playing together early that day, but that he left Dalbert at a local store and then went on to play without him at a nearby park.

Six days later, four sailors on the Navy's destroyer ship, USS Dobbin, found Dalbert Aposhian's body floating in the waters of the San Diego Bay. Dr. Frank Toomey of the San Diego Coroner's Office released the grisly results of the boy's autopsy. Toomey concluded that Dalbert had been sodomized, murdered, and mutilated. He reported the boy's lips, eyes, fingers, and genitals had been removed. Toomey believed the implement used was a knife, in fact, the kind of knife typically issued to sailors. The physician said no water was found in the lungs, so the boy couldn't have drowned, but rather was dumped in the Bay. He also reported traces of semen in the boy's distended rectum.

The press ran with the story, calling the perpetrator "a maniacal killer" that had to be stopped. The community was up in arms, insisting that the police find the vicious sex fiend before he struck again. One local women's civic group wrote a letter to the Sheriff's Office demanding they clear the city of its "underworld gangsters, prostitutes, immoral dance halls, burlesque theatres, burglars and degenerates." They set forth that "all known degenerates, feeble-minded, congenital perverted, and all irresponsible criminals should henceforth be kept in humane but permanent confinement."

The media, not police, dubbed Dalbert's case as the sixth victim in a string of crimes they were trying to link to a suspected sexual predator who was believed to be operating in the San Diego area. By the way, the first victim in that series was a 13-year-old girl who was killed a couple of years earlier; her case still remains the oldest open cold case in the San Diego Police Department's files. And in the Dalbert Aposhian case, like so many other high-profile incidents, like the Black Dahlia murder and the killing of young JonBenet Ramsey, there was at least one false confession, that of a mentally-imbalanced 19-year-old man who apparently copped to Aposhian's death to impress his girlfriend, but later recanted his confession.

In the midst of all the publicity, police again returned to the last person known to have seen Dalbert alive, his friend Jackie Confar, but now Jackie changed his story. He told them the boys were playing that day in the downtown area, near where their parents worked, as they always did with permission, but, at one point that day, the two boys decided to sneak out onto the San Diego pier to try to fish, using some discarded materials they found. Jackie said Dalbert slipped and fell into the water off the pier, but that he had been too afraid to tell anybody what really happened, because the boys had gone beyond their designated play area.

Some of the investigators, especially those in the San Diego Sheriff's Office, tended to believe this version of Jackie's story, but still, Toomey stuck to his autopsy findings. Unlike what's often shown in today's TV shows, the medical examiner does not usually interrogate witnesses, but Toomey insisted on interviewing nine-year-old Jackie Confar himself, in the hopes of extracting the full truth, or at least getting some new leads.

Two years later, in 1935, when the crime was still unsolved, a Los Angeles pathologist performed a second autopsy on the body of Dalbert Aposhian, which I suppose had to have been kept in cold storage. This pathologist disputed the original findings and believed the cause of death was drowning. But Dr. Toomey and numerous members of the San Diego Police, as well as the local community, they refused to believe that second opinion. Dalbert's anguished parents, likewise, thought their son had been brutally murdered and mutilated, until they died many years after their boy's untimely end.

Over 70 years later, the San Diego Police Department received a grant to open cold cases using DNA technology. There was no DNA left from the Aposhian case, since whatever tissue or semen samples existed were long since discarded, but the San Diego Medical Examiner, Dr. Jonathan Lucas, wanted to reopen Dalbert's case anyway. After reading the two conflicting autopsy reports and looking at lots of photographic evidence, Lucas determined the death to be the result of drowning and the mutilation to the boy's body simply to have been caused by marine life. He called it "A classic case of crustacean and fish activity." The body parts that were missing—lips, eyes, fingers, genitals—are all areas either close to body orifices or that project out from the body and are readily and easily consumed by smaller forms of sea life.

Lucas also stated it's now well-known that not all drowning victims have water in their lungs, particularly if they lose consciousness and don't struggle, such as when hitting the water from a height or from breath-holding. Lucas also concluded it was highly unlikely that sperm could survive in open water for the six days the boy's body was in the bay, so no one really knows how the original autopsy report concluded that there was semen present. Lucas said in many cases of submersion, body orifices dilate as muscles relax, making it appear as though the victim may have been sexually violated. And as a result of this re-examination of the evidence, in December of 2005, Dalbert Aposhian's case was officially closed as an accidental death.

Fortunately for this case, most records and images were maintained, which is definitely not always the case in too many other coroner's offices, including some agencies with which I've worked. In fact, the San Diego Medical Examiner's Office claims they never discard anything from open homicide cases. If I had one wish for the forensic and medicolegal community, it would be that no records, no tissue samples, no X-rays, no photographs, or any other evidence ever be discarded for any reason. The Dalbert Aposhian case, like many others we'll examine in this series, shows that if data are kept, insights from cutting-edge research or new technology can sometimes be applied to old cases and result in fresh conclusions.

For instance, through modern studies, we now know much more about the decomposition process and the effects of animal activity than was understood or documented back then, and that's one of the things that allowed Dr. Lucas to come to a new conclusion in the Aposhian case. Conducting experiments and documenting the results are really the cornerstones of scientific understanding, and today's forensic community takes its role in the advancement of our field very seriously. But to my surprise, even back in the 1930s, using experimentation to try to solve a forensic case wasn't completely unheard of.

At the time of his second autopsy, just two years after young Dalbert Aposhian's death, investigators conducted an experiment using an animal model, in this case, rabbits, to try to simulate the conditions and attempt to show that sperm could not have survived in that water, as Toomey had first reported. Now this sounds a little disturbing, but researchers inserted semen into the rabbits' rectums, wrapped the animals in corduroy, which is what Dalbert had been wearing, then placed the rabbits in crab cages, and submerged them in water. Their results documented the unlikelihood of finding semen after six days, but no one at the San Diego Coroner's Office, or the press, apparently paid any attention to the conclusions of this unconventional experiment in the 1930s.

Now, since that time, there's been much more groundbreaking, and sometimes also controversial, research into the science known as *taphonomy*. Taphonomy is the study of what happens to living organisms once they die and are deposited in the environment. That includes everything from studies on the effects of marine life on bodies, as ultimately helped solve the Aposhian case, to the effects of different climate variables on bodies deposited on land, and even all the way to how fossilization gives us dinosaur bones.

The Anthropology Research Facility, or so-called "Body Farm" at the University of Tennessee, is a human decomposition research facility. It was started in the 1970s and was the prototype for formal taphonomic studies of humans. Today, several other research facilities are devoted to human decomposition, and they exist around the United States. They include Western Carolina University, Texas State University at San Marcos, Sam Houston State University in Texas, Colorado Mesa University, and a few others, with more such facilities planned for the future.

Donated bodies are subjected to a number of settings in different parts of the country so that forensic anthropologists and other investigators can gain new insights into the science of decay. In addition to those, many other experiments use pig models to study decomposition, since pigs have a relatively hairless skin, like humans. In fact, at an American Academy of Forensic Sciences meeting I recently attended, I saw a presentation about research undertaken in 2012 to document and intensively examine what happens to remains in underwater environments. This study was conducted in a bay off British Columbia, Canada. Researchers put pigs on pallets covered by wire cages and set up time-lapse cameras to record the types and actions of the marine life, the organism that colonized and consumed the pig bodies. Incidentally, the presenter told us the wire cages were actually added to the project at the last minute once she discovered that sharks demolished the first pig within a matter of minutes.

The most amazing part of this presentation was that they showed the accelerated videotape of the experiment, which incidentally, I later found posted on the Internet. In this video, they capture eight days of activity and condense it into about just six minutes. By the end of the first day, critters called sea lice pretty much covered the pig, and some shrimp and crabs occasionally visited too. The sea lice remain as the main feeders, but by the end of the fourth day, it's clear they're no longer interested in the remains, so they move to colonize the wire cage for a while before disappearing on the fifth day.

At times in the video, the water is almost too murky to see what types of organisms are actually feeding, but by day six, the body is pretty much completely reduced to bones. A cool surprise was that a large octopus, quite a lot bigger than the original pig itself, came in on day eight and hugged onto the cage for a while to check out what was in it.

Now, this kind of thing might seem disturbing to some members of the general public, but let me give you just one important reason why this kind of research is crucial to forensic science, and how, like in the Aposhian case, it helps demonstrate that things aren't always what they seem. I know a woman named Pat Peden who conducted decomposition research studies using pigs of approximately the same weight as adult humans. Now, she wanted to examine the role that insects and other animals may have in disturbing the clothing on deceased individuals. Behind the scenes, many of us referred to her study as Patty's Pigs in Pink Panties.

Now, essentially, what Pat did was dress the pigs in ladies clothing, including bras, panties, and pantyhose, and then she put the carcasses out to decompose in the environment. What she discovered was that the insect larvae, particularly fly maggots, could get into such a rolling and writhing mass, they could literally peel the pantyhose and underpants about halfway down the thighs of the pigs.

You can easily figure out why this kind of discovery is so important to investigators. If a badly decomposing body is found with the undergarments pulled down well below the waist, the natural assumption would be there had been a sexual assault. But Pat's studies, as well as others conducted by other forensic scientists since then, have better documented insect activity and the effects it has not only on the body, but also on artifacts associated with a victim.

Now I'd like to tell you about one of my own cases. It happened in 1990, but still stands out to me as one of the most unusual in my almost 30 years of practice. It involves atypical decomposition and some pretty unexpected evidence. But I'm going to relay the story as it unfolded, rather than how I got involved in it. Two women, a mother and a daughter, were living in a very run-down area near downtown Cincinnati. The mother, Susan, was near 80; her daughter, Ann, was around 60, and the two lived together in an apartment.

Every month Ann would bring rent checks to the landlord, and he would ask how her mother was doing. Ann would always reply something like, "Oh, mom's fine." Occasionally, the landlord would go to the apartment where the ladies lived to fix something, or whatever, and he would ask Ann where her mom was that day. Ann's reply was always something like, "Mom's not home right now; she's out." As time went on, the landlord got increasingly suspicious, and he finally told Ann he would not accept another rent check signed by her mother unless Susan delivered it herself.

So, no kidding, the next month, in Norman-Bates fashion, Ann actually dressed like her mother, purported herself to be Susan, and walked into the landlord's office with the rent check. The man accepted payment, but as soon as Ann left, he contacted the local police and told them the unbelievable story. So that same morning, officers went to the apartment where the two women lived to question Ann and also ask to search the house. I didn't go to the scene, but I heard the condition of the place was pretty horrendous; it was one of those hoarder-type situations.

There were a couple-dozen cats living in the house and no litter box in sight, and the roaches were said to be so thick on the walls, it looked like the wallpaper pattern was moving. Now, other than being generally foul, there were no signs of foul play. Officers asked Ann where her mother was, and when she replied her mom was visiting relatives in Detroit, they asked for the contact information where Susan was staying.

Back at the police station, officers got in touch with the family in Detroit who said they hadn't seen Susan for four years. Investigators returned to the ladies' apartment that same afternoon. When they got there, everything was exactly as it had been that morning, with one exception; this large, freestanding metal locker, you know, the kind of thing you might keep in your garage to hold brooms or cleaning supplies. Well, that cabinet, which had been in the kitchen, of all places, that morning was now thrown down a ravine in the backyard. They opened the locker and saw a large cloth-covered bundle inside of it. Based on their suspicions, police transported the locker and its contents back to the coroner's office. That's when I got involved.

Now, before I talk about the contents of the bundle, which I'm sure you've already guessed at, I need to mention that the officers did ask Ann if the cabinet did, in fact, contain the remains of her mother, and Ann admitted it did. When they asked what happened, Ann replied that her mother had fallen down the stairs and was badly hurt. She claimed she helped her mom into bed and attempted to nurse her back to health. Ann tried to get her mother to eat or drink, but with no success.

Eventually her mother began to smell, so Ann said she bathed her mom's body with chlorine bleach; that was what they used around the house when so ever-infrequently they cleaned anything. When Ann finally realized her mother had died, she wrapped the body in the bedclothes, which consisted of a sheet and thin bedspread, and she put Susan's body in that locker in the kitchen. When police asked why Ann didn't report the death or bury her mother, she said they were descendants of the Blackfoot Indian tribe, and burial was not their custom.

Now, I studied Native American burial practices in graduate school as an anthropology major. And while the Blackfoot did typically wrap the body, there was no mention of bleach. They sometimes put the remains in a lodge, or up on a hilltop, or maybe on a platform in a tree. But it definitely was not a common Blackfoot practice to use a storage locker in the kitchen to inter the dead.

So what do you think would be the steps we'd use in examining that bundle at the morgue, and establishing that the body was that of Ann's missing mother, let alone figure out what happened? Well, the first thing we did, even before unwrapping the remains, was to run the whole bundle through the X-ray machine. That allows a radiographic view of the body before disturbing anything, and, any dense, radiopaque objects, whether part of the body or not, will show up on X-ray.

Well, when we looked at the films of the body, which was kind of curled up in a fetal position, we saw quite a few small, round metallic objects scattered throughout the chest and neck area, and even some embedded in the victim's skull. At that point, I looked at the pathologist and I said, "What do you think? Buckshot?" He said, "Naw, it's too small; I think it's birdshot." And we started to speculate that maybe mom might have had a little help falling down those stairs at home.

Now, the next step was to unwrap the remains and perform an examination with two main objectives. The pathologist's job was to try to establish the cause or manner of death, as best he could, in other words, how and why the victim died. While my role as an anthropologist was to look at any anatomical features that might help us figure out who this person was. Of course, we had what we call a presumptive ID, in that we all suspected this was the body of Susan, Ann's missing mother. But I needed to use skeletal features to assess the sex, the age, ancestry, height, and any other characteristics that would help substantiate, in a scientific way, that the remains in the footlocker really belonged the missing woman.

When we removed the bedclothes, we discovered a completely mummified corpse, or as I sometimes like to say, we found Mommy the Mummy. Now, all of the soft tissues, and I guess I should put "soft" in quotation marks here, since they were as dry as shoe leather. Well, those soft tissues were completely desiccated. We figured out that the wrapping, combined with the relatively constant environment within the cabinet, and maybe even something about the bleach, caused a modern-day case of mummification.

And that might surprise you, but in my nearly 30 years of practice, I've actually encountered probably a dozen cases where atypical decomposition resulted in mummification. It's generally a product of circumstances where there are relatively constant temperatures, like we'd find inside a house, and limited access to insects and other organisms that would normally consume the remains. The wrappings and the locker apparently provided such an environment. And not to be disrespectful, but the analogy I use with my students is, "mummy is to human, what raisin is to grape."

Given the dried-up skin and muscle, we had a difficult time looking at the body to see if there were wounds on the surface. We found no openings to indicate either entrance or exit wounds, and were left wondering maybe if Susan lingered long enough while in her daughter's care for some significant healing to have occurred before she eventually died.

Now, in terms of assessing the biological profile, in other words, who this person was, anatomically, both the skeletal features and external anatomy suggest that the body was that of a woman. The victim was also edentulous, meaning she had no teeth remaining, and Susan was known to have no teeth. Next I had to peel the dried tissue from the skull in order to more thoroughly examine the cranium and measure it. The skull measurements were used in mathematical equations, and that corroborated she was likely female, and interestingly, it suggested a mix of Native American and Caucasian ancestry. So, although none of those things positively proved the woman's identity, they were all consistent with Susan, Ann's missing mother.

Also when we removed the tissue from the skull, we were able to see several of the small, metallic pellets; some were partially embedded in the bone. The pathologist removed one of them so we could better examine it, and we found it to be the consistency of almost like putty. Now, lead is a very soft metal, and it takes little heat to work it, but shot doesn't squish like putty. Birdshot, in particular, is increasingly being made of other compounds, like steel or tungsten, which are not only more durable, but also less toxic to us and to the environment. After all, a bird that doesn't die after being shot, might wind up elsewhere in the food chain, or a human hunter may accidentally consume toxic lead. Anyway, it was very curious that the shot pellets, embedded in the surface of the skull, were so soft.

Okay, let's get back to the investigative end of the story. While we were examining Susan's body, law enforcement officers had returned to further question her daughter, Ann. We had already told them our findings, so the first question they asked Ann was, "Was your mother shot?" And Ann, who had clearly been a bit scrambled throughout this entire investigation, and before, became completely lucid and told the officers that yes, her mother had been shot, back in 1926, in of all places, Bullitt County, Kentucky. How appropriate is that!

Now, if you remember your American history, 1926 was during what was known as the Prohibition Era. That was the time from 1920 to 1933 when the manufacture and sale of drinking alcohol was illegal in the United States. Now, the way the story was told to me was that in 1926, Susan, who was at that time a young girl, lived in a one-room cabin with her brother and their mother, in other words, Ann's grandma. Well, grandma found out the "neighbors was runnin' a moonshine still, and makin' corn whiskey," so she called the Revenuers. Now, that was the general term used for the authorities who had once been responsible for taxing alcohol, but were later charged with enforcing its prohibition. Of course, the neighbors didn't take too kindly to this situation, and they decided to retaliate.

So, one night, in typical Hatfield and McCoy fashion, the neighbors set fire to the one-room cabin the family lived in. And when Susan, her brother, and their mother ran out to escape the flames, the neighbors pelted them with birdshot. Grandma's injuries were so serious that she died, and as a result,

Susan and her brother were sent to be raised by other relatives, who lived near the big city of Cincinnati. So Ann clearly knew the history of how her mother had come from Bullitt County to grow up near Cincinnati up north, and, how her mom had come to be full of shot pellets. In fact, Ann told investigators that she would wash her mother's hair and Susan would say, "Be careful of my bumps now; you know it hurts me when you press on my bumps."

OK, should the authorities have just taken Ann's word for all this? Well, they can't, so to try and confirm the story, they made contact with Bullitt County, and although I've not seen the article, I was told that the front page of an issue of a 1926 Bullitt County Gazette told the whole story of the shootout at the cabin, including granny's death and all. And not only was Susan full of birdshot, but she had a brother who was living in a nursing home in Lexington in 1990, and he was also full of birdshot too.

So, in this case, there really was no bad guy, at least not in the present. The authorities decided not to charge Ann with desecration of a corpse, or anything along those lines, since it was clear she really wasn't of sound mind in many ways. But, they did end up charging her with social security fraud, because she was cashing her mother's government checks, right along with her own, for the whole four years she had hidden Susan's body. Now, that may have been the real reason Ann never reported that her mother had died. Now, that was definitely a forensic case with quite a history!

I'd like to close this lecture with one more forensic matter where things were not what they initially seemed. A friend and colleague of mine, Dr. John Williams, presented this case at one of the American Academy of Forensic Sciences meetings, maybe it was about a decade back. A couple came home from vacation to find a hole, about a foot in diameter, in the middle of their wooden deck in their backyard. The homeowners called the local authorities, and I think they got someone from the local university or museum involved, too, because the original thought was that a meteorite had crashed through the deck.

Now, the height of the deck was just a couple of feet off the ground, and it was dark and hard to see underneath it. But the investigators were able to determine that what crashed through the deck was not a solid mass, but rather a plastic bag. Now, how could a plastic bag have possibly busted through a wooden deck? Well, the bag had broken open during impact, and in and around it was a powdery substance that was scattered under the deck. Clearly, the bag had to have fallen from a pretty great height in order to damage the deck like that.

Any idea what happened or what the powdery substance was? Consider that they eventually called in my friend, who is a forensic anthropologist; that might help you figure it out, if you haven't already. The bag was filled with cremated human remains that had been dropped from a small airplane. The cremains were supposed to be scattered in the wind during flight, but the bag was accidentally let go before it was opened and literally, pardon the pun, hit the deck. The police got in touch with the local airport, and they were able to find the pilot involved in the incident, and eventually, the identity of the person whose ashes were supposed to have been scattered to the wind. The family recovered what was left of their loved one's cremains and were then able to give him a second chance to fly away.

Lizzie Borden and the Menendez Brothers
Lecture 5

Some sources say that a newspaper peddler created the famous "Lizzie Borden took an axe" rhyme to try to generate sales, but no one really knows for sure. Still, this case from the late 1800s is a good example of how a relatively unknown person can become infamous—and plead not guilty at trial—despite a mountain of evidence. In this lecture, we'll look at the Lizzie Borden case in detail, as well as a more modern-day family feud—the case of Lyle and Erik Menendez. The two cases—more than 100 years apart—share some striking similarities.

Background on the Bordens

- Lizzie Borden was born on July 19, 1860, in Fall River, Massachusetts, to Andrew Borden and Sarah Morse Borden. She had a sister, Emma, who was about 10 years older. When Lizzie was just 2, her mother died, and in 1865, her father remarried a woman named Abby Durfee Gray, who was already well into her 30s.

- Andrew was a successful entrepreneur, bank president, and landlord. He was admired for his business sense but not very well liked, though he was said to treat his wife and daughters well. Despite his wealth, Borden was known to be tight with money.

- Both Lizzie and Emma were well-respected spinsters, and despite not getting along with their stepmother, they both still lived at home. Lizzie was a Sunday school teacher and was involved in the women's temperance movement. She was said to envy her richer relations and friends.

- A burglary took place in broad daylight at the Borden house in 1891, when only the two sisters and their live-in maid, Bridget Sullivan, were at home. The only things taken were about $50 and some jewelry. The police were called, but a couple of weeks after the incident, for unknown reasons, Mr. Borden asked to have the investigation stopped. The family may have suspected Lizzie as the burglar.

- Lizzie didn't have much of a social life outside of church, but she was an animal lover and kept pigeons in the loft of the family's barn. Sometime in May or June of 1892, Mr. Borden got upset by Lizzie's hobby, saying that it drew neighborhood boys who were shooting at the birds. He beheaded Lizzie's pigeons with an axe. Things became so tense in the household that by July, Emma and Lizzie went on a week's vacation to New Bedford.

The Murders

- In early August of 1892, Mrs. Borden went across street to visit a neighbor, Dr. Seabury Bowen, saying that the family and their maid had been sick for several days. Mrs. Borden feared that they might have been poisoned, but the doctor believed it was probably something they'd eaten. Lizzie told her friend Alice Russell that she thought someone—perhaps an enemy of her father—had poisoned their milk; Lizzie claimed to have seen a strange man hanging around the house and barn.

- The next day was August 4, 1892, and while Emma was away at a friend's house party, 32-year-old Lizzie screamed for Bridget to get the doctor, saying that her father had been killed. Although there was no sign of a struggle, Mr. Borden's body lay on the sofa in the family's sitting room; his face had been smashed in by a sharp weapon. A short time later, Mrs. Borden was found dead upstairs.

- Autopsies were conducted at the Borden home at 3:00 that afternoon. Based on the victims' body temperatures and the condition of the blood at the scene, the doctor thought that Mrs. Borden had been killed between 9:00 and 9:30 a.m. Mr. Borden's face had been struck with a large, sharp implement—possibly an axe—from someone probably standing above and behind him. Many blows had crushed the back of Mrs. Borden's skull.

The Investigation

- Lizzie and Bridget had been the only ones home at the time of the murders, and the two were questioned for most of the day. Their stories didn't exactly agree, but a composite timeline could be constructed.
  - Mr. Borden had left the house earlier and returned about 10:45 a.m. He asked Lizzie where his wife was, and she replied that her stepmother had gone to visit a sick friend. At about 10:55, Mr. Borden told Bridget that he was going to rest in the sitting room, and the maid went up to her attic room, as well.
  - Lizzie said that she had gone out to the barn and returned to the house around 11:10. She saw her father's bloody body on the couch and called for Bridget to go to Dr. Bowen's house. He arrived at 11:30. The police were called at 11:15 and arrived by 11:45.
- A search of the Borden property is said to have turned up two axes. Also found was a claw-hammer hatchet, a regular hatchet, and the head of a standard hatchet with its handle broken off. No bloody clothing was found at the scene, except the clothes on the victims.
- The funeral was held two days after the Borden's deaths, but investigators delayed the burial because they wanted a more thorough autopsy. The heads of the couple were removed and defleshed, just as they would be in an examination today. The medical examiner, Dr. Dolan, estimated that Mr. Borden's face had been struck 10 times with a hatchet (smaller than an axe), while Mrs. Borden's skull had sustained about 18 blows.
- On August 7, three days after the murders, Lizzie's friend Alice Russell saw Lizzie burning a blue dress in the kitchen stove. Lizzie claimed that the dress had a paint stain on it. Two days later, a judge held a closed inquest and ordered Lizzie, Bridget, and a few others to answer questions. Bridget told the magistrate that Lizzie had been wearing a blue dress on the morning of the murders, but she was wearing something else when police arrived.

- During her four-hour interview, Lizzie's story seemed suspect on a number of counts, and she was arrested two days after the inquest. At her arraignment hearing, she was charged with the double murder, to which she pled not guilty. At a grand jury hearing in November, it looked as if Lizzie might not be indicted until Alice Russell testified about the dress burning. Lizzie was jailed from arrest to trial, which took place in June of 1893 in New Bedford.

The Trial

- Lizzie had some support from members of her church, as well as an excellent defense team. The defense managed to have Lizzie's statements at the initial inquest ruled inadmissible because her lawyer had not been present. Further, the defense convinced the panel of three judges to exclude the testimony of witnesses that Lizzie had attempted to buy prussic acid (hydrogen cyanide) at a pharmacy the day before the murders.

- The prosecution emphasized motive—the inheritance Lizzie would receive, combined with her hatred of her stepmother—and opportunity. There were no prime suspects other than Lizzie, but the bulk of the prosecution's case was built on circumstantial evidence. No physical evidence could definitely be tied to Lizzie.

- After 10 days, the prosecution rested, and the defense spent only a day presenting its case. Emma testified that her sister and stepmother had a fairly cordial relationship and that Emma herself had been the one to tell Lizzie to burn the blue dress that was stained with paint. The defense also tried to poke holes in the prosecution's timeline and introduced witnesses who said they had seen strangers in the area on the day of the murders.

- The jury quickly found Lizzie Borden not guilty of killing her parents. Lizzie was released and returned to Fall River, where she lived for the rest of her life. She and Emma sold the family house and moved to a mansion. Emma later left town, changed her name, and stopped all contact with her sister. Lizzie died in 1927 at age 67; both she and Emma are buried next to their parents.

The Menendez Brothers

- The idea for killing their parents came to the Menendez brothers while they watched a movie in which a group of wealthy Los Angeles hipsters commits murder. Following the movie, 21-year-old Lyle and 18-year-old Erik discussed their overbearing father, who had pressured them to achieve and had cheated on their mother. The brothers decided that if they murdered their father, they would also have to kill their mother; not only would she be a potential witness, but she couldn't live without her husband.

- On August 20, 1989, José and Kitty Menendez were watching television in the den of their 23-room mansion. Apparently, their sons made sure that the security system was off as they entered the house, each with a shotgun.
  - According to the autopsy results, the boys most likely fired one round into José from the front, then walked behind him and fired again.

  - Kitty must have jumped up with the first shot to her husband and, as she tried to get away, was shot 9 or 10 times. The boys ran out of ammunition before realizing that Kitty was still alive, and one of them went back to the car to reload.

  - The brothers also shot both parents in the kneecaps in an attempt to make the murder look like an organized crime hit, a theory they later proposed to the police. They then picked up the shell casings and called the police.

- Lyle and Erik were not immediately suspects, but within six months after their parents' murders, the brothers had spent about $1 million, and police interviewed them again. A break in the case came when police were tipped off about audiotapes recorded by Erik's therapist, Dr. Jerry Oziel, in which the young man confessed to killing his parents. The Menendez brothers were arrested in March of 1990.

Doctor/patient communications are usually considered privileged and are not admissible in court, but confidentiality does not apply if a patient threatens the doctor.

- It took three years before a trial began. During that time, the judge ruled some of the audio recordings admissible but not the one in which Erik described the crime in detail. The brothers pled not guilty, claiming that the act of killing their abusive parents had been a form of self-defense. Prosecutors pointed to the massive amount of money the Menendez brothers had spent after killing their parents and to their elaborate plan for the murders.

- Although there was a single trial, there were two juries, one to hear the case against Erik and another to try Lyle, but neither jury was able to reach a conclusion. The district attorney requested another trial, which was granted. The result was two guilty verdicts and life in prison for the brothers, without the possibility of parole.

Suggested Reading

Brown, *Lizzie Borden.*

Davis, *Bad Blood.*

Douglas and Olshaker, *The Anatomy of Motive.*

Pietras, *Unanswered Evidence.*

Porter, *The Fall River Tragedy.*

Schroeder and Fogg, *Beverly Hills Confidential.*

Soble and Johnson, *Blood Brothers.*

Spiering, *Lizzie.*

Questions to Consider

1. Do you think Lizzie did it? What is the evidence you would give for or against that opinion?

2. Women have married each of the Menendez brothers while they serve life in prison without parole (and are not allowed conjugal visits). What are your thoughts on the motivations of those involved?

3. Reportedly, the Menendez brothers have not spoken in a decade or more. What do you think may be behind their separation? Does their estrangement potentially tell us anything about the case?

Lizzie Borden and the Menendez Brothers

Lecture 5—Transcript

Hard to believe a twisted little jump-rope skipping rhyme originated out of the story of alleged parent-killer Lizzie Borden:

> Lizzie Borden took an axe,
> And gave her mother forty whacks.
> When she saw what she had done,
> She gave her father forty-one.

A lesser-known second verse goes:

> So, close your door,
> And lock and latch it,
> Cause here comes Lizzie
> With her hatchet.

Some sources say a newspaper peddler created the saying to generate sales, but no one really knows for sure. But this case from the late 1800s is a good example of how a relatively unknown person can become infamous, and how someone can plead not guilty at trial, despite a mountain of evidence.

Lizzie was born on July 19, 1860 in Fall River, Massachusetts to Andrew Borden and Sarah Morse Borden. She had a sister Emma who was about 10 years older; another sister between the two girls had died while a baby. When Lizzie was just two, her mother died, and several years later, in 1865, her father remarried a woman named Abby Durfee Gray. She was already well into her 30s, so previously considered a spinster.

Andrew was a successful entrepreneur; he was a bank president, and landlord, admired for his business sense, but not very well liked, though he was said to treat his wife and daughters well. Despite his wealth, Mr. Borden was tight with money. The family had a modest home, without indoor plumbing, and used kerosene lamps, while other wealthy folks had gas lighting. Andrew Borden had lots of business dealings, including with both his second and first wife's relatives. Some of these exchanges produced tension in the family, in particular, between Mr. Borden and his daughters. He's alleged to have made quite a few enemies in the community during his rise to wealth.

Like their stepmother had, both Lizzie and her sister, Emma, grew into well-respected "spinsters" in their own right. Despite not getting along with their stepmother, even calling her Mrs. Borden, rather than "mother," both still lived at home. In those days, unmarried women just didn't move out on their own. Emma was said to be more like a mother to Lizzie than a big sister, having cared for her from the time their mother died. Lizzie was a Sunday school teacher and was involved in the women's anti-alcohol, or so-called temperance movement. She was said to envy her richer relations and friends who lived up on the hill in Fall River, rather than down by the mills as her family did.

In 1891, there was a burglary at the Borden house in broad daylight, when only the two sisters, Emma and Lizzie, and their live-in maid, Bridget, were at home. The only things taken were about $50 and some jewelry from Mrs. Borden's desk drawer. The police were called, but a couple of weeks after the incident, for unknown reasons, Mr. Borden told them to stop the investigation. Apparently, Lizzie may have been the family's suspect. She was known to be a shoplifter in town; the clerks were just told to bill her father if they saw Lizzie take something. After the burglary, the family always kept all the house doors locked.

Lizzie wasn't a very good student; she dropped out of high school in junior year. And she didn't have much of a social life outside of church since, other than a trip to Europe once, her father didn't go for the fancy activities of other wealthy families. Instead, Lizzie kept pigeons in the loft of the family's barn; she was quite the animal lover. But, sometime in either May or June of 1892, Mr. Borden got fed up with Lizzie's hobby, saying it was drawing neighborhood boys who shot at the birds. So Mr. Borden beheaded Lizzie's pigeons with an axe. The household got so tense that by July, Emma and Lizzie went on an extended vacation to get away for about a week.

One day, in early August of that year, Mrs. Borden went across the street to visit their neighbor, Dr. Seabury Bowen, reporting the family and their maid,

Bridget Sullivan, had all been sick for several days. Mrs. Borden told Dr. Bowen she feared they might have been poisoned, but the doctor believed it was probably just something they'd all eaten. Lizzie told her friend, Alice Russell, that she thought someone, maybe an enemy of her father, had poisoned their milk, and that she had seen a strange man hanging around their house and barn.

The next day was August 4, 1892, and while her sister, Emma, was away at a friend's house party, 32-year-old Lizzie screamed for the maid, Bridget, to run and get the doctor, saying her father had been killed. Although there was no sign of a struggle, Andrew's body was on the sofa in the family's sitting room, with his feet still on the ground and his face smashed in by a sharp weapon. A short time later, Mrs. Borden was found on the floor in an upstairs guestroom with blows to the back of her head.

Autopsies were conducted, believe it or not, right there on the family's dining room table at 3:00 that afternoon. Based on the victims' body temperatures and the condition of the blood at the scene, the doctor thought Mrs. Borden had been killed between 9:00 and 9:30 that morning, even though she had been found after her husband. His face had been struck with a large, sharp implement, like an axe, from someone probably standing above and behind him. Many blows had crushed the back of Mrs. Borden's skull. The doctor stated that the first blow would have been enough to kill either of the victims, but the perpetrator didn't stop there.

Lizzie and the maid, Bridget, had been the only ones home at the time of the murders, and the two ladies were questioned for most of day. Their stories didn't exactly agree on several points, but here's a composite timeline of the morning's events. Mr. Borden had gone out, having been asked by Lizzie to mail some letters. He came back about 10:45 that morning, and Bridget had to let him in the front door; the back door, which he usually used, was bolted from the inside. As Mr. Borden walked in the front door, Bridget claimed she heard Lizzie laughing on the landing of the front stairs just above them. Mr. Borden asked where his wife was, and Lizzie told him her stepmother had received a note that a friend was ill, so she went to visit her. Remember, the doctor ruled Mrs. Borden was likely already been dead by this time.

At about 10:55, Andrew Borden told Bridget he was going to rest on the sofa in the sitting room. Bridget went up to her room in the attic to relax, as well and said she recalled hearing a nearby clock tower chime 11:00. Lizzie told authorities she'd gone out for about a half an hour around then to go up into the loft of the barn, the same barn loft where her father had killed her pigeons. When Lizzie came back around 11:10, she saw her father's bloody body on the couch and called for Bridget to come down and run to Dr. Bowen's house; someone had killed her father.

Now, Doctor Bowen wasn't home, but after someone found him, he got to the Borden house at around 11:30. The police had been called in between, around 11:15, and they got there by 11:45. During questioning, when an officer referred to Mrs. Borden as Lizzie's "mother," she replied, "She is not my mother; she is my step-mother. My mother died when I was a child."

Now, a search of the Borden property is said to have turned up two axes, one of which, according to one source, had hair and blood on it that turned out to be from a cow. Now, I couldn't find anything to support how this was known, since serology was in its infancy at that point, and even human blood typing was not yet understood. They also found what is termed a claw-hammer hatchet, a regular hatchet, and the head of a standard hatchet with its handle broken off. No bloody clothing was found anywhere at the scene, except the clothes on the victims. Fingerprint comparison was not standard in forensics at that time and was not used in the case.

One initial suspect was Mr. Borden's brother-in-law from his first wife; he had spent the prior night at the family home and had breakfast with the Borden couple that morning. But when police located him, they learned he had spent the rest of the morning across town with other relatives and had a clear alibi. The maid, Bridget, was also briefly considered a suspect, but that was dismissed. Nothing had been taken, so robbery wasn't a motive. And although Andrew Borden had many business enemies, why would Mrs. Borden have been killed, too, and first?

The funeral was held two days after the Borden's deaths, but police delayed the actual burial, because they wanted another, more thorough, autopsy. So a second examination occurred a week after the couple's deaths in the receiving vault of the Oak Grove Cemetery by the Medical Examiner who was named Dolan.

Now, this is really interesting to me, because, this is the way I would handle skull trauma analysis today. They removed the heads and defleshed them, and that's when they examined Mr. Borden's face and estimated it had been struck about 10 times with a hatchet, which is smaller than an axe, while Mrs. Borden's skull sustained about 18 whacks. Dr. Dolan's report documents each blow in detail, including its position and its measurements.

So, the nursery rhyme gets it wrong on both counts with regard to the number of blows; and the implement was not an axe. Mr. and Mrs. Borden's heads were not returned to their coffins for burial, but rather retained as evidence for a potential trial. Oh, and the couple's stomachs had earlier been taken to Harvard Medical School, and no evidence of toxins were found to substantiate Mrs. Borden's allegations that maybe they were being poisoned. But lots of possible poisons could probably have been tested for at that time, and quite a few others could not.

Now, I mentioned that no bloody clothing had been found, but on August 7, three days after the murders, a friend of Lizzie's, Alice Russell, who was staying with the Borden sisters, saw Lizzie burning a blue dress in the kitchen stove and reportedly said, "If I were you, I wouldn't let anybody see me do that." Lizzie said the dress had gotten a paint stain on it. Two days later, a judge held a closed inquest and ordered Lizzie, the maid Bridget, and a few others to answer questions. Bridget told the magistrate that Lizzie had been wearing a blue dress on the morning of the murders, but she was wearing something else when police arrived.

During her four-hour interview, when Lizzie was asked where she was when her mother was murdered, Lizzie said she was in the kitchen ironing, but this is not a huge house; there were only five rooms on the second floor and five on the first, if you counted the front entry hall. When asked why she went out to the barn, Lizzie said she had gone up into the loft to look for lead to make sinkers for a planned fishing trip. But the barn was searched that day, and it was said the loft was so hot no one in their right mind would have gone up there, plus, there were no footprints in the loft, only cobwebs. It was also reported that Lizzie had been to the drug store the day before the murders to try to buy prussic acid, which is what they called cyanide in those days. She said she wanted it to clean her sealskin coat, but then she denied it, even though the druggist and two customers said they saw her at the pharmacy.

Lizzie was arrested two days after the inquest for the double murder. She had to be jailed in a nearby town because Fall River didn't even have a place to house female prisoners, that's how unheard of it was for a woman to be arrested and held for something. At her arraignment hearing the next day, Lizzie Borden was charged with the double murder of her parents, to which she pled not guilty.

A preliminary hearing was held in late August, and the judge found probable cause to move ahead with the court proceedings. At a Grand Jury hearing that happened in November, it was looking like Lizzie might not be indicted, until Alice Russell testified about the dress-burning incident. Lizzie Borden was jailed from arrest to trial, which didn't happen until almost a year later in June of 1893 in New Bedford. The State of Massachusetts had a mandatory death sentence at that time; she would be hanged if found guilty.

Lizzie did have some support from her pastors and friends from her church. She also had an excellent defense team of three lawyers—Andrew Jennings, who had been her father's attorney; a young Boston attorney named Melvin Adams; and importantly, the former Governor of Massachusetts, George Robinson. Now, while Robinson had been Governor, he had appointed Justin Dewey to a judgeship. Guess who wound up being the primary judge on the three-judge panel in the Borden case? Dewey—who in my opinion should probably have recused himself once Lizzie hired former Governor Robinson to defend her.

Anyway, the prosecution wasn't too shabby, either. One of the prosecutors, Hosea Knowlton, went on to become the Massachusetts State Attorney General; the other was William Moody, who after the trial became U.S. Secretary of the Navy and then U.S. Attorney General under Theodore Roosevelt, as the two had been classmates at Harvard. Roosevelt ultimately nominated Moody to the U.S. Supreme Court.

The case was big news that was intensely followed by the media. Some of the defense's pre-trial strategies were very obvious. Lizzie and her sister Emma posted a $5,000 reward for any information leading to the real killer. The attorneys put ads in the paper to try to find the person who allegedly delivered a note about a sick friend to Mrs. Borden the morning of the murders. But the defense managed to land two big blows. First, they got the judges to rule Lizzie's statements at the initial inquest as inadmissible because her lawyer wasn't present, so that evidence was never heard by the jury. Also not allowed by the judges was the testimony about the attempted prussic acid purchase at the Fall River drugstore. Surprisingly, even though Lizzie's visit to the pharmacy happened only the day before the murders, the judge said it was too remote in time to be considered by the jury and that any attempt to buy poison wasn't really relevant to the crime, since the couple had been killed with a hatchet.

So, the prosecution went with motive and opportunity. Their first motive was the inheritance that would allow Lizzie to spend some of the cash her father held so dear. Second, they presented Lizzie's hatred of her stepmother. Their third main point was opportunity. I mean, there was no sign of forced entry, and the family kept all doors locked after that 1891 burglary. There were just no other prime suspects than Lizzie, but the bulk of the case was built completely on circumstantial evidence.

The prosecution did have some direct evidence in the form of testimony, for example, Alice Russell, who testified about the burning of the dress. They also had some physical evidence, but none of it could be definitively tied to Lizzie. The hatchet with the broken handle found in the Borden basement—that was presented as the murder weapon.

Forensic scientists, though they weren't called that at the time, said there was some kind of powdery white substance on its blade that had either been used to clean it or maybe make it look like it hadn't been used in a long time. And when Medical Examiner Dolan took the stand to try to match the hatchet's blade to the wounds in her parent's skulls, Lizzie swooned and fainted right there in the court room. I've testified myself with a skull in hand while explaining my findings to the jury. You can imagine it provides some pretty compelling evidence.

After 10 days, the prosecution finally rested. The defense, though, spent only one day presenting its case. Although Lizzie didn't take the stand, as was her right, her sister, Emma, did. Emma testified that her sister and stepmother had a fairly cordial relationship, and that Emma herself had been the one to tell Lizzie to burn that blue dress that was stained with paint. The defense also tried to poke holes in the prosecutor's timeline, and they brought on witnesses who said they had seen strangers in the area on the day of the murders. They also delivered up both a plumber and a gas-line workman who said they had been up in the Borden's barn loft just days before the murders, despite the prosecution's claim that the dust was undisturbed.

When both sides rested, the judges delivered a jury charge that some have called nothing more than "directions to acquit." It's said the jury of 12 men took just 10 minutes to decide the case, but waited another hour, just so it didn't seem like they were acting in haste. They found Lizzie Borden not guilty of killing her parents. The courtroom cheered. Lizzie cried and was released, and she headed home to Fall River.

Surprisingly, that's where she lived the rest of her life, even though people generally stayed away from Lizzie Borden. She and Emma sold the family house, bought a mansion on the hill, and hired several servants. They took on that lifestyle Lizzie always envied, but Emma wasn't very pleased with it. She finally left town, changed her name, and stopped all contact with Lizzie, though Emma never admitted Lizzie killed their father and stepmother, or even that she thought Lizzie had.

Media followed Lizzie the rest of her life, until she died in 1927 at age 67 due to complications from surgery. Her estranged sister, Emma, died just nine days later. Both are now buried next to their parents, whose skulls were buried at the foot of their graves after the trial was over. In her will, Lizzie left nothing to her sister, but money to the city for the upkeep of her father's grave in perpetuity. She also left funds to former friends, a cousin, and servants who had stood by her. But the bulk of her estate, $30,000, went to an animal shelter in Fall River.

Now, one last bit of Borden trivia. There was a piece of artwork, a painting on porcelain, that Lizzie had at her house. She gave it to one of her few remaining friends, but the woman accidentally broke it and took it to an art shop to have it repaired. The store recognized it as being stolen from them years ago and asked the woman where she got it. Lizzie was charged with the theft, but settled out of court. Now, shoplifting is a far cry from murder, but maybe that wasn't the only crime that Lizzie committed and got away with, at least for a while. Perhaps the discovery of the journals of Lizzie's attorney, Andrew Jennings, in 2012, will shed new light on the case when their details are released.

Let's contrast the Lizzie Borden case with another, more recent, crime in which a middle-aged couple was killed, this time in Beverly Hills, California, and in 1989, just a little over a hundred years after the Lizzie Borden case. Some of the similarities are remarkable. The family was well-to-do; the parents had only two children, but sons, not daughters. The boys stole from others to provide extravagances their father wouldn't allow. After the deaths of their parents, the two sons used their inheritance to finance the kind of lifestyle they wanted. The murders were brutal examples of overkill; the case was highly publicized; and the sons, but both of them this time, were accused of murdering their parents. Like Lizzie Borden, they pled not guilty, and the courtroom testimony the judges ruled admissible had a great bearing on the case. You may have already guessed I'm talking about the Menendez brothers—Erik and Lyle.

Apparently, the idea for killing their parents came about while the brothers watched the *Billionaire Boys Club*, a movie in which a group of wealthy L.A. hipsters ends up committing murder. Following the movie, 21-year-old Lyle and 18-year-old Erik discussed how their father's pressure was just so overbearing; they couldn't imagine living with it the rest of their lives. He wanted them to work for a living. Can you imagine that? He wanted them to go to prestigious schools and be tennis champions. He had cheated on their mother and threatened to change his will to leave them out. The brothers decided, though, if they murdered their father, they would also have to kill their mother, because not only would she be a potential witness, she just couldn't live without her husband. How kind.

On August 20, 1989, José and Kitty Menendez had given the maid the night off and were watching television in the den of their 23-room mansion, well, actually, Kitty had dozed off with her head in José's lap, as he watched TV with his feet up on the coffee table. Apparently, their sons made sure the security system was off as they entered the house, each with a shotgun. According to the autopsy results, the boys most likely fired one round into José from the front, then walked behind him and put the shotgun against the back of his head and fired again. Kitty must have jumped up with the first shot to her husband, and as she tried to get away, slipping in her own blood, and crawling on her hands and knees, wound up enduring a total of nine or 10 shotgun blasts, four of those to her head. She had defense wounds on her hand from trying to block the blasts. Can you imagine her last thoughts as her two only sons fired shotguns at her?

What's worse, the boys ran out of ammunition before realizing Kitty was still alive, so one of them went back to their car to reload; they were out of buckshot, but still had birdshot. She wound up being finished off by a blast of birdshot through a gun placed against her left cheek. Then José and Kitty were each shot in the kneecap with the birdshot. This was to make it look like an organized crime hit. The theory the brothers later proposed to police was that their parents had been killed by the mob.

Next the boys picked up the shell casings, except for one, which Erik later admitted he spotted on the floor while the police were questioning the boys later that night. He managed to bend over and recover it before the officer even saw it. How did the police get there? Lyle had called them. And unlike the cry for help in the Lizzie Borden case, the 911 call was recorded, with Erik screaming in the background and Lyle yelling not to touch the bodies.

Lyle and Erik were not immediately suspects, so officers didn't do a gunshot residue test on them, which would have easily shown they were the killers. But within the six months after their parents' murders, the brothers had spent about a million dollars. That's when police began to suspect they had committed the killings and started to use some additional interrogation to try to rattle the brothers and get them to turn against each other.

But the case really broke when a woman called police to say that her lover, Dr. Jerry Oziel, was Erik Menendez's therapist. She knew the boy had confessed to killing his parents about two months after their deaths. She claimed that

Lyle Menendez had found out about Erik's confession and threatened to kill Dr. Oziel. The therapist had asked his girlfriend to come to his office so she could secretly listen in on and tape a subsequent conversation with both brothers, in case anything happened to him. Police got a search warrant that led to the audiotapes being recovered from Oziel's safe-deposit box. The Menendez brothers were arrested in March of 1990, just over six months after the murders of their parents.

You've probably heard about doctor-patient confidentiality and other so-called privileged communications, like between clergy and confessor, or attorney and client. Those are usually not admissible in court. But you may not know that doctor-patient confidentiality doesn't apply if the patient threatens the doctor. Once Lyle said he would kill Oziel if he told anybody about his brother's confession, it should have removed the legal restrictions on what would normally be considered confidential communication.

It took three years before the trial started, and during the typical legal wranglings, the judge ruled some of the audio recordings admissible, but not the one in which Erik described the crime in detail. Court TV filmed and broadcast the six-month trial that began in 1993. It was easily as much of a media circus as the Borden trial. The brothers pled not guilty, but a key difference between this trial and that of Lizzie Borden was that the Menendez brothers' attorneys acknowledged the young men did kill their parents but said decades of severe abuse by both parents—physical, psychological, and even sexual—had caused the murders. The brothers claimed they feared that their parents would end up killing them, and based on that, the act was a form of self-defense for which they should be found not guilty.

The prosecution pointed to the massive amount of money the Menendez brothers spent in the six months after the killings and said greed was the simplest motive. The prosecution also showed that the brothers had hired a computer expert about 10 days after the murders to wipe the family computer hard drive in an effort to erase the planned changes to their parents' will that would have left them out of the estate. The prosecution also pointed to the elaborate plan the brothers had developed. Shotguns bought out of town with a fake ID just two days before the murders, which were later thrown into a canyon on the night of the killings. And how the boys went out to the movies after dumping their bloody clothes and shell casings at a gas station, only to come home later and pretend to find their parents. Can you imagine them sitting at a movie after just committing a double murder, and of their parents, no less?

Although there was a single trial, there were two juries, one to hear the case against Erik, and another to try Lyle. But neither jury was able to reach a conclusion. The district attorney requested another trial, which the judge granted. The judge in the second trial, which began in 1995 and lasted seven months, did not allow cameras in the courtroom and was much more judicious about what he allowed the defense to present at this time, especially about the alleged sexual abuse for which there was not a shred of evidence. The result was two guilty verdicts and life in prison for the Menendez brothers, without the possibility of parole.

The Tylenol Murders
Lecture 6

On September 29, 1982, in a Chicago suburb, 12-year-old Mary Kellerman woke and complained to her parents that she was coming down with a cold. They gave her some Extra Strength Tylenol. Shortly afterward, they found Mary on the bathroom floor and rushed her to the hospital. She was dead just a little while later. Some of you might clearly remember the media attention in this terrible case. The Tylenol poisoning was on the national news, and panic quickly spread. In this lecture, we'll look at the mobilization of forensic scientists and the Illinois Department of Health for this unprecedented case of product tampering.

Untimely Deaths

- When 12-year-old Mary Kellerman died in a Chicago hospital on September 29, 1982, after her parents found her on the bathroom floor, doctors suspected that she might have had a stroke, but that's rare in someone so young.

- One the same day as Mary's death, in another Chicago suburb, paramedics were called to the home of Adam Janus, a 27-year-old postal worker. Janus, found unconscious, was rushed to the hospital but died soon after of a suspected heart attack. That evening, two of Janus's family members also collapsed, and both died shortly thereafter. Forensic investigators suspected poisoning, and blood samples were taken from all three for toxicology.

- When someone dies unexpectedly, forensic investigators routinely examine the scene and question witnesses, family members, or coworkers. To help pathologists establish cause and manner of death, these investigators try to find links among circumstances, events, and materials surrounding the loss of life.
    - When these Chicago-area investigations led to the discovery that Mary Kellerman and all three of the Janus family members had

taken Extra Strength Tylenol before their deaths, police rushed back to the scenes and confiscated the bottles of acetaminophen.

  - Forensic chemistry indicated that several of the remaining capsules were tainted with high levels of potassium cyanide—thousands of times the lethal amount.

Response to the Crisis

- Johnson & Johnson, the maker of Tylenol, was alerted and issued an extensive product recall. But in the meantime, three more women in the Chicago area had also taken the drug and were dead. The lot numbers on the tainted bottles indicated that they hadn't all come from the same manufacturing plant, which meant that Johnson & Johnson probably wasn't the source of the poison, either through accident or sabotage.

- Chicago police cars took to the streets, using loudspeakers to warn the public not to take Tylenol. Johnson & Johnson offered a $100,000 reward for information leading to the capture of those involved. Investigators removed the product from store shelves all around the nation for forensic testing.

- Only eight adulterated packages of Tylenol were ultimately recovered, but many more may have been thrown away by consumers after hearing about the recall. A combination of store receipts and surveillance videos helped investigators determine that the eight bottles of tainted medication came from seven different locations.

- No usable fingerprints were recovered from any of the packaging or its contents. Investigators believed that someone was taking Extra Strength Tylenol bottles from stores—either stealing or buying them—opening the gel caps, adding cyanide, then returning the products to store shelves to be purchased by unsuspecting victims.

- Illinois Department of Health officials used a rapid test involving a small slip of chemically treated paper that could be put into a bottle and would turn blue if cyanide was present. These types of quick assays are called *field tests* or *screening tests* because they allow rapid presumptive identification. If a substance tests positive, it's then subjected to *confirmatory tests*, which involve more sophisticated chemical analysis and high-tech lab equipment.

The Search for Suspects

- Because potassium cyanide is used in some industrial settings and in the mining industry, investigators looked for leads in those directions. They also considered people who might have been disgruntled with Johnson & Johnson.

- The first major suspect was a do-it-yourself chemist named Roger Arnold. He worked on the loading dock at one of the stores involved and had a weak connection to another store where tainted capsules were found. The father of one of the victims was also a coworker of Arnold. He was ultimately cleared of suspicion but was later convicted of the murder of man he thought had steered the police to him.

- The investigation turned to a more promising suspect after Johnson & Johnson received an anonymous letter with a New York postmark.
    - The handwritten letter—although not technically claiming to be from the killer—described how perfectly the poisoning had gone and how simple it had been to put the cyanide into the gel caps. The

© JuNi Art/iStock/Thinkstock.

A letter sent to Johnson & Johnson bragged that putting cyanide in the Tylenol gel caps had required less than 10 minutes per bottle and cost less than $50.

letter instructed Johnson & Johnson executives to deposit $1 million to a specific account at the Continental Illinois Bank in Chicago if they wanted the deaths and the bad press to stop.

- The bank account listed in the letter was closed but had previously belonged to Frederick Miller McCahey, one of the heirs to the Miller Brewing Company. McCahey had run a Chicago travel agency that had recently gone bankrupt. But why would someone put his own bank account number in an extortion letter?

- What investigators didn't yet know was that one of McCahey's former employees, Leann Lewis, and her husband, James Lewis, were furious with him. They had even tried to sue McCahey in the summer of 1982—just months before the Tylenol poisonings—because one of Leann's paychecks had bounced.

Background on the Lewises

- James Lewis was in his mid-30s in 1982. His history included a difficult upbringing and a diagnosis of schizophrenia. In his late teenage years, Lewis allegedly chased his adoptive mother with an axe and beat up her husband before trying to kill himself with an overdose of over-the-counter pain reliever. He recovered at a mental health facility and later that claimed the attacks and suicide attempt had been a ploy to evade the draft.

- After his release, James met and married Leann, and soon after, the two had a baby, who had Down syndrome. The couple opened their own accounting and tax service in Kansas City so that they could keep their daughter with them at work. One day, a retired bachelor named Raymond West stopped into the storefront; he befriended the family and became one of the Lewises' tax clients.

- Some three and a half years later, in the summer of 1978, Raymond West disappeared. A friend who was suspicious contacted police, and after about three weeks, West's dismembered body was found hidden in his own home.

- James Lewis was linked to this gruesome discovery through a number of avenues, including a check from West's bank account made out to Lewis for $5,000 and dated on the last day West had been seen alive. After a search of Lewis's car yielded additional evidence, he was charged with murder. Unfortunately, the charges were later dismissed as a result of several legal missteps.

- Although Lewis was free, police continued to investigate his tax business and other ventures. By the end of 1981, authorities had amassed enough evidence to get a search warrant for the Lewises' home, but James and Leann went on the run. They moved to Chicago and assumed the names Robert and Nancy Richardson; Leann went to work for Frederick Miller McCahey's doomed travel agency.

Search and Arrest

- When the travel agency went bankrupt, James Lewis tried to mobilize the former employees to take action with the Illinois Department of Labor. He also had Leann contact one of her former supervisors to get a list of McCahey's bank account numbers, possibly to find out if he had any remaining money. When the claim ultimately went nowhere, James and Leann bought one-way train tickets to New York and left on September 4, 1982.

- Regarding the investigation of the killings and the extortion, authorities hadn't yet connected James and Leann to anything. When questioned, McCahey quickly identified James and Leann as people who might hold a grudge against him, but no one knew where they had gone.
    - Investigators initially thought that the extortion letter was just a hoax to embarrass McCahey, but they reconsidered after they learned of James's connection to the Raymond West case and the suspected business and tax fraud.

- By this time, the couple's faces and aliases had been highly publicized; Leann quit the job she'd just gotten in New York, and the two dropped out of sight.

- James then sent to the press a package of material related to the travel agency payroll default, indicating that there was good reason for framing McCahey but stating that he and his wife had not committed the Tylenol murders. In another letter, Lewis mocked police for reopening the Raymond West case. He began signing his real name and even put his right thumbprint on one letter.

- About two months after the Tylenol killings, a Manhattan Western Union surveillance camera captured Leann picking up a money order sent by her father. The big break came when the FBI received a call from a librarian who said that she believed James Lewis was sitting right then in the New York Public Library. Agents closed in and arrested Lewis, but they had enough evidence only to charge him with extortion.

- Leann turned herself in but refused to turn against her husband. About a year after the Tylenol killings, James Lewis was convicted of attempted extortion. He was sentenced to 10 years in prison, to be served consecutively with another 10-year sentence for other charges. Lewis was released in 1995 and has lived in the Boston area.

Continuing Investigation

- In 2009, the FBI searched the Lewises' home as part of an "ongoing investigation," citing the possibility of new evidence and technological advances that might help solve the Tylenol killings. This may have been a reference to *touch DNA testing*, a technique in which only a small number of human skin cells can be used to create a DNA profile. No new charges have been filed.

- In 2012, the year that marked the 30th anniversary of the Tylenol murders, media reports indicated that a grand jury might be convened to indict James Lewis in the seven murders, but that never occurred. Even lead investigators on the case disagree about whether or not Lewis is actually responsible for the killings. Today, some describe the case as one of the earliest examples of what we would now call domestic terrorism.

- In the fall of 2013, the FBI announced that it could no longer serve as lead investigator in the killings. Still, Chicago-area police have not given up; several agencies have started a new task force to continue to pursue the Tylenol killer.

Suggested Reading

Bartz, *TYMURS*, books 1 and 2.

Blum, *The Poisoner's Handbook.*

Douglas and Olshaker, *The Anatomy of Motive.*

Emsley, *Molecules of Murder*.

Questions to Consider

1. Do you remember the news reports of the Tylenol murders when they occurred? If so, how did they make you feel?

2. Do you think James Lewis is responsible for the poisonings? Why or why not, based on the evidence presented in this lecture?

3. Can you think of other crimes that have prompted specific changes in the way we do things in society?

The Tylenol Murders
Lecture 6—Transcript

On September 29, 1982, in a Chicago suburb, 12-year-old Mary Kellerman woke and complained to her parents she was coming down with a cold. They gave her two Extra Strength Tylenol acetaminophen capsules. Shortly after that, they found Mary on the bathroom floor, and rushed her to the hospital. She was dead just a little while later. Doctors suspected she might have had a stroke, but that's rare in someone so young.

The same day, in another Chicago suburb, paramedics were called to the home of Adam Janus, a 27-year-old postal worker. By the time they arrived, Adam was unconscious. They rushed him to the hospital, but he died soon after of a suspected heart attack. That evening, Adam's family gathered at his home to make plans for his untimely funeral. Adam's 19-year-old sister-in-law, Theresa, had a headache. Her husband, 25-year-old Stanley Janus, saw a bottle of Extra Strength Tylenol on his brother's counter. Stanley and his wife each took a couple of the capsules. A little while later, both collapsed, and paramedics were again called to Adam Janus's home.

Authorities evacuated the rest of the family immediately, thinking there might be some kind of toxic gas leak or something. Stanley died that night, and his wife, Theresa, two days later. Obviously, the deaths of three young people from the same family, at the same location, were very suspicious. Forensic investigators suspected poisoning, and blood samples were taken from all three of them for toxicology.

When someone dies unexpectedly, forensic death investigators routinely examine the scene and question witnesses, family members, or co-workers. This helps pathologists establish cause and manner of death. These investigators try to find links between circumstances, events, and materials surrounding the loss of life.

When these Chicago-area death investigations led to the discovery that Mary Kellerman and all three of the Janus family members had taken Extra Strength Tylenol before their deaths, police rushed back to the scenes and confiscated the bottles of acetaminophen. Forensic chemistry indicated several of the remaining capsules were tainted with very high levels of potassium cyanide, thousands of times the lethal amount.

Johnson & Johnson, the makers of Tylenol, were quickly alerted and very rapidly issued an extensive product recall of 31 million bottles. But by that time, three more women in the Chicago area had also already taken the drug and were dead; 27-year-old Mary Reiner, who had just given birth just days before, 35-year-old flight attendant Paula Prince, and 31-year-old Mary McFarland, who collapsed in her office at the phone company.

Comparing the locations of the victims' homes, investigators thought it highly unlikely that they'd all purchased the medication from the same store. Investigators used the lot numbers on the bottles to see if they'd all come from the same manufacturing plant. They didn't. So Johnson & Johnson probably wasn't the source, either through accident or sabotage.

Some of you might clearly remember the media attention in this terrible case; I know I do. The Tylenol poisoning was all over the national news, and panic was spreading. But it was so important to get the word out. Chicago police cars traveled up and down the streets using loud speakers to warn the public not to take Tylenol. Johnson & Johnson offered $100,000 in reward money for information leading to the capture of those involved. Investigators removed the product from store shelves across the whole nation for forensic testing.

Only eight adulterated packages were ultimately recovered in this hideous death lottery, but who knows how many poisoned packages consumers could have thrown away, after hearing about the recall. The eight tainted packages included the five involved in the deaths, two others returned to stores, and one found on a store shelf. A combination of store receipts and surveillance videos helped investigators figure out the eight bottles of tainted medication came from seven different locations. In those days store surveillance equipment wasn't nearly as common as it is today.

No usable fingerprints were recovered from any of the packaging or contents, and each of the eight bottles had from 3 to 10 contaminated capsules in it. The best investigators could figure was that someone was taking Extra Strength Tylenol bottles from stores, either stealing or buying them, then opening the gel caps, adding cyanide, and returning the products to store shelves to be purchased by unsuspecting victims.

Now, you can imagine how quickly forensic scientists and the Illinois Department of Health had to mobilize for this massive test of product tampering. They developed a rapid test that was a small slip of chemically-treated paper that could be put into a bottle. The strip would turn blue if cyanide was present. We call these types of quick assays, *field tests*, or sometimes *screening tests*; they allow rapid presumptive identification. If a substance tests positive, it's then subjected to what we call *confirmatory tests*, which involve more sophisticated chemical analysis and high-tech lab equipment.

Something investigators looked into was who could have gotten their hands on potassium cyanide? It is used in some industrial settings and in the mining industry, so investigators looked for any leads in those directions. They also considered people who might have been disgruntled with Johnson & Johnson and maybe wanted to punish the company by ruining its reputation. The FBI and several Chicago police agencies came up with suspects, and about 400 people were interviewed.

The first major suspect was a do-it-yourself chemist named Roger Arnold. He worked at the loading dock at one of the stores involved and had a weak connection to another store where tainted capsules were found. Tylenol victim, Mary Reiner's, father was one of Arnold's coworkers, so initially, police wondered if Arnold had some kind of grudge against Mary's father and maybe used the other killings to try to disguise his one real target.

Investigators couldn't pin the crimes on Arnold, but he did end up having a mental breakdown from all the media attention. He thought a guy he knew from a local tavern had steered police to him, and Arnold decided to shoot the man. But one night outside the bar, Arnold actually shot a different guy that he mistook for his intended target. So that innocent man was really another casualty of the Tylenol killings. Arnold was convicted of second-degree murder, was paroled after 15 years of a 30-year sentence, and died in 2008.

Anyway, while Arnold was being cleared of suspicion in 1982, the Tylenol investigation quickly turned to a more promising suspect; Johnson & Johnson received an anonymous letter with a New York postmark. The handwritten letter, while not technically admitting to be from the killer, described how perfect the poisoning had gone and how simple it was to put the cyanide in the gel caps. The person claimed it required less than 10 minutes per bottle to complete the task and hadn't even cost $50.

The letter said the whole plan was a "beauty." This was because the bitter poison was housed in the capsules, so it couldn't be tasted. Plus, cyanide required such a small amount to be fatal, and acted so quickly, that realistically nothing could be done to save the victims. The letter told Johnson & Johnson executives to deposit a million dollars to a specific account at the Continental Illinois Bank in Chicago, if they wanted the deaths and the bad press to stop.

Was this letter really from the Tylenol killer, or was it a prank? Or could it have been from an unrelated extortionist who wanted to capitalize on the poisonings? The bank account listed in the letter was closed, but had previously belonged to Frederick Miller McCahey. He was one of the heirs to the Miller Brewing Company and had run a Chicago travel agency that recently went bankrupt. But why on Earth would a man put his own bank account number in an extortion letter like that? Could someone be trying to frame him?

What investigators didn't yet know was that one of McCahey's former employees, his bookkeeper Leann Lewis, and her husband, James Lewis, well, they were furious with him. They had even tried to sue McCahey in the summer of 1982, just months before the Tylenol poisonings, because they'd been stiffed on a bad paycheck the travel agency bounced.

Now, to get the full picture here, we need to go back into the couple's lives a bit. Although he was in his mid-30s in 1982, James Lewis's history included a really difficult upbringing and a diagnosis of schizophrenia. He had earned good grades in school and had been pretty involved there, but was known to have some serious problems at home. In his late teenage years, Lewis allegedly chased around his adoptive mother with an axe and beat up her husband before trying to kill himself with an overdose of Anacin, an over-the-counter pain reliever. He recovered at a mental health facility and later claimed the attacks and suicide attempts were just ploy to get him out of the draft, since this was during the Vietnam War.

Later, while attending the University of Missouri in Kansas City, he met and married Leann, and soon after, they had a baby with Down syndrome. So the couple opened their own accounting and tax service in Kansas City; that way they could have their daughter with them at work each day. Supposedly, the little girl would sit in the store window and wave at people on the street. One day, an older man named Raymond West saw the child, stopped into the store, and befriended the family. West was a retired bachelor who lived in the Kansas City neighborhood where the Lewises had their business. He quickly became a tax client and grew even closer to the couple after their little girl died following heart surgery when she was just five years old.

Some three-and-a-half years later, in the summer of 1978, Raymond West made his usual Sunday stop at the local florist, and among other things that were discussed, he complained of not feeling well. That's the last time anyone saw him alive. Now, the death of Raymond West is a puzzle in itself within this whole Tylenol story, but I'll summarize it by saying that his decomposed and dismembered body was found in his home three weeks later in the Kansas City summer. Now, let's see how James Lewis connects to that.

While West was missing, a concerned friend had tried several times to contact him. He went to West's house, and after seeing the man's unmade bed through an open window shade, he called police who agreed to look into it. The same friend then returned a little later to West's house, only to find the same window shade closed and a note on the door. The note, on Lewis and Lewis tax firm letterhead, stated that Raymond West was out of town for a few days, and anybody who had questions could call James Lewis.

Now, this made West's friend really suspicious, so he got police to do a cursory search of the house; they actually had to break the door lock to enter, but they found nothing really unusual, except, West wasn't there. Later, when this concerned friend was putting in a new lock at West's house, James Lewis showed up to ask what was going on, and the two men talked for a while. But when three weeks passed with no word from Raymond West, his friend used the key to the new lock he'd just installed. When he opened the door, the smell of decomposition almost knocked him over.

Police came back and found things weren't exactly the way they'd been during the first search. Besides the smell, there was a bloody lawn chair in the basement and a garbage bag that held bloodstained sheets and West's hairpiece and glasses. But the basement didn't seem to be where the odor was coming from. So they went back upstairs and saw a bloody stain oozing through the drywall on the bedroom ceiling.

The dismembered body parts of a man assumed to be Raymond West had been hoisted up into the attic, using some kind of a pulley system. As for positive ID, this was prior to forensic DNA testing. The remains were too decomposed for fingerprints, and West wore dentures made by a lab that was no longer in operation. The body was determined to be his through a hair match between the little hair West had left and a hair found in one of his hats, plus, of course, West was missing.

Now, how is James Lewis further linked to this gruesome discovery? By a check from Raymond West's bank account made out to Lewis for $5,000 and dated on the last day the old man was seen alive. When police brought Lewis in, he said the money was a loan that the bank didn't even honor anyway, since they couldn't reach West for verification. Officers point-blank asked Lewis if he had any of Raymond West's checks, and he said no. Lewis willingly submitted to a handwriting sample and fingerprints, and police let him go, while they continued to investigate.

The next day, officers went to Lewis's house and requested to search his home, his tax agency, and his vehicle. In Lewis's car, investigators found garbage bags and rope, like those at the crime scene; they also found some of West's personal papers and a pack of his blank checks. Lewis still denied any involvement in the man's death, but they charged him.

Lewis's trial for the 1978 murder of Raymond West was set to begin in the fall of 1979, but the charges ended up dismissed due to some legal missteps along the way. For one thing, Lewis hadn't been read his Miranda rights, which rendered everything that was collected as evidence inadmissible in court. There were also errors in the legal filings. Besides, West's body was so decomposed that the county coroner couldn't provide either a cause or manner of death. So even if the prosecution could show that Lewis dismembered West's body and even tried to steal or extort money from the old man, there was likely no way they could prove Lewis killed him.

Although James Lewis was free, police were determined to try to find some way to put him behind bars. They continued to investigate his tax business, as well as other ventures Lewis partnered in, one of which, coincidentally or not, involved the import of machinery used to manufacture pills. By the end of 1981, authorities had amassed enough evidence to get a search warrant for Lewis's home. They collected files as well as typewriter ribbons that I assume they were going to use for ink comparisons on some questioned documents. But this time, rather than waiting around to see what happened, James Lewis and his wife went on the run. They left Kansas City for Chicago, where the couple assumed the names Robert and Nancy Richardson. That's where "Nancy," who was really Leann Lewis, went to work for Frederick Miller McCahey's doomed travel agency.

Now that was quite a sidetrack in the investigations of the 1982 Tylenol killings, but I think it's an important one. It sets us up to know much more about James Lewis, who you can probably already guess becomes the second major suspect in the poisonings. It also explains how he and his wife came to Chicago using false names following which Leann's employment unraveled to the point where the couple wanted to get even with McCahey—all for a bounced paycheck in the amount of $511 and change.

Regarding the Lewis's attempt to punish McCahey, James Lewis was instrumental in trying to mobilize the travel agency's former employees to take action with the Illinois Department of Labor. Lewis also had Leann contact one of her former supervisors to get a list of McCahey's bank account numbers, I assume maybe to find out if they had any money left to go after. If you remember, the Tylenol extortion letter provided one of McCahey's bank account numbers. James Lewis also helped the former employees get a hearing with the Illinois Department of Labor. James was said to be the most outspoken of the group, even though he had never even worked for the travel agency, nor was any type of official representative for them. In fact, the person at the hearing finally told Lewis he needed to just sit in the corner and listen. When the meeting was almost over, McCahey showed up. At that point, witnesses said that he and the Lewis couple began yelling at each other, ending with McCahey threatening Leann Lewis.

Now, when their claim against McCahey went nowhere, James and Leann, or Robert and Nancy Richardson, as they were called the whole time they were in Chicago, well they decided to leave town. Using the names William and Karen Wagner now, the couple bought one-way train tickets for New York City and left on September 4, 1982. That was 25 days before 12-year-old Mary Kellerman died from the cyanide-laced Tylenol. Surprisingly, when the pair hit New York, they took back their Chicago names, as Robert and Nancy Richardson.

Authorities hadn't connected James and Leann Lewis to any of this yet, I was just giving you their backstory. As for the investigation of the Tylenol murders, all police knew early on was that McCahey's account number was in that extortion letter and that it was postmarked from New York, just days after the poisonings. During the forensic analysis of the envelope, when investigators scratched the postmark away, they found an office postage meter stamp from McCahey's travel agency, dated April of 1982. Although police didn't know it yet, that was because Leann Lewis had metered some blank envelopes when she still worked for McCahey; she took them home, and then months later used one of them to send the extortion letter from New York.

So there were several lines leading to McCahey, but he was kind of a prominent guy in the area. Police just didn't think he'd be foolish enough to implicate himself like that, in either the extortion or the Tylenol killings. Among the first things they asked McCahey was who might have been upset with him for some reason. He immediately said, "Robert and Nancy Richardson." It didn't take long for police to figure out the couple's legal names were James and Leann Lewis. But they had left Chicago for New York by this time, and no one really knew where they'd gone.

Investigators initially thought the extortion letter was just a hoax to bring McCahey into the limelight and embarrass him. But after police learned of

James Lewis's connection to the Raymond West dismemberment case back in Kansas City, as well as the business and tax fraud that caused him to leave Missouri, well then they reconsidered. By this time the couple's faces and aliases were all over the media, and they needed to really drop out of sight quickly. Leann quit the job she'd just gotten in New York; they moved out of their apartment; and stopped using the names Robert and Nancy Richardson.

People who knew the couple at one time or another started coming forward to the press and police at this time. Some said James Lewis was a near genius that could talk on almost any subject, but others said he was mentally unstable, cold, and vindictive. Things really got crazy when it was reported that Lewis had also written a threatening letter to President Ronald Reagan, alleging it was from McCahey. Handwriting analysis showed the same person had written both letters, and they found Lewis's fingerprint on the extortion letter.

Then Lewis began writing to the press, just like some other killers have done, including Ted Kaczynski, the Unabomber; and Dennis Rader, the BTK killer. Lewis sent a package of material to a Chicago newspaper. The papers related to the travel agency payroll default, I suppose to try and indicate that framing McCahey was with good reason. But Lewis's letter stated that he and his wife did not commit the Tylenol murders. How could they, after moving to New York weeks before the deaths?

Co-workers saw Leann daily at her New York job and also saw James meet her for lunch and at the bus daily. Authorities never did find any public transportation records to indicate either of them had returned to Chicago during the Tylenol killings. In another letter, Lewis mocked police for reopening the Raymond West case, and again trying to connect him to it. Lewis began signing his real name. He even put his right thumbprint on one letter after the media reported the FBI had recently discovered Lewis's right thumbprint on the pulley system used in Raymond West's attic.

Since Leann, who generally supported them, had quit her job, they needed money. About two months after the Tylenol killings, a Manhattan Western Union surveillance camera captured Leann picking up a 140-dollar money order sent by her father, so authorities knew the couple was still in the area. But the big break came next when the FBI received a call from a New York Public librarian who said she thought she had just given the wanted man, James Lewis, two references he had requested. One was a listing of newspaper outlets and their addresses. And he was presently sitting at a table in the library. Agents closed in and arrested Lewis, but really, all they had enough evidence to charge him with was the extortion.

Leann turned herself in, and although investigators bargained with her to turn against her husband, she refused. The only thing they could charge her on was social security fraud, for using identification that wasn't hers during her prior employment. About a year after the Tylenol killings, James Lewis was convicted of attempted extortion. He got 10 years in prison to be served consecutively with 10 years he got for other charges, including tax fraud. While in prison, Lewis wanted to work with authorities to try to solve the poisonings. They agreed in the hopes he would slip up. He didn't, but neither would he take a lie detector test. Lewis was a model prisoner for 13 years until his release in 1995. Leann was still waiting for him.

Since then, Jim Lewis has been living in the Boston area. He maintains he was a political prisoner, who served as a scapegoat for an inept investigation. Lewis says that while he sat in jail, and since he remains a prime suspect, the real Tylenol murderer has gone free. In 2004, Lewis was arrested for the rape and kidnapping of a woman in Cambridge, Massachusetts. Despite being held for three years awaiting trial, the case was dropped on the eve of the court proceedings, because the victim decided she wouldn't testify.

In 2009, the FBI searched Lewis's home as part of what they called a, quote, "ongoing investigation," citing the possibility of new evidence and technological advances that might finally help them solve the crime. My guess is they were referring to what's called "touch DNA" testing, developed just about that time, in which only a handful of human skin cells, maybe as few as six to eight, can create a DNA profile. Agents were seen leaving Lewis' home with boxes of papers and an Apple computer, they also searched a storage facility in the Cambridge area. But no charges were filed as a result. In 2010, the FBI asked James Lewis and his wife to submit DNA samples and fresh fingerprints, and they complied.

Unbelievably, also in 2010, Lewis released a self-published novel that he had the audacity to title *Poison!* He gave a 45-minute interview to a Boston public access show to try to promote the book. You can find video clips of the interview on the Internet, and the whole thing is very bizarre. Lewis said the book is totally fictional and is set in the kind of rural area he grew up in. The storyline is about a well that is intentionally contaminated with mercury, causing people in the area to die. He said there is no message about the Tylenol case intended by his work, and still maintains he had nothing to do with it. He has also said that if he had any idea of what the extortion letter did to the victims, by directing the investigation onto him, instead of the real killer, well, he'd never have written it.

Another strange twist: In 2011, the convicted and imprisoned so-called Unabomber, Ted Kaczynski was also asked by the FBI to submit a DNA sample for evaluation in the 1982 Tylenol killings. He grew up in the Chicago area and at the time of the Tylenol case, Kaczynski's parents still had a home there that he occasionally visited. Four of his attacks occurred in Chicago, between 1978 and 1980. Kaczynski tried to bargain with the FBI, filing a motion saying he would give the sample if they would stop the pending auction of his belongings, including items seized from the Montana cabin where he had lived. The auction was an attempt to raise money for his victims. The Unabomber said the 30,000 pages of his journals, in particular, would show he couldn't have committed the Tylenol poisonings. But the auction went on.

2012 marked the 30th anniversary of the Tylenol murders. That year, media reports indicated that a grand jury might be convened to indict James Lewis in the seven murders. That never happened. Even lead investigators on the case disagree on whether or not Lewis is actually responsible for the killings. Today, some describe the case as one of the earliest examples of what many would now call domestic or homegrown terrorism. Either James Lewis, or someone else, randomly targeted the general public to cause fear, and I'm quoting here from one of the men involved in the investigation from its beginning, and "bring the country to its knees."

Since then, America, among other places in the world, has sadly seen all too many examples of this kind of random violence—cases in which targets aren't hand-selected by madmen who single out victims for their own demented reasons. Instead, we've seen the indiscriminate killing of groups of fellow Americans, such as in the Sandyhook and other school shootings, the Boston Marathon bombings, the shootings at Fort Hood Texas, or the Oklahoma City bombing. Some experts disagree with calling the Tylenol killings terrorism, because by their definition terrorism involves a clear motive or goal that's politically or religiously motivated—and often involves a person or group claiming responsibility. At this point, the 1982 Tylenol deaths themselves are no mystery. In a span of just three days, seven people were killed using cyanide-laced acetaminophen capsules. What remains unknown is who did this terrible thing and why?

In the fall of 2013, the FBI announced it could no longer serve as lead investigator in the killings. The case has just grown too cold. Despite ever-advancing forensic technology, there are just no fresh clues and all existing leads have been exhausted. I guess that means they still found no definite connection to James Lewis, Ted Kaczynski, or anyone else at this point. Still, Chicago-area police have not given up; several agencies have started their own task force to continue to pursue the Tylenol killer.

Copycats and Hoaxes

Lecture 7

The Tylenol murders of 1982 gave rise to changes in product packaging, federal anti-tampering laws, increased in-store surveillance, and new insurance protections for manufacturers. They also prompted a rash of copycat attacks and related hoaxes. Within four weeks of the last death in the Tylenol series, 270 cases of suspected product tampering were reported to the U.S. Food and Drug Administration (FDA). Some of the incidents were thought to be a form of mass hysteria, but the FDA was able to substantiate that in 36 of the cases, a product had been altered. In this lecture, we'll explore a number of other famous copycats and hoaxes.

Stella Nickell

- In 1986, a copycat named Stella Nickell of the Seattle area used the Tylenol case as inspiration to kill her husband, Bruce. First, she bought life insurance on him, then decided to commit additional tamperings to disguise him as her intended target. Nickell put cyanide into packages of Extra Strength Excedrin and Anacin-3; as a result, not only did her husband die, but six days later, so did Susan Snow. In Snow's death, the cyanide was discovered at autopsy and linked to the product found at her home.

- Amazingly, Nickell herself came forward to claim that her husband—whose cause of death had been ruled as emphysema—had taken Excedrin of the same lot number publicized by the media after Snow's death. Bruce Nickell had already been buried, but the hospital still had a sample of his blood, and tests found cyanide in it.

- Stella Nickell, along with Snow's husband, instigated wrongful death lawsuits against Bristol-Meyers, the company that manufactured Excedrin. An FBI investigation found cyanide in the two packages Nickell turned in, the bottle from Snow's home, and two other tainted packages on store shelves.

- In those five packages, investigators also found green flecks that they determined were from an algaecide commonly used in fish tanks. Agents knew from a visit to her home that Nickell had an aquarium. They canvassed local pet stores and found a clerk who identified Nickell as a customer.

- The FBI's suspicions grew even greater after Nickell told agents that she had bought the two tainted packages she had on two separate occasions and from different stores. Digging further into Nickell's background, agents discovered the recent insurance policies on Bruce Nickell and found that his signatures on them had been forged.

- The break in the case came in early 1987 when Nickell's daughter from a prior relationship told police that her mother had spoken about wanting to kill her current husband. The daughter knew that Nickell had even researched poisoning methods at the public library.

- Why did Stella come forward if all she wanted was to get rid of her husband? The insurance carried on Bruce Nickell paid an additional $100,000 if his death was ruled an accident, plus there was the chance of cashing in on a lawsuit against Bristol-Meyers. Nickell was sentenced to 90 years in prison in the first case prosecuted under the new federal laws prompted by the Tylenol murders.

The Pepsi Hoax

- There is a technical difference between a fraud and a hoax. Fraud nets the perpetrator some financial or personal gain, whether by harming someone else or not. A hoax is an act that's humorous, malicious, or both but not intended to gain any real benefit for the originator. A case from 1993 involving reports of hypodermic needles found in Pepsi cans serves as an example of a hoax.

- The first report came from an 82-year-old man in Tacoma, Washington. Within the next week or so, more than 60 reports of hypodermic needles found in Pepsi cans spanned 24 states, and the story became headline news. In addition to hypodermic syringes, also reported were screws, a sewing needle, a bullet, and a crack cocaine vial.

- Pepsi brought the media into its bottling plants to demonstrate how unlikely it was that the reports were true. The plants produced 2,000 cans a minute, each one open for just 0.9 seconds. At that rate, tampering was literally impossible. Pepsi refused to issue a recall, and eventually, it was determined that not one of the claims was true; the allegations represented a vicious copycat cycle.

- Pepsi's CEO went on television with the head of the FDA to emphasize that the penalty for fraudulent product-tampering claims was five years in prison and a $250,000 fine. The CEO promised that Pepsi would seek out and prosecute hoaxers. In the end, 53 people were arrested, and the incident cost the company an estimated $35 million.

Piltdown Man

- The story of Piltdown man is one of the most famous scientific hoaxes in history. In 1912, an amateur archaeologist named Charles Dawson reported on five thick skull fragments of what appeared to be a primitive hominid discovered in the Piltdown quarry in Sussex, England. Dawson was affiliated with the British Museum and was delighted with the find, which seemed to show that a cradle of humanity was in England.

- The specimen was given the scientific name *Eoanthropus* (meaning "dawn of man"), followed by Dawson's self-tribute—*dawsoni*. It showed both human and ape characteristics and was quickly touted as the missing link scientists had been seeking for nearly half a century. Other hominid fossils had been found in Germany, France, and Asia.

- Two main discoveries were made: One at Piltdown in 1912 and one allegedly found two miles away in 1915. Dawson brought the British Museum two partial skulls, half of a lower jaw, and one canine tooth, all attributed to *Eoanthropus dawsoni*, along with a tool made of carved elephant bone and the fossil teeth from a number of other animals from the Pleistocene Epoch.

- From the start, both American and French paleontologists criticized the Piltdown find, but objections quieted down a bit after papers were published about the second skull in 1917. Piltdown man took his place in fossil history. It wasn't until 1953 that the scientific community finally learned that Piltdown man was a complete hoax.

- Existing cracks in the Piltdown story were significantly weakened in 1949 when the specimens were analyzed by British physical anthropologist and paleontologist Kenneth Oakley, one of the pioneers of dating fossils using fluorine content. Oakley determined that the two Piltdown skulls were from humans of the medieval period, probably about 620 years old. The jaw was about 500 years old and from a Borneo orangutan.

- Later testing showed that a weak acid solution had been used to remove some of the mineral content from the skull fragments, and a stain containing iron had been applied to age them. The molars from the orangutan jaw had been filed down flat to appear more humanlike. Microscopic examination showed that the canine tooth had also been filed down and covered with brown paint; a piece that had been broken was reattached with chewing gum covered with sand.

- To this day, no one knows for sure who tainted and planted the specimens—whether it was Dawson alone or in collaboration with others or whether Dawson was given the finds and simply took credit for them. It has been suggested that the doctoring of the evidence was done at the same time, after which the quarry was salted with the specimens.
    - The perpetrator had to have a certain amount of paleontological and anatomical knowledge to pull off the scam, which is why some suspect that Dawson, an amateur, was a pawn in a game created by others. Still, some of Dawson's other finds have been determined to be hoaxes in modern times.

- Among the other suspects in the Piltdown hoax are Sir Arthur Conan Doyle, the creator of Sherlock Holmes; a zoologist named Martin Hinton who worked at the British Museum; and Arthur Smith Woodward, the head of the museum's Natural History Department.

Sir Arthur Conan Doyle may have been involved in the Piltdown hoax; he was an amateur bone hunter and actually dug at Piltdown with Charles Dawson.

- In 2012, the 100th anniversary of the Piltdown discovery, scientists made new attempts to conduct sophisticated testing on the specimens. DNA, however, has been difficult to extract. Modern radiocarbon dating puts the estimated age of the skull fragments at about 1,000 years, twice the 500 years that Oakley's fluorine analysis suggested but nowhere near old enough to contribute to our understanding of human evolution.

The Disappearance of John Stonehouse

- During a business trip to Florida's Miami Beach on November 20, 1974, British cabinet minister John Stonehouse told companions that he was going for a swim. When he didn't return to the hotel, investigators found a pile of his clothing on the beach. Apparently, Stonehouse had drowned or, perhaps, was attacked by a shark.

- Stonehouse, who was 49 at the time of his disappearance, had held several political posts for his country and had aspirations to become prime minister. But a string of failed business ventures in the early

1970s had tarnished Stonehouse's political future, and by the time of his disappearance, bankruptcy was looming. Stonehouse had been married for more than 25 years and had three children.

- Some of Stonehouse's friends were suspicious that members of organized crime or opposing political factions might have killed him, but a month after his disappearance, Australian police found Stonehouse living in a Melbourne resort community.
    - He had entered the country from Hawaii one week after disappearing, using a forged passport in the name of J. D. Norman. Once in Australia, Stonehouse called himself Clive Mildoon and opened bank accounts.

    - Further investigation into his movements showed that Stonehouse had left Australia the day after he first arrived there and traveled to Lebanon, Singapore, and Denmark (where he stayed for a time with his 28-year-old secretary, Shelia Buckley) before returning to Australia alone on December 10.

- Prior to his disappearance, Stonehouse had obtained passports and credit cards in several names and a birth certificate for Clive Mildoon. With those, he had opened Australian and Swiss bank accounts using overdraft loans from his U.K. banks. With that knowledge, Australia deported Stonehouse to face British authorities.

- Stonehouse claimed that he had suffered a mental breakdown and that his alternate names were akin to multiple personalities. At his trial, the prosecution asserted that the breakdown was moral, rather than mental. In the summer of 1976, Stonehouse was sentenced to prison for seven years on various counts, including fraud, deception, and theft. After his release, he married his former secretary and wrote spy thrillers. He died in 1988 at the age of 62.

Suggested Reading

Farquhar, *A Treasury of Deception.*

Innes, *Fakes and Forgeries.*

Spencer, *Piltdown.*

Weiner, *The Piltdown Forgery.*

Questions to Consider

1. What were some of the slip-ups Stella Nickel made that allowed police to follow her trail of evidence and apprehend her?

2. What's the key difference between a fraud and a hoax?

3. Given the nature of science, what are some of the dangers in scientific hoaxes, such as that of the Piltdown man?

Copycats and Hoaxes

Lecture 7—Transcript

The Tylenol murders of 1982, in which cyanide-tainted medication killed seven people in Chicago, give us a noteworthy example of the interactions between crime, forensic science, and society. That incident gave rise to significant changes in product packaging, resulting in the types of tamper-proof containers we've become accustomed to today. Federal anti-tampering laws, enacted in 1983 and '84, made crimes like that a federal offense. Stores increasingly began to install surveillance cameras. In addition, insurance companies started offering manufacturers protection against tampering, whether intended or accidental, to cover the costs of product recalls, or related lawsuits, if an incident occurred.

The second phenomenon the Tylenol murders prompted was a rash of copycat attacks, as well as related hoaxes. The last death in the Tylenol series occurred on October first, and within the four weeks following, 270 cases of suspected product tampering were reported to the U.S. Food and Drug Administration, known as the FDA. Some of the incidents were thought to be a form of mass hysteria, in which anybody who felt sick, after eating or drinking anything, wondered about having been poisoned. But the FDA was able to substantiate that 36 of those 270 cases were legitimate acts of small-scale terrorism, in which a product had been altered.

For instance, at the end of that October, several Halloween trick-or-treaters discovered pins and needles in candy items in areas as diverse as New York, Connecticut, Florida, and Chicago. In Detroit, razor blades and nails were found in packages of hot dogs, a Minneapolis 14-year-old was sickened by chocolate milk tainted with lye, and a 27-year-old Florida police officer drank insecticide-poisoned orange juice and began vomiting immediately. Luckily, both of them recovered.

In 1986, a copycat named Stella Nickell, of the Seattle area, used the Tylenol case as inspiration to kill her husband, Bruce. First, she bought a bunch of life insurance on him, and then decided to commit additional tamperings to disguise him as her intended target. Nickell put cyanide into packages of both Extra Strength Excedrin and Anacin-3, and as a result, not only did her husband die, but six days later, so did Susan Snow, who picked up one of Nickell's deadly packages at a local store. The cyanide was discovered at Snow's autopsy, linked to the product found at her home, and then the media released an alert.

Amazingly, Nickell herself came forward to claim that her husband, whose cause of death had been ruled emphysema, had taken Excedrin, of the same lot number publicized by the media, after Snow's death. Bruce Nickell had already been buried, but since he was a registered organ donor, the hospital still had a sample of his blood. Toxicology results revealed cyanide, and luckily, no one received any of his donor tissues. Stella Nickell, along with Snow's husband, instigated a wrongful death lawsuit against Bristol-Meyers, the company that manufactured Excedrin. An FBI investigation found cyanide in the two packages Nickell turned in and the bottle from Snow's home, and two other tainted products discovered on store shelves. But in those five packages, investigators also found mysterious green flecks. They determined these were from an algaecide commonly used in fish tanks. The FBI's chemical analysis even named the specific brand, a compound marketed as Algae Destroyer.

Agents knew from a visit to her home that Nickell had an aquarium and figured she could have used the algaecide container to crush the cyanide crystals in, so the next thing they did was canvass local pet stores, using a photo array that included Nickell's picture. One clerk identified Nickell's image, saying the woman stood out because she had a bell jingling from her purse. The clerk said the lady was a regular purchaser of fish supplies, including the type of algaecide found in the tainted capsules.

The FBI's suspicions grew even greater after Nickell told agents she bought the two tainted packages on two different occasions and from two different stores. Now, the odds of her picking up two of only five poisoned bottles discovered, just seemed incredibly unlikely. At an FBI interview, where Nickell refused a polygraph, the agent noticed a bell dangling from her purse. Digging further into Nickell's background, they discovered the recent insurance policies, and FBI handwriting experts discovered that Bruce Nickell's signatures on them had been forged.

The real break in the case came in early 1987, when Nickell's daughter from a prior relationship, told police that her mother had spoken about wanting to kill her current husband. The daughter knew Nickell had even researched poisoning methods at the public library. Agents went to the local library and asked about a patron named Stella Nickell. Librarians showed them an overdue notice for a book Nickell never returned—its title: *Human Poisoning*. She'd also checked out other books on the subject, and in them, FBI agents found no less than 84 of Nickell's fingerprints, most on pages related to cyanide, including in a book titled *Deadly Harvest* and the C volumes from several of the library's encyclopedias.

But why did Stella come forward, if all she really wanted was to get rid of her husband? Well, the insurance carried on Bruce Nickell paid an additional $100,000 if his death was ruled an accident. Investigators learned that Nickell had questioned her husband's doctors in the week following his death about their ruling he died from emphysema. Plus she had the chance of cashing in on the lawsuit against Bristol-Meyers. But Nickell was sloppy. At trial, the tainted packages were described with seals clearly cut and obvious re-gluing on the boxes. Sadly, that was not enough to dissuade Susan Snow, the other casualty, from taking the medication. Nickell was sentenced to 90 years in prison in the first case prosecuted under the new Federal laws prompted by the Tylenol murders.

If there can be a lighter note to product tampering, this is it. In 2010, a Long Island couple, 68-year-old Alexander Clement and his wife Christine, who was 64, they were charged with petty larceny and product tampering. Their target was Jell-O brand pudding mix. Alexander drove Christine to four different stores, where she bought 10 boxes of the pudding mix each time; their favorites were apparently pistachio and butterscotch. The couple brought the packages home, presumably ate the pudding, and then filled plastic sandwich bags with aquarium sand and salt, put them in the Jell-O boxes, resealed the packages, and returned them to the stores for the $1.40 refund. The fraud came to light when other customers bought the tampered packages and complained to the stores. Surveillance tapes from security cameras identified the Clements. They were charged, but police say they didn't really think the couple meant any real harm. They were very apologetic, and Christine suffered from what they called an age-related mental issue.

Now, technically there's a difference between fraud and what's considered a hoax. Fraud nets the perpetrator some type of financial or personal gain, whether by harming somebody else or not. A hoax is something that is humorous, malicious, or maybe both, but not intended to gain any real benefit to the person who originates it. Let's look at an example. In 1990, an Ontario, Canada grocery store clerk found a hypodermic syringe in a sealed Pepsi bottle. Although authorities never got to the bottom of it, they eventually assumed the likely source was a disgruntled employee from the bottling company. But three years later, in June of 1993, in the span of just a couple days, reports of hypodermic needles being found in Pepsi cans cropped up around the United States in what was a media frenzy.

The first report came in from an 82-year-old man in Tacoma, Washington, who said he poured himself a Pepsi one day, and then, the next day, retrieved the can because of a contest he heard the soft drink company was running. Inside the can the man saw a syringe, with its needle bent, so he contacted his attorney, who reported the incident to the police. The story quickly found its way to the media, and within the next week or so, over 60 reports of hypodermic needles being found in Pepsi cans spanned 24 states. The story was headline news. Products included Pepsi, Diet Pepsi, Caffeine-Free Diet Pepsi. In addition to hypodermic syringes, also reported were screws, a sewing needle, a bullet, and a crack-cocaine vial.

Now, Pepsi brought in the media on their end to demonstrate how unlikely it was these reports were true. Their automated plants produced 2,000 cans a minute, each one open for just nine-tenths of a second. At that rate, tampering was literally impossible. They refused to issue a recall. Eventually, it was determined that not one of the claims was true; the allegations were just a vicious copycat cycle in which more and more gold diggers and attention seekers jumped on the tampering bandwagon to discredit Pepsi or get their five minutes of fame.

The FBI arrested at least 20 people who had planted foreign objects in soft drink containers. Now, at that point, to try to stop the madness, Pepsi's

CEO went on NBC's *Nightline* with the head of the U.S. Food and Drug Administration. They emphasized the penalty for fraudulent product tampering claims was five years in prison and a quarter-million dollar fine, adding they would continue to seek out and prosecute hoaxers. Now people started coming forward to admit false claims, because they didn't want to be prosecuted. Shortly after that, Pepsi placed full-page ads and discount coupons in national newspapers, thanking the American public for believing in their product. But between that hoax and the cashed-in coupons, the incident cost Pepsi an estimated $35 million. All in all, 53 people were arrested in the nationwide hoax.

Even the elderly man who first reported the incident was discredited after authorities discovered one of his relatives injected insulin and probably had placed the syringe in the can to protect others; that's a practice recommended by the American Diabetes Association. Also, the can the man had brought to his attorney was a Diet Pepsi, while the 24-pack he claimed it came from was regular Pepsi. Perhaps he had just gotten confused and picked up an empty can he thought was from the day before but really had grabbed one his diabetic relative placed a syringe in, say the last time she visited.

A major thread of forensic evidence throughout the case was that investigators could find no correlation among the allegedly contaminated cans. They came from multiple plants around the U.S., and some cans had been filled days prior to being reported, others weeks before, while others were months after they left the plant. One source stated the saddest part of this story is that many people remember the needles in Pepsi cans as fact, rather than as a nationwide scam intended to sabotage a major soft-drink manufacturer. So hopefully we've dispelled that in telling this tale.

As an anthropologist, I have to say that one of my favorite false claims, if one may have a favorite, is the story of Piltdown Man, one of the most famous scientific hoaxes, or, some might say frauds, in all of history. In 1912, an amateur archaeologist named Charles Dawson, reported on five thick skull fragments of what appeared to be a primitive hominid discovered in the Piltdown quarry in Sussex, England. Dawson was affiliated with the British Museum and was delighted with the find, particularly, since it showed that a cradle of humanity was in England. The specimen was given the scientific name *Eoanthropus*, which translates to the "dawn of man," followed by Dawson's self-tribute, *dawsoni*. *Eoanthropus dawsoni* showed both human and ape characteristics and was quickly touted as the big missing link scientists had been searching for for nearly half a century.

Neanderthals had been discovered as early as 1856 in the Neander Valley of Germany, just three years before Charles Darwin published *On the Origin of Species* in England. By the late 1860s, Cro-Magnons had been found in France, and in 1871, Darwin published *The Descent of Man*. But after that, the fossil focus shifted to Asia, where Java Man was discovered in 1891, then back to Germany for Heidelberg man in 1907. As tensions between Britain and Germany increased, in the prelude to World War I, perhaps Dawson just grew tired of there being no great hominid-fossil finds in England.

There were two main Piltdown discoveries, one at Piltdown in 1912, and a second cache in 1915, allegedly found some two miles from the quarry at an undisclosed location. Between those two finds, Dawson brought the British Museum two partial skulls, half of a lower jaw with molar teeth in it, and one canine tooth that fit the jaw, all attributed to his *Eoanthropus dawsoni*. Plus, he had a tool made of carved elephant bone, fossil teeth from a number of other animals from the Pleistocene epoch, which lasted from about 1.5 million to 10,000 years ago.

From the start, both American and French paleontologists criticized the Piltdown find, saying the first skull, and its supposedly related jawbone, could not possibly be from the same individual. But objections quieted down a bit after papers were published about the second skull in 1917, and Piltdown man took his place in fossil history. Some scientists simply chose to ignore the Piltdown fossils as an aberration. In fact, physical anthropologist Sherwood Washburn stated, "I remember writing a paper on human evolution in 1944, and I simply left Piltdown out. You could make sense of human evolution if you didn't try to put Piltdown into it."

It wasn't until 1953, over 40 years since the initial discovery, that the scientific community finally learned that Piltdown man was not only an incredible contribution to paleontology, it was a complete hoax. Already-existing cracks in the Piltdown story were significantly weakened in 1949, when the specimens were analyzed by British physical anthropologist and paleontologist Kenneth Oakley. He was one of the pioneers of dating fossils using fluorine content. Oakley determined that Piltdown skulls one and two were both from modern humans of the medieval period, likely about 620-years old. The jaw that many scientists said did not belong with the skull was about 500 years old and from a Borneo orangutan; the canine was from a Pleistocene chimpanzee; the elephant fossils likely from Tunisia; and the fossil hippo tooth was probably from Malta or Sicily.

Later testing showed that Dawson, or someone else, used a weak acid solution to remove some of the mineral content from the skull fragments and then applied some kind of a stain containing iron to age them, as well as the jawbone, to make the specimens look consistent with fossils discovered in gravel, where iron is common. The molars from the orangutan jaw had been filed down flat to appear more human-like, and the joint surfaces where the jaw would fit onto the skull were broken off before the staining, so it wouldn't be so obvious that the jaw didn't anatomically fit the skull. Microscopic examination showed the canine tooth was also filed down and actually covered with brown paint. A piece of the tooth that had been broken was attached with, are you ready for this, chewing gum covered over with sand.

To this day, no one knows for sure who actually tainted and planted the specimens, whether it was Dawson alone, him in collaboration with others, or he was handed the finds, maybe even for money, and simply just took credit for them. It's been suggested, though, that despite the two main discoveries being a couple of years apart, the doctoring of the evidence was done at the same time, after which the quarry had either been salted with the specimens on a single occasion, taking a real chance that the alleged fossils would ever be found, or that Dawson returned to the quarry several times to plant and then allegedly discover them. The thing is, though, other than the jawbone, which Dawson himself dug up, anonymous workers at the quarry actually made the discoveries, but, maybe in exchange for payment from Dawson or others.

In any event, the perpetrator had to have a certain amount of paleontological and anatomical know how to pull the scam off, and Dawson was just an amateur. That's why some sources suspect he may have been a pawn in a bigger game created by others. Dawson was famous, though, for his amazing discoveries in the world of antiquities and fossils. He was known locally as the Wizard of Sussex. It's since come to light that some of those finds were other hoaxes of his day. In 2003, Bournemouth University analyzed some of Dawson's famous antiquities and found at least 30 of them were fakes.

Interestingly, Sir Arthur Conan Doyle, of Sherlock Holmes fame, he's also been suggested as having been involved, or even being the main perpetrator. He was an amateur bone hunter in his day; he was also a neighbor of Dawson's, and he actually dug at Piltdown with him. Doyle was an avowed spiritualist, and some authors have suggested he promulgated that Piltdown hoax to get back at science for its attack on psychics and other aspects of his beliefs.

It's been argued that Doyle's 1912 work, *The Lost World*, published the same year as the first Piltdown discovery, provides a veiled literary account of the hoax. Others argue there's no way he could have planted the bones and realistically expected them to be found by either Dawson or the quarry workers. Another theory suggests the perpetrator was a zoologist named Martin Hinton who worked at the British Museum at the time. In the 1970s, workers found an old, dusty trunk with his initials on it in the museum's loft. Inside the trunk they found teeth and bones that were stained the same way as those from the Piltdown hoax. They also found stained pieces of both elephant and hippo fossil teeth that looked as though they came right out of the Piltdown quarry with the others of their exact same species. Scientific analysis demonstrated the staining methods on the items in Hinton's trunk matched those in the Piltdown case.

However, in 1899, years before the Piltdown discoveries, Hinton had published a paper about nature's effects on bones and teeth when they fossilize in gravel. So, is it possible that the specimens in Hinton's trunk were created in the lab, maybe to try and duplicate the natural processes he was seeing in the field? If so, could Dawson have used some of them to salt the quarry? In addition, Hinton had done work for Oakley, the scientist who exposed the Piltdown fraud using fluorine dating, so could Hinton have been trying to replicate the hoax in the lab, maybe to help Oakley discredit the

Piltdown hoax? None of that seems likely since Hinton died in 1961, years after the hoax was revealed, yet never said a word about any confirmatory studies he had done to try to expose the fraud or any other lab experiments related to staining bones or teeth.

Another problem with Hinton as a suspect is he really didn't have the access to the Piltdown site that others from the British Museum had. So, if it was Hinton, he had to have an accomplice in the field. Another theory is that someone else planted a handful of the never-used Piltdown fakes in Hinton's storage at the Museum; otherwise, if Hinton made the decoys, why wouldn't he have destroyed any possible evidence after the hoax was revealed? As for Dawson, he died of blood poisoning at the age of 52, just a year after the second Piltdown skull fragments were discovered. Fortunately, no one has said anything suspicious about Dawson's death. I mean, it's one thing to fake a man, it's another thing to kill one.

The last prime contender worthy of mention in this case is Arthur Smith Woodward, the head of the British Museum's Natural History Department at the time. Woodward, who was then vying for the position of Museum Director, would have benefitted greatly from the Piltdown discovery. He was the only person who had consistent and unlimited access to all the Piltdown specimens and the one who, following Dawson's death, reported the second skull as having been discovered by Dawson a year earlier. Also suspicious is that Woodward was highly reluctant to allow any real scientific testing on the materials discovered. He essentially kept the original specimens locked up for nearly 40 years and provided access to very few researchers, preferring they work with plaster casts, instead. Some say Woodward was just incompetent, not really malicious, since he did return to Piltdown year after year to continue the search.

It's hard to imagine how on Earth those makeshift fossils weren't exposed long before they were. Some say both British nationalism and cultural racism influenced people to believe in Piltdown man's authenticity. Europeans, and some Americans, wanted to cling to the theory that Europe was the cradle of mankind, despite concurrent finds of true hominids being made in Asia and Africa.

In 2012, the 100th anniversary of the Piltdown discovery, scientists made new attempts to conduct sophisticated testing on the specimens. DNA has been difficult to extract, though, likely due to the acid used on the remains before staining. Researchers have said the skull is pathologically thicker than normal and may have been chosen for the hoax precisely for that reason. Modern radiocarbon dating puts the estimated age of the skull fragments at about a 1,000 years; that's twice the 500 years that Oakley's fluorine analysis suggested, but nowhere near old enough to contribute to our understanding of human evolution.

Now, since we've been talking about famous dead Britons, whether real or not, let's continue that theme with the story of British Cabinet Prime Minister John Stonehouse. During a business trip to Florida's Miami Beach on November 20, 1974, Stonehouse told companions he was going for a swim. When he didn't return to the hotel, investigators found a pile of his clothing on a nearby beach. Apparently, Stonehouse had either drowned or maybe was attacked by a shark.

Stonehouse, who was 49 at the time of his disappearance, had held several political posts for his country, including Aviation Minister in the '60s, and the UK's 100th Postmaster General. At one time, he had been projected to be a future Labour Party leader and had aspirations to become Prime Minister one day. But a string of failed business ventures in the early 1970s had tarnished Stonehouse's once-bright political future. His ventures had been called a money-go-round, where he would pass unsecured fraudulent loans from one of his companies to another; by the time of his disappearance, bankruptcy was looming. Stonehouse had been married for over 25 years to his wife, Barbara, and the couple had three children. The Parliament Member's obituary was released the same month as his disappearance, despite never finding his body.

Some of Stonehouse's friends and allies suspected that organized crime mobsters or opposing political factions might have killed him. Back in the 1960s, allegations had been made that Stonehouse was a spy for the Czech government and may have been associated with the U.S. Central Intelligence Agency. The U.K.'s Military Intelligence Section 5 had investigated those charges in 1969 after a Czech defector informed on Stonehouse, but MI-5 ultimately decided the spying allegations were false.

A month after his disappearance, on December 24, Australian police found Stonehouse living in a Melbourne resort community. He had entered the country from Hawaii, one week after disappearing, using a forged passport in the name of J. D. Norman. Once in Australia, Stonehouse was calling himself Clive Mildoon and opening bank accounts, but he'd also created accounts in the name of Joseph Markham. Further investigation into his movements showed Stonehouse as having left Australia the day after he first arrived there, traveling to Lebanon, Singapore, and Denmark, where he stayed with his 28-year-old secretary, Sheila Buckley for a while, before coming back to Australia, alone, on December 10.

Australian police had been watching him, sitting in restaurants reading British newspapers, but not because they thought the man was John Stonehouse; it was because they suspected he was Richard Bingham. Bingham. Now Bingham, otherwise known as Lord Lucan, was a professional gambler, once considered for the role of James Bond. Lucan had disappeared from England the month prior after his children's nanny was found bludgeoned to death in his estranged wife's basement. When they apprehended Stonehouse, investigators asked him to pull down his pants, since murder-suspect Bingham had a six-inch scar on his right thigh. The man wasn't Bingham, so who was he?

Prior to his disappearance, Stonehouse had obtained passports and credit cards in several names and a birth certificate for Clive Mildoon. With those, he had opened Australian and Swiss bank accounts, using overdraft loans from his U.K. banks. With that knowledge, Australia deported Stonehouse to face British authorities. While out on bail, he had the nerve to return to his seat in Parliament, even though it had already been turned over to somebody else.

Apparently, Stonehouse's disappearing act was intended to create a new life for himself and his secretary, Sheila Buckley, or at least that's what he told her. Stonehouse claimed he had suffered a mental breakdown that he called psychiatric suicide, and his alternate names were akin to multiple personalities. Turns out his aliases were the names of some of Stonehouse's dead constituents. Less than a year before disappearing, Stonehouse had called local hospitals to gather names of men around his age who had recently died. He then visited the men's widows, under a ruse, to find out whether they held passports at the time of their deaths.

At his trial, the prosecution claimed it was moral, rather than mental breakdown, that was behind Stonehouse's actions. After serving as his own attorney in the summer of 1976, Stonehouse was sentenced to prison for seven years on various counts, including fraud, deception, and theft. His secretary got two years in prison, but that was suspended. After serving about half his sentence, Stonehouse was set free, having suffered three heart attacks during as many years in prison and undergoing heart surgery. He and Buckley married in 1981, and the couple had a son. Stonehouse began writing, including spy thrillers, which Buckley typed for him, but in 1988, Stonehouse died of his final heart attack, at age 62.

In 2010, the British National Archives declassified papers confirming Stonehouse had, in fact, served as a paid spy for the Czech government, making him the only U.K. politician to be proven a spy, while a minister of parliament. Records showed that in 1980, Prime Minister Margaret Thatcher herself agreed to keep the knowledge confidential, and not prosecute Stonehouse, who was just getting out of jail at the time. Even MI-5 thought nothing would be gained from going after him.

Returning to the notion of copycats, two years after Stonehouse's disappearance, the U.K. television program *The Fall and Rise of Reginald Perrin* was released. It's a sitcom about a guy who starts a new life after leaving his clothes on a beach. One final note: Suspected killer Lord Lucan, the man Interpol thought Stonehouse was when they tracked him to Australia, he's never been found.

Frauds and Forgeries
Lecture 8

There are three main ways to perpetrate art fraud: (1) Create a piece of artwork and sell it as an original by someone famous, (2) resell a known fraud as an original, and (3) credit an existing piece of artwork to a known artist and sell it for a high price. In this lecture, we'll look at two interesting cases that fall into the first category. We'll also explore the hoax perpetrated by the writer Clifford Irving, who claimed that he had been authorized by Howard Hughes to write the billionaire's biography.

Elmyr de Hory

- In 1946, in Paris, France, a Hungarian named Elmyr de Hory sold a Picasso drawing to Britain's Lady Malcolm Campbell for a little less than $100. Shortly thereafter, de Hory began selling other Picassos to art galleries around Paris. He explained that his parents had been aristocratic Jews killed in the Holocaust, and the remarkable pieces of artwork were remnants of their vast estate prior to imprisonment.

- Using some of his proceeds, de Hory traveled, selling other pieces of artwork for even larger sums. He ultimately settled in Miami, Florida, in 1950 and worked as an art dealer. He sold works from Picasso, Matisse, Renoir, and Modigliani.

- As you might guess, de Hory created all of these masterpieces himself. A key to his success was that he didn't reproduce existing works but drew and painted new pieces in the style of great artists. He also used pseudonyms and sold artwork through the mail, enabling him to stay one step ahead of authorities for many years.

- At times, de Hory struggled to become a celebrity artist in his own right, but he continued to find more success in forgery. His life was marked by stormy personal relationships, including with associates in his con game who cheated him out of millions of dollars. He suffered through occasional legal battles when deceptions were detected and was essentially on the run throughout Europe.

- Eventually, de Hory decided to go to Spain and face the music. Though the courts there tried to convict him of numerous charges, de Hory wound up with a sentence of only two months in jail because no one could prove that he had created any of the fraudulent artwork within Spanish borders. There were also questions about the legality of calling his works forgeries because de Hory rarely signed the works with the name of the artist he was imitating.

- His crimes made de Hory infamous, and once he was released from prison, the value of his forgeries increased. Not only that, but his celebrity created public interest in his own artwork. He appeared on television shows, and the author and investigative reporter Clifford Irving wrote his biography in 1969.

- But in late 1976, when de Hory found himself faced with criminal extradition to France, he took an overdose of sleeping pills and ended his life. It's unknown how many forgeries he created or how many are still in circulation and presumed authentic.

John Myatt

- In 1985, John Myatt was a struggling art teacher in England when a friend offered to pay him to reproduce a famous painting. Later, Myatt's friend remarked that the copy was so good it fooled many people who saw it.
  - To raise cash, Myatt then put an ad in a London magazine indicating that he would create 19th- and 20th-century fakes on request for a cost of $240. He didn't intend to defraud anyone; he just needed money.

  - A con man named John Drewe ordered some masterpieces by Myatt and was so impressed that he persuaded Myatt to work with him to try to pass the fakes off as originals.

- Drewe's ability to fake the provenance of the works allowed the pair to sell more than 200 fake masterpieces. Drewe was able to gain entry to famous art museums and art houses, where he would steal dealers' stamps and gallery labels and change auction records. This strategy allowed the pair to fool dealers, collectors, and even experts at several of London's best art museums.

- When Myatt heard how much his fakes were selling for, he ramped up his game by using an emulsion that helped the paint dry faster and allowed him to produce even more quickly. Even that didn't catch the experts' eyes for several years. But eventually—after almost 10 years of scamming the art world—authorities became suspicious.

- Ultimately, Myatt confessed and implicated Drewe. The two were arrested, tried, and convicted in February of 1999. Myatt served only four months and Drewe, two years.

Authenticating Art

- An important consideration in authenticating any piece of art is that the methods used must be as noninvasive as possible. Considering the analysis of paintings, for example, there are three main steps in forgery investigation, usually undertaken in the following order:
  - The exam begins with the surface of the painting. *Optical microscopy* or *light microscopy* allows forgery experts to view the details of tiny cracks that naturally form on paint over time. Through this examination, experts judge whether the cracks are authentic, were accelerated using solvents, or were even merely drawn on the surface. When ultraviolet light is applied to a painting's surface, old layers of varnish will dramatically fluoresce, while newer paint on a retouched or forged painting will not.

  - If the surface exam doesn't easily weed out a fake, investigators dive deeper into the background, or *underpainting*, of a piece. Artists commonly reuse canvases over time, whether masters or frauds. Conventional medical X-rays and infrared analyses can see through layers of paint and detect earlier artwork created on the same background. If the styles of the surface and the underpainting don't match, forgery may be suspected.

- The last step undertaken is an examination of the mediums used in the body of the surface painting. Polarized light microscopy is a basic way to analyze pigments in paint or even ink. X-ray diffraction studies, which look at how a given medium bends X-rays, can reveal the crystalline structure of components within pigments. Some types of crystalline pigments were only discovered and incorporated into paint during certain eras. Thus, if they're used to fake an old master, X-ray diffraction studies can expose the fraud.

One of the easiest ways to detect an art forgery is to show that anything used to create it—such as paint or other natural material—was not available in the place or time the artwork was allegedly created.

- Techniques called *X-ray fluorescence* and *neutron activation* can take authenticators down to the most basic chemical level. In X-ray fluorescence, the object being examined is bathed with radiation, and the pattern of X-rays it emits reveals its elemental breakdown. With neutron activation, the sample is bombarded with energized neutrons, causing some of the atoms in the sample to take up a neutron and become radioactive isotopes. As these isotopes decay, they send out gamma rays that identify which elements are present in the paint or other medium.

- *Radiocarbon dating*, which also relies on the decay of radioisotopes, can be used to estimate the age of organic constituents, whether canvas, paint, or other carbon-containing media. But part of the sample is destroyed during the dating, and this technique works only in the range of about 50,000 to 400 years in the past.

- When analysis of three-dimensional artwork, such as a sculpture, is required, *CT scanning* (computed axial tomography) can allow internal views. *Magnetic resonance imaging* (MRI) can also reveal internal aspects of ceramics, bronzes, and wooden sculptures. A new method, *molecular Raman spectroscopy*, relies on lasers to identify both nonorganic and organic pigments, as well as binders used in paint. This technique is overtaking X-ray diffraction studies because it can be performed in the field and is completely nondestructive.

Clifford Irving and Howard Hughes

- The year after his biography of the master forger Elmyr de Hory was published, Clifford Irving told his editors at McGraw-Hill that he had been contacted by the reclusive billionaire Howard Hughes. According to Irving, after reading the de Hory biography, Hughes had decided to ask Irving to serve as his biographer.

- At the invitation of McGraw-Hill, Irving flew to New York and presented three letters from Hughes. In these letters, Hughes stated there were far too many rumors and inaccuracies in circulation regarding his unusual life. The reclusive billionaire wanted these misconceptions cleared up before he died, but he wanted the project to remain a secret until publication, with all contact passing through Irving only. McGraw-Hill had the letters authenticated by a nationally recognized firm and issued two contracts: one for Irving and one for Hughes.

- As most people know, Howard R. Hughes Jr. was born in 1905, the only child of a man who became a millionaire developing technology used in oil well drilling.
  - At age 18, Hughes was orphaned and inherited three-fourths of his father's company and wealth. He went on to become a successful movie producer, created the Hughes Aircraft Company, and invested in various casinos and hotels in Las Vegas.
  - Throughout the 1950s, Hughes became increasingly reclusive. Among other rumored maladies, he was thought to have suffered from obsessive-compulsive disorder. From the 1960s until his death in 1976, Hughes lived in seclusion with a small entourage, moving from one hotel penthouse to another in Las Vegas, Beverly Hills, Boston, London, and elsewhere.
- Of course, as we now know, Irving had concocted the biography scheme in conjunction with another author, Richard Suskind. Both men were counting on the assumption that Hughes would not come out of seclusion and probably would never learn of the pending publishing deal. Irving had Hughes's signature forged on the contract from McGraw-Hill.
- Irving got his hands on the unpublished memoir of a former accountant and friend of Hughes, as well some material about Hughes that had been donated to the Academy of Motion Picture Arts. Using these factual sources, he built his phony biography, filling in what he didn't know with lies. By the fall of 1971, Irving and Suskind had a manuscript of more than 1,000 pages. In December of that year, McGraw-Hill made a public announcement about the project.
- Shortly after the announcement, a journalist named Frank McCulloch—the last person known to have interviewed Hughes, back in 1958—received an angry phone call from a man claiming to be the billionaire. The caller said that the book was a hoax

and that he had never even met Clifford Irving. When Irving was questioned, he claimed that the caller was clearly an imposter.

- Following the coverage of these events in the news, Hughes surfaced and announced that he would hold a teleconference to be recorded by seven journalists and aired at a later date. Within weeks of the teleconference, Irving and his wife admitted to the hoax, and Suskind soon followed. Ultimately, the three faced several state and federal charges, including mail fraud, forgery, and perjury. In the end, Clifford pled guilty and served 17 months, and Suskind served 5 months.

Suggested Reading

Farquhar, *A Treasury of Deception*.

Innes, *Fakes and Forgeries*.

Irving, *Fake*.

———, *The Hoax*.

Nilsen, *Art Fraud Detective*.

Shelton, *Forensic Science in Court*.

Warden and Drizin, eds., *True Stories of False Confessions*.

Questions to Consider

1. Can you identify some other famous examples of art frauds or document forgeries throughout history?

2. What kinds of items or materials are forged and why?

Frauds and Forgeries

Lecture 8—Transcript

In 1946, in Paris France, a Hungarian by the name of Elmyr de Hory sold a Picasso drawing to Britain's Lady Malcolm Campbell for a little less than $100. Shortly thereafter, de Hory began selling other Picassos to art galleries around Paris. He explained that his parents had been aristocratic Jews killed in the Holocaust, and the remarkable pieces of artwork were remnants of their vast estate prior to imprisonment.

Using some of his proceeds, de Hory traveled around, selling other pieces of artwork for even larger sums. At times he worked with sales partners; other times de Hory sold on his own. He moved to the United States, ultimately settling in Miami, Florida in 1950, where de Hory worked as an art dealer. He sold Picassos, but also works from Matisse, Renoir, and Modigliani. Not just drawings, but expensive oil paintings, too. Some of his clients included Texas oil baron Algur H. Meadows, the Fogg Art Museum at Harvard University, the National Museum of Western Art in Tokyo, and actress Zsa Zsa Gabor.

There was just one major problem with business; Elmyr de Hory created all of those masterpieces himself. A big key to de Hory's long success was that, unlike most other art forgers, he didn't reproduce existing works, but rather drew and painted brand new pieces in the style of the great artists he was imitating. Reproducing an existing masterpiece means easy exposure if potential clients know the whereabouts of the original or ever come across it. But another part of de Hory's success was that he often used pseudonyms and sold artwork through the mail. That's why he was able to continue peddling forgeries, even after the Museum at Harvard University figured out they had a fake, as did other of de Hory's customers. Authorities tried to investigate the forger's trail, but for many years de Hory stayed just one or more steps ahead of them.

Obviously, in addition to being a consummate con man, de Hory was an amazing artist in his own right. While running from the law, he expanded his repertoire to styles of other sought-after artists, like Chagall, Toulouse-Lautrec, and Degas, among others. Many of de Hory's fakes were even featured in art books, in which they were attributed to the great masters he copied. Brokers and galleries that unknowingly purchased de Hory's forgeries were selling them to others at top dollar, far in excess of what he had been paid for them. The forger began to consider how little he had been paid for his fakes, versus how much they could command.

At times, de Hory struggled to become a celebrity artist in his own right, but continued to find more success in forgery. His life was marked by stormy personal relationships, including his associates in the con game, who cheated him out of millions of dollars, leaving him nearly broke. He suffered through occasional legal battles when deceptions were detected, and ultimately a failed suicide attempt. In 1962, de Hory moved to the island of Ibiza, off the coast of Spain. There he continued to dabble in art forgery, but the quality of his fakes declined to the point where they were much less passable as authentic.

While living in Ibiza, de Hory and his criminal associates were more or less outlaws. On and off, as various law-enforcement agencies, including Interpol, investigated him, de Hory was on the run throughout Europe. But eventually he decided to return to Spain and face the music. Even though the Spanish courts tried to convict him of numerous charges, de Hory wound up with only two months in jail. That was because no one could prove he actually created any of the fraudulent artwork within Spanish borders. There were also questions about the legality of even calling his works forgeries, because de Hory rarely signed them with the name of the artist he was imitating. It's not illegal to duplicate the style of another artist; de Hory merely peddled them as authentic.

His crimes made de Hory infamous, and once he was released from prison, the value of his forgeries actually increased. Not only that, but his celebrity created a public interest in de Hory's own artwork, something he had longed for his entire life. He appeared on television shows, hobnobbed with the likes of Marlene Dietrich, Ursula Andress, Marilyn Monroe, Liz Taylor, Montgomery Clift, and Rita Hayworth. The author and investigative reporter, Clifford Irving, wrote de Hory's biography in 1969. It was titled *Fake! The Story of Elmyr de Hory, the Greatest Art Forger of Our Time*. In 1974, Orson Welles created a documentary about de Hory called *F is for Fake*.

But the noose tightened around his decades of forgery in international circles, and when de Hory found himself faced with criminal extradition to France in late 1976, he took an overdose of sleeping pills and ended his life. It's unknown how many forgeries the man created or how many are still out there in circulation and presumed authentic. Some estimate his total sales exceeded $100 million, but he died nearly broke, due to unfair business partners combined with de Hory's own propensity for a lavish lifestyle.

First and foremost, de Hory's works were not readily identified as forgeries because he was that good of an artist. Plus, as I mentioned earlier, de Hory didn't reproduce existing works; he made his own. But just as importantly, forensic science itself was not nearly as state of the art, pun intended, in the middle of the 20th century, like it is now. Even if stylistic qualities of a forgery pass muster, today's sophisticated chemical and physical testing should easily expose a fake. For instance, the ability to test paints and canvases for their chemical constituents and their age, well that's one of the prime means of identifying materials that are inconsistent with the time or place in which a fake was allegedly produced.

There are three main categories of art fraud. One is what de Hory routinely did, make a piece of artwork and play it off as created by someone famous. Another category is what some of the galleries or investors did once they found out they had a de Hory, and that is to know a piece of artwork is a fraud, but sell it as an original. Galleries and art investors need to turn a profit, so if they discover a ruse, but then pretend a forged piece is authentic, well they might easily get away with resale, especially if they act fast. The gallery owners might sell a forgery for far less than what they paid for it, just to unload it before they're caught or stuck with it, since having the public find out a gallery was duped could ruin its reputation. The third main art forgery category is when someone finds an existing piece of artwork and decides to credit it to someone they know was not the artist in order to sell it for a high price.

There's a long history of what could be thought as art forgery, and the legalities are not often clear. The first known examples were ancient Romans who copied artwork from Greece. Later, during the Renaissance, master artists had numerous apprentices working under them, and all the collective works were signed and attributed to the master. There really wasn't anything fraudulent about that. But, as time has gone on, that practice has sometimes made it difficult for authenticators to know for certain which, among a group of signed works, was actually from the master and not of his apprentice.

By the end of the Renaissance, though, plenty of artists were creating pieces they claimed were from the great masters, and that was the true beginning of illegal art fraud. Here's an interesting twist, though. Salvador Dali signed his name to thousands of pieces of paper and then gave them away, allowing other people to create artwork on them. In fact, because of this unusual practice, some art investigators believe an unsigned Dali is more likely authentic than one with his signature on it.

Here's another good story. In 1985, John Myatt, a struggling art teacher in England, was raising his two young children alone, when a friend offered to pay him to reproduce a famous painting. Later, Myatt's friend remarked that the copy was so good that it was fooling many people who saw it. So to raise cash, Myatt put an ad in a London Magazine indicating he would create 19th- and 20th-century fakes, on request, for a cost of $240 each. He wasn't intending to defraud anyone; he just needed money. A con man named John Drewe put in an order for some masterpieces by Myatt, and he was so impressed that he persuaded Myatt to work with him and try to pass the fakes off as originals.

Myatt's artwork wasn't really that great, I mean, he was no de Hory. But Drewe's ability to fake the provenance of the works was what allowed the pair to sell over 200 fake masterpieces. Drewe was able to gain entry to famous art museums and art houses, where he would steal dealer's stamps and gallery labels, and change auction records. The strategy allowed the pair to fool dealers and collectors, including at several of London's best art museums.

Drewe even used fake names, or names of dead people, as previous holders of the paintings, to show a chronology of ownership, even bribing an old friend who needed money to pose as an art dealer and sign false documents for Drewe. In the mid-1980s there was a huge interest in buying artwork as investment, so the pair's timing was perfect. They were even able to fool experts at the famous Christie's and Sotheby's auction houses.

When Myatt heard how much his fakes were going for, he ramped up his game by using an emulsion that helped the paint dry faster, so he could produce even more quickly. Even that didn't catch the experts' eyes for several years. But eventually, after almost 10 years of scamming the art world, authorities got suspicious. When they visited Myatt at his home, he confessed; he implicated Drewe, as well. Myatt and Drewe were arrested, tried, and in February of 1999, were convicted. Myatt served only four months, and Drewe two years, but the damage was done. Of the over 200 paintings the pair sold, less than half have been recovered. Myatt ended up a veritable celebrity, hosting a couple of his own TV series about art. He also has worked with Scotland Yard on methods to help detect art forgery.

An important consideration in authenticating any piece of art is that the methods used be as non-invasive as possible. Considering the analysis of paintings, for example, there are three main steps in forgery investigation, usually undertaken in order. The exam begins with the surface of the painting. *Optical microscopy*, often called *light microscopy*, in the magnification range of 5 to 50 times, well that allows a forgery expert to view the details of tiny cracks that naturally form on paint over time. This will let them judge whether the cracks are authentic, were accelerated using solvents, or were even merely drawn on the surface. When ultraviolet light is applied to a painting's surface, old layers of varnish will dramatically fluoresce, while newer paint on a retouched or forged painting will not.

Now, if the surface exam doesn't easily weed out a fake, investigators dive deeper into the background, or what's known as *underpainting*, of a piece. Artists commonly reuse canvases over time, whether masters or frauds. Conventional medical X-rays and infrared analysis can see through layers of paint and detect earlier artwork created on the same background. When using X-rays, the film is placed in front of the painting, and the X-ray is shot from behind to reveal any underpainting. Infrared light can also be applied to the surface and then reflected back to a light-sensitive camera that can capture subsurface painting, or even prior charcoal drawings. If the styles don't match, especially if the background art is of a type that came into fashion later than the foreground, they can suspect forgery.

The last step undertaken is an examination of the mediums used in the body of the surface painting. *Polarized light microscopy* is a basic way to analyze pigments in paint or even ink. On the other hand, X-ray diffraction studies, which look at how a given medium bends X-rays, can reveal the actual crystalline structure of components within pigments. Some types of crystalline pigments were only discovered and incorporated into paint during a certain era. So, if they're used to fake an old master, X-ray diffraction studies can expose the fraud.

Techniques called *X-ray fluorescence* and *neutron activation* can take authenticators down to the most basic chemical level. In X-ray fluorescence, the object being examined is bathed with radiation, and the pattern of X-rays it emits reveals its elemental breakdown. Neutron activation relies on the use of energized neutrons to bombard the sample, causing some of the atoms in the analyzed substance to take up a neutron and become radioactive isotopes. As these newly-created isotopes then decay, they send out gamma rays that identify which elements are present in the paint or other medium.

Radiocarbon dating, which also relies on the decay of radioisotopes, can be used to estimate the age of organic constituents, whether that's canvas, paint, or other carbon-containing media. But, part of the sample has to be destroyed during the dating, and it only works in the range of about 50,000 to 400 years ago.

When analysis of 3-D artwork is required, *CT scanning*, more formally known as computed axial tomography, can allow internal views of objects like sculptures. *Magnetic resonance imagery*, commonly called MRI, can reveal the internal aspects of ceramics, bronzes, and wooden sculptures. But, one of the problems with nearly all of these high-tech methods, except maybe the optical microscopy, is that the painting or other object has to be brought to the lab or to a medical facility. A new method, called *molecular Raman spectroscopy*, relies on lasers to identify both non-organic and organic pigments, as well as binders used in paint. This technique is overtaking X-ray diffraction studies, since it can be performed in the field, and is completely non-destructive.

One of the easiest ways to detect a forgery is to show that anything used to create it was not available in the place or the time the artwork was allegedly created. If a given type of wood, clay, paint, or other natural material was not available in the region or period the artwork is claimed to be from, it's a fake. Just like other criminal enterprises, though, people intent on forgery will try to find a way around any known investigative tools.

Can you believe in 1496, Michelangelo himself buried a sculpture he made of Cupid to allow the acidic soil to age it? Even during the Renaissance, ancient artwork was still hugely popular and could command more money. The Cardinal he sold it to found out about the artificial aging technique, but he let young Michelangelo keep some of the profit, because he recognized the man's talent. And Elmyr de Hory, our master forger, well he deliberately took blank pages from vintage sketchbooks to use for his drawings. He also visited flea markets to purchase old, low-cost paintings so he could scrape off the paint and use the aged canvases to create make own oil paintings.

In 2011, the New York Times reported that the man who had been living with de Hory when he died, named Mark Forgy—a name ironically only one letter away from the word "forge"—well he was working with a Paris art historian to uncover more about de Hory's life. Forgy had kept boxes of de Hory's documents, including legal paperwork and false passports, but now decided to share them with the world. Birth records and other documents from Budapest, Hungary, were found that dispelled some of the myths set forth in Clifford Irving's biography of de Hory.

The research also led to the discovery of new works from the master forger located in an attic in western France. The team plans to send some of de Hory's pieces for high-tech analyses to create what they have called a "forensic footprint" of his work. Now that may help define de Hory's methods, in then that forensic footprint can be used for comparison to other suspected forgeries of his that are probably still in circulation.

De Hory's story doesn't even end there. Beginning in 2014, some of his artwork, along with pieces from dozens of other forgers, became part of a traveling exhibition called "Intent to Deceive: Fakes and Forgeries in the Art World." The exhibition was created by the organization International

Arts and Artists, headquartered in Washington, DC. The curator of the exhibit, Colette Marvin, is the same art historian who worked with de Hory's companion, Mark Forgy, to document the artist's deception.

Interestingly, Marvin is also a trainer for the U.S. Department of Homeland Security's Cultural Heritage Protection Program. In that regard, she helps teach federal agents the more esoteric, in other words, non-scientific and stylistic ways that art historians look for forgery. That way, they can still combat art fraud within or coming into or out of, the U.S. without resorting to costly high-tech methods.

Now, let's look at what I think is a really interesting twist off the Elmyr de Hory story. Remember I said that the American news reporter and author, Clifford Irving, wrote de Hory's biography? This was because the pair were neighbors in 1962 on the Spanish island of Ibiza. That's where Irving lived with his third wife, and then later his fourth. After de Hory was exposed, the forger asked Irving to write a biography of his exceptional life, and in 1969 McGraw-Hill published Irving's book. Well, Clifford Irving has a relevant story of his own I'd like to tell.

The year after his biography of the master forger was published, Irving contacted his editors at McGraw-Hill. He told them that in an unbelievable stroke of luck, the reclusive billionaire, Howard Hughes, had recently contacted him. Irving said after reading the de Hory biography he'd written, Hughes had decided to ask Irving to likewise serve as his biographer. At the enthusiastic invitation from McGraw-Hill, Irving flew to New York and presented three letters from Hughes.

Within these letters, Hughes stated there were far too many rumors and inaccuracies in circulation regarding his unusual life, both before and after he'd gone into his self-imposed seclusion. The reclusive billionaire wanted these misconceptions cleared up before he died, but wanted the project to remain a secret until publication, with all contact passing through Clifford Irving only.

McGraw-Hill trusted Irving. He had been a successful writer with their organization for over 10 years. But just to be sure that Irving wasn't being scammed by somebody, the publishing house decided to have a forensic document examiner look over the letters from Hughes. They took the paperwork to The Osborn Associates, a nationally-recognized company that specialized in the authentication of handwriting and documents. Their experts judged the signature to be from Hughes. The assessment satisfied McGraw-Hill, and the Hughes' biography project went forward. Now, most people know plenty about Howard Hughes, but just in case you don't, let me give a few highlights of his remarkable life.

Howard R. Hughes Jr., was born in 1905 near Houston, Texas, the only child of a man who became a millionaire developing technology used in oil-well drilling. Hughes Jr., was said to be introverted, creative, and very talented. At age 18, he was orphaned and inherited three-fourths of his father's company and wealth. A year later, Hughes married and moved to Hollywood, where his uncle was a screenwriter. Hughes eventually became a successful movie producer, launching the careers of numerous Hollywood stars. Affairs with various actresses ultimately ended his marriage. While in Hollywood, Hughes hired a man named Noah Dietrich as his accountant and personal business manager, building a relationship that lasted more than 30 years. Dietrich's name will come up again shortly.

Hughes got his pilot's license, and in the early 1930s created the Hughes Aircraft Company. He became a world-renown aviator, and his engineering led to many innovations in the aircraft industry. In 1942, Hughes was commissioned to design and build a wooden seaplane for use in World War II. It was nicknamed the Spruce Goose. Unfortunately it wasn't finished until 1947, after the war was over, and Hughes flew it only once.

Hughes moved to Las Vegas, where, throughout the 1950s, he lived and conducted business on the top floor of the Desert Inn Hotel. He began buying up various casinos and hotels in Vegas, and continued in other ventures, like RKO Pictures and TWA Airlines, but he was becoming more and more reclusive. Among other rumored maladies, Hughes was thought to have suffered from obsessive-compulsive disorder, and as time went on, severe psychosis and mental deterioration, some say possibly from syphilis. Throughout the 1960s, until his death in 1976, Hughes lived in seclusion with a small entourage. He moved from one hotel penthouse to another between Las Vegas, Beverly Hills, Boston, London, Acapulco, Managua, Nassau, and other cities, typically only leaving the hotel to move elsewhere.

Given that spectacular life and all of his accomplishments and idiosyncrasies, Hughes was definitely someone of the public interest, especially after his seclusion began. Hughes became even more of an enigma as rumors circulated about his increasingly bizarre behaviors throughout the 1960s and 1970s. So it was no wonder that once Hughes' handwriting had been authenticated, McGraw-Hill jumped on the biography proposal.

They gave Irving a contract for himself and another one for Hughes to sign. The publisher offered Irving a decent advance, but he bargained for even more—$100,000 for himself and $750,000 for Hughes. But as I've said before in this lecture, there was just one big problem; Irving had completely concocted this whole scheme, in conjunction with fellow author Richard Suskind. The guy who had just finished writing the biography of one of the greatest forgers of all time decided to perpetrate his own artistic fraud in the form of a book. Maybe de Hory's story was Irving's inspiration.

Irving had gotten another artist friend, not de Hory, to forge all three of the letters he presented to the publisher. The artist designed the documents using a sample of the recluse's writing that had appeared in a 1970 Newsweek article that was titled "The Case of the Invisible Billionaire." Both Irving and Suskind were banking on the assumption that Hughes would not come out of seclusion, and probably wouldn't learn of the pending publishing deal. So Irving took the McGraw-Hill contract for Hughes and forged the recluse's signature on it.

Next, with the help of his wife, Irving figured out how to get the money he negotiated for Hughes. They got Mrs. Irving a false passport in the name of Helga R. Hughes, and instructed McGraw-Hill that checks for the project were to be made out to H. R. Hughes. Edith Irving then signed as Helga R. Hughes and deposited the funds into a Swiss bank account.

Irving and Suskind began their research and hit paydirt in Palm Springs, California, the resort town Hughes had frequented for nearly 30 years before his seclusion. There, Irving ran into a friend of his, Hollywood producer

Stanley Meyer, who coincidentally happened to be trying to get a memoir of Noah Dietrich published. Remember, Dietrich was Hughes' accountant and friend. Now, Dietrich had been working with a young journalist on the project, but Irving's friend, Meyer, suggested Dietrich get a more seasoned author to really, "polish it up." While Dietrich was searching for his new writer, Meyer secretly copied the manuscript and gave it to Irving. The Dietrich memoir was full of biographical information about Howard Hughes.

Also while in Palm Springs, Irving and Suskind got their hands on three boxes of material about Hughes, which his former publicity agent had donated to the Academy of Motion Picture Arts. So, Irving had plenty of factual information to work from to build his phony biography, and he filled in the rest with lies. He told McGraw-Hill he got the information by interviewing Hughes at several undisclosed locations. Later, Irving confessed the trips were really excuses to rendezvous with his lover.

Irving and Suskind rushed to get their manuscript to McGraw-Hill before Noah Dietrich was any the wiser. By fall of 1971 they had a manuscript of over 1,000 pages. In December of that year, McGraw-Hill announced to the public what they had. Shortly after the announcement, journalist Frank McCulloch from *Time-Life*, who was the last person known to interview the recluse in 1958, well he got an angry phone call from a man claiming to be Hughes. The caller said the book was a hoax and that he never even met Clifford Irving. McCulloch was a Hughes' expert, and when he read the manuscript, McCulloch found the information to be pretty factual. When Irving was questioned about the contact allegedly from Hughes, he said the caller was clearly an imposter. McGraw-Hill asked Irving to take a lie detector test, and although the results were deemed inconclusive, they showed no, "concrete lies."

Following the coverage of all these events in the news, Hughes surfaced. He announced that on January 7, 1972, he would hold a teleconference to be recorded by seven journalists and then aired at a later date. The reporters, who were all familiar with Hughes, questioned him for 20 minutes. Anticipating more Irving denials, NBC hired voice-print expert Lawrence Kersta, and he was to forensically compare the recording of Hughes's 1972 call to previously-recorded audio of him. Kersta acquired a 30-year-old tape that was known to be of Hughes' voice. That recording was converted to a spectrogram that could measure tone, pitch, and other voice elements. Using this technology, Kersta concluded the voice on the 1972 teleconference belonged to Hughes and no one else.

In the meantime, Irving's story continued to unravel. He tried to play the whole affair off as an experiment he had drummed up, more or less as an adventure in fraud that he could write about. Within weeks of the teleconference, though, Irving and his wife admitted to both the hoax and the check forgery, and Suskind wasn't far behind. Ultimately, they faced several state and federal charges, including mail fraud, forgery, and perjury for lying in a sworn affidavit. Clifford pled guilty and was sentenced to two and a half years, but served only 17 months. Suskind ended up serving only 5 months. For her part, Edith Irving served just 2 months in America, but also wound up with a bank fraud sentence in Switzerland that was suspended.

Questioned documents, like letters, wills, ransom notes, and manuscripts, have a long forensic history, similar to the analysis of art forgeries. The examination of handwriting styles has been used to authenticate documents for centuries, as have been the brush strokes of the great masters, or, those who try to imitate them. Chemical and physical laboratory testing of inks and papers mirrors those used on paints and canvases. And media found to postdate the time a document was allegedly produced is a dead giveaway to a fraud. The perpetration of art and document forgeries also shares something else in common; both are creative efforts of con men and women determined to make money by deceiving someone else.

Blood Doping and Other Sports Scandals

Lecture 9

More than 100 years before competitive cyclist Lance Armstrong used performance-enhancing drugs, other professional cyclists were already doing the same thing. In particular, English cycling coach Choppy Warburton encouraged the use of racing drugs as early as 1891, when the 350-mile Bordeaux-to-Paris cycling race first took place. Competitors were known to use stimulants, such as caffeine and cocaine, as well as nitroglycerine, which dilates vessels to help deliver more blood to body tissues, and even strychnine, which is excitatory to the nervous and muscular systems. In this lecture, we'll look at Armstrong's case and a number of other well-known sports scandals.

Lance Armstrong

- America's most well-known cyclist of the recent past is Lance Armstrong. Born in Texas, Armstrong turned pro in 1992 at age 21 and, within the next few years, won major competitions. In the 1996 summer Olympics, he cycled for the United States, coming in 6th in time trials and 12th in the road race. But Armstrong's athletic career had to be put on hold in October of 1996, when he was diagnosed with testicular cancer.

- Armstrong underwent surgery and successful treatment. Incredibly, by the late 1990s, he was not only back in competitive cycling but began a series of seven consecutive first-place wins in the Tour de France between 1999 and 2005. He announced his retirement from the sport in the summer of 2005, only to make a comeback in 2009 and take third place in that year's Tour de France.

- In early 2011, Armstrong retired again but this time under a cloud of shame. After years of denying claims that he used performance-enhancing drugs, in 2012, Armstrong was accused by the U.S. Anti-Doping Agency (USADA) of being part of "the most sophisticated, professionalized and successful doping program that sport has ever seen."

- Although allegations had been made for years, the first highly publicized suspicion dates to 1999, when Armstrong's urine tested positive for corticosteroids during preliminary drug testing for the Tour de France. To explain the positive result, Armstrong's team doctor produced a prescription for a steroid ointment he said was used to treat a saddle sore Armstrong had developed. It's now known that the physician backdated the prescription.

- Another hormone that has been used by athletes is erythropoietin (EPO). This substance is made by the kidneys to increase red blood cell production in bone marrow. Because red cells deliver oxygen to muscles, the use of EPO ultimately enhances endurance. During Armstrong's hearings, many of his teammates admitted to using EPO and testified that he did, too.
    - Teammates also claimed that Armstrong helped supply them with testosterone, human growth hormone, and other banned substances that he pressured them to use.

    - Further, teammates described undergoing reinfusions of their own blood and disposing of drugs when they thought police were about to raid their van.

- Ultimately, the forensic evidence used by the USADA in the summer of 2012 included lab reports from blood samples taken during Armstrong's 2009–2010 comeback, as well as e-mails, photographs, and financial documents. Armstrong was indicted by the agency for using banned substances and for possessing and trafficking in drugs. Law enforcement authorities did not file criminal charges.

- Armstrong refused to respond or even request a hearing to challenge the charges. But in 2013, he finally admitted that he had been involved in the use of performance-enhancing drugs. The cyclist alleges that he stopped doping in 2005 and that his comeback wins in and after 2009 were unaided by banned substances.

Major League Drug Use

- In 1998, baseball fans were caught up in the race between Mark McGwire and Sammy Sosa for the most homeruns in a single season; both were on track to surpass the 1961 record of 61 homeruns hit by Roger Marris. By the end of the season, McGwire had hit 70 and Sosa, 66.

© Design Pics/iStock/Thinkstock.

By the 1960s and 1970s, according to many professional players, both amphetamines and steroids were widely used in baseball.

- Seven years later, in 2005, both record breakers, as well as many other stars, were called to testify before Congress on the use of performance-enhancing drugs in baseball. In that same year, outfielder and power hitter José Canseco admitted to drug use in his book, *Juiced*. Canseco claimed that the vast majority of other Major League Baseball players also used steroids.

- In 2006, baseball commissioner Bud Selig asked Congress to institute an investigation into the use of steroids and human growth hormones in his sport. As a result of the investigation, baseball enhanced its drug-testing policies and made penalties for users more severe.

Nancy Kerrigan and Tonya Harding

- Even before they had reached their teenage years, both Nancy Kerrigan and Tonya Harding were promising figure skaters. Tonya was a strong, athletic performer, and Nancy was tall, elegant, and graceful. Both began entering and winning competitions, and each earned a spot at the 1992 winter Olympics, where Nancy won the bronze medal, while Tonya came in fourth.

- In January of 1994, the U.S. Figure Skating Championships were scheduled for Detroit. The day before the competition began, as Kerrigan was leaving the practice ice, a man ran up behind her and hit her just above the right knee with some type of club, knocking her to the ground. The suspect ran off; witnesses reported that he got into a waiting car and sped away. The next day, Harding won first place, while Kerrigan nursed a bruised right thigh.

- A man named Shawn Eckhardt confessed to his minister and friend that he was part of a conspiracy to attack Kerrigan. Eckardt was Tonya Harding's personal bodyguard and the lifelong friend of her ex-husband, Jeff Gillooly. The minister reported the incident to two of his college professors: an ethics professor who was also an attorney and another professor who was a private detective. Eckardt had an audiotape of the four conspirators discussing their plan.

- The attorney/professor agreed to represent Eckhardt in any subsequent legal actions against him if the minister would take the allegation to the FBI. Ultimately, the FBI picked up Eckhardt for questioning, and the bodyguard confessed.
    - Eckhardt claimed that it had been Gillooly's idea to eliminate Kerrigan from the competition to promote Harding's chances of winning the U.S. Championships and going on to the Olympics.

    - Because both Eckhardt and Gillooly were known to be closely associated with Harding, they had brought another man, Shane Stant, into the plan. Stant agreed to damage Kerrigan's right knee for $6,500.

    - The conspirators initially planned to attack Kerrigan at her home rink in Massachusetts, but by the time Stant reached

Massachusetts, Kerrigan was already on her way to the championship in Detroit.

- Stant and his uncle, Derrick Smith, cased the arena in the days before the assault. The weapon Stant used was a collapsible police baton, and after hitting Kerrigan, he tried to run to the exit door he had scoped out the day before, but it was now chained shut. He rammed through the Plexiglas panel to make his escape, threw the baton away, and jumped into his uncle's car. Nobody got a good look at him because all eyes and cameras were on Kerrigan.

- After winning the championship, Harding—along with Gillooly, Eckhardt, and Smith—returned to Portland, Oregon. But one week after the attack, Eckhardt and Smith were arrested and charged with conspiracy to commit assault. FBI agents arrested Stant the next morning in Phoenix. Both Gillooly and Harding denied involvement, but evidence discovered later solidly implicated them in the plot.

- In February 1994, Jeff Gillooly pled guilty to racketeering; he received two years in prison and a $100,000 fine. The other three conspirators were indicted and all received jail sentences. Both Kerrigan and Harding skated in the winter Olympics, but the following month, Harding pled guilty to conspiracy to hinder prosecution. She was sentenced to three years' probation, 500 hours of community service, and a fine of $165,000. She still alleges that she had no prior knowledge of the plan to sabotage Kerrigan.

The 1919 World Series

- The 1919 World Series was a best-of-nine series, rather than the usual best-of-seven. It was to be a contest between the Cincinnati Reds and the Chicago White Sox, but the Sox's first baseman, Arnold "Chick" Gandil, apparently convinced seven of his teammates to deliberately lose the series. Their incentive was payment promised from organized crime members, who would bet against them. Gandil apparently wanted to get back at the White

Sox owner, Charlie Comiskey, who had a reputation for paying low salaries.

- The Sox players managed to throw four of the first seven games. On the night before the eighth game, Claude "Lefty" Williams, a Sox pitcher, was visited by some mobsters, who threatened to kill him and his wife if he didn't lose the next day; Williams lost the series for the Sox.

- During the following season, rumors of the fix spread, and in September of 1920, two players, Eddie Cicotte and outfielder "Shoeless" Joe Jackson, confessed to a grand jury. In October of 1920, the grand jury handed down indictments naming eight Chicago players. Shortly after the defendants were arraigned, the confessions of Cicotte and Jackson went missing. At trial in 1921, despite eyewitness testimony from teammates, all eight players were acquitted, but the commissioner of baseball banned them from the game for life.

"Mr. Baseball"

- Born and raised in what is now a rundown Cincinnati neighborhood, Pete Rose eventually became baseball's all-time hit king. His first 15 professional years were spent in Cincinnati, but he moved to Philadelphia in 1979, when the Phillies offered him a four-year contract for $3.2 million. Following that, Rose played a year for the Montreal Expos before coming back to his hometown in 1984 to finish his career. Rose was a player/manager for two years, then managed full time until 1989.

- Back in Cincinnati, Rose was known to have problems with gambling; he was often seen at the horse track, but suspicions also emerged that he had been betting on baseball. In early 1989, the baseball commissioner appointed a lawyer to investigate, and by the spring of that year, a 225-page report had been compiled showing evidence of Rose betting on 52 games.

- Although there was no evidence that Rose bet against the Reds, any betting on baseball is strictly forbidden by anyone associated with Major League Baseball. In the summer of 1989, Rose agreed to a voluntary lifetime ban from the sport and a settlement with the league. His compromise had three provisions: (1) The league would stop its investigation and make no finding of fact; (2) Rose would neither admit nor deny the charges; and (3) Rose could apply for reinstatement after one year.

- In 2004, just before the release of his autobiography, *My Prison without Bars*, Rose publicly admitted to betting on baseball. He has applied for reinstatement twice and has been refused.

Suggested Reading

Albergotti and O'Connell, *Wheelmen.*

Canseco, *Juiced.*

Finley, Finley, and Fountain, *Sports Scandals.*

Macur, *Cycle of Lies.*

Prouse, Torke, and Strozier, *The Tonya Tapes.*

Rose and Hill, *My Prison without Bars.*

Questions to Consider

1. Can you think of any professional sports in which performance-enhancing drug use is far less likely than others?

2. Some have said that blood doping should be permitted as just another way for athletes to get ahead, to complement anatomical and physiological differences that already naturally exist among people. Do you agree or not?

3. Do you think Pete Rose belongs in the Baseball Hall of Fame?

Blood Doping and Other Sports Scandals

Lecture 9—Transcript

Over a hundred years before competitive cyclist Lance Armstrong used performance-enhancing drugs, other professional cyclists were already taking advantage. In particular, cycling coach Choppy Warburton, from England, encouraged the use of racing drugs, at least as early as 1891, when the 350-mile Bordeaux to Paris cycling race first took place. That was the precursor to the modern Tour de France, the world's most famous multiple-stage bicycling competition, which began in 1903.

Competitors were known to use stimulants, like caffeine and cocaine, as well as nitroglycerine, which dilates vessels to help deliver more blood to body tissues. Racers even tried strychnine, which, although a known poison, is excitatory to the nervous and muscular systems due to its effects on a particular neurotransmitter. Amphetamines eventually became the more common drugs of choice, especially by the late 1880s, when pharmaceutical labs started synthesizing them. At that point, there really wasn't anything illegal about so-called doping, but it did lead to several deaths, including at least three members of Choppy's cycling teams.

At the time of these early races, people more or less accepted, or at least thought, it would be near impossible to compete for such long distances under all kinds of conditions without some kind of chemical assistance. So, performance-enhancing drugs didn't become illegal until June of 1965. In July of the following year, the Tour de France began drug testing its athletes.

America's most well-known cyclist of the recent past is, no doubt, Lance Armstrong. Born in Texas, Armstrong turned pro in 1992 at age 21, and within the next few years, won major competitions, including stages within the Tour de France. In the 1996 Summer Olympics, Armstrong cycled for the U.S., coming in 6th in time trials and 12th in the road race. But Armstrong's athletic career had to be put on hold in October of 1996 when he was diagnosed with stage-three testicular cancer.

Armstrong underwent surgery and successful treatment, despite the cancer already having traveled to his brain, lungs, and abdomen. And, incredibly, by the late '90s, he was not only back into competitive cycling, but began a series of seven consecutive first-place wins in the Tour de France between 1999 and 2005. Armstrong announced his retirement from the sport in the summer of 2005, only to make a comeback in 2009, taking third place in that year's Tour de France.

In early 2011, Armstrong retired again, but this time under a cloud of shame. After years of denying claims he had used performance-enhancing drugs, in 2012 Armstrong was accused by the U.S. Anti-Doping Agency of being part of "the most sophisticated, professionalized and successful doping program that sport has ever seen."

Let's look at some of the forensic evidence that led up to that charge. Now, while allegations had been bantered around for years, the first highly-publicized suspicion involving Armstrong dates back to 1999 when his urine tested positive for corticosteroids during preliminary drug testing for the Tour de France. Although made naturally by our adrenal glands, the subclass of corticosteroids known as glucocorticoids are banned by the World Anti-Doping Agency. Glucocorticoids are natural, anti-inflammatory hormones that also help us regulate protein and carbohydrate metabolism, as well as electrolyte balance. Glucocorticoid hormones are banned if suspected to be from an exogenous source, in other words, in levels higher than the body itself could normally produce.

To explain the positive test result, Armstrong's team doctor produced a prescription for a steroid ointment he said was used to treat a saddle sore Armstrong had developed. Though most of Armstrong's associates didn't fall for the story, the governing body for international cycling races, known as the UCI, did believe the doctor's claim and let Armstrong continue with the race.

It's now known, through sworn testimony at the Anti-Doping Agency's hearing, that the team physician backdated the prescription. Armstrong's personal assistant and massage therapist also stated she never knew anything about a saddle-sore ointment, and if that treatment had been ordered, she would have been the one to apply it. In her testimony, this assistant also described having to dispose of used syringes and once having to use her makeup to cover an injection bruise on Armstrong's arm.

Another hormone that's been used by athletes is erythropoietin, also known as EPO. Erythropoietin is made by the kidneys to increase red blood cell production in bone marrow. Since red cells deliver oxygen to muscles, the use of EPO ultimately enhances endurance. Athletes got away with EPO use for years because there was no blood screen capable of detecting the pharmaceutical versions, that is, until 2000, when the French National Anti-Doping Agency developed a test for synthetic EPO.

During the hearings that Armstrong later faced, many of his teammates admitted to using EPO and testified that he did too. Riders were taught by the team doctor to inject it directly into their veins, not under their skin, and to get rid of their syringes in empty soda cans. Teammates also claimed Armstrong helped supply them not only with EPO, but also testosterone, human growth hormone, and other banned substances that he pressured them to use. Armstrong and other team members even engaged in blood transfusions, or rather, re-infusions of their own blood that they had previously extracted from each other. They hung blood bags from picture hooks in hotels or in the team van along the roadside while the driver faked engine difficulty. Some cyclists recalled the icy chill of the cold blood as it entered their veins or dumping drugs into the commode on the team's camper when they thought police were about to raid them.

On one occasion, Armstrong suddenly dropped out of a race in Spain; that was because a teammate texted him a heads up that testing officials were waiting at his hotel. There were so many near misses when Armstrong wasn't caught, like the time a team physician diluted the cyclist's tainted blood using an IV bag of saline the doctor smuggled under his coat and into Armstrong's bedroom, walking right past the drug inspector setting up his blood-testing station in the apartment living room.

Eventually, the forensic evidence used by the U.S. Anti-Doping Agency in the summer of 2012 included lab reports from blood samples taken during Armstrong's 2009 to 2010 comeback, as well as emails, photographs, and financial documents, including forensic accounting records detailing payments of over a million dollars to one of the team's physicians. The agency ultimately amassed more than 1,000 pages of testimony from 26 different witnesses, including 11 of Armstrong's cycling teammates who decided to break their code of silence and come forward. Armstrong was not only indicted by the agency for using banned substances, but also drug possession and trafficking, among other offenses. They also brought charges against the team's director, trainer, and three doctors.

The U.S. Anti-Doping Agency issued what they called a "reasoned decision in the matter and submitted it to the UCI international cycling organization. By the fall of 2012, Armstrong was stripped of all medals he had accumulated between the years 1998 and 2010, which included all of his Tour de France victories. He was permanently banned from participation in any future professional competitions.

Law enforcement authorities didn't file criminal charges, though, probably because, in the words of the United States Anti-Doping Agency, the UCI didn't have any official record of Armstrong failing drug tests. Their position was that this also implicated the UCI for aiding in what the agency believed was a long-standing cover-up. In other words, that Armstrong should have been caught long ago, and probably was, but his blood doping went ignored by nearly all of cycling, especially the international union.

Armstrong refused to respond or even request a hearing to challenge the charges. Eventually, though, in early 2013, during an interview with Oprah Winfrey, Lance Armstrong finally admitted to the world that he had been involved in the use of performance-enhancing drugs. However, the cyclist alleges he stopped doping in 2005 at the time of his first retirement and that his comeback wins, in and after 2009, were unaided by banned substances. I don't think many people believe that. Some have even wondered whether taking substances like human growth hormone either caused or accelerated his cancer, since cancer is a form of uncontrolled cell growth, but clinicians haven't demonstrated any type of link at this point.

As we know, the use of banned substances is not particular to cycling. After years of denial, seven-time Mr. Olympia, world-champion bodybuilder, Arnold Schwarzenegger, admitted to using anabolic steroids, saying you do whatever it takes to get to the top. After all, they were still legal at the time, he did them under a doctor's care, and Schwarzenegger says he doesn't regret it. Doping has undoubtedly touched virtually all professional sports. Athletes have probably looked for magic potions to increase speed, strength, and endurance, for as long as competitions have been held. Baseball's mighty Babe Ruth is even said to have administered himself an injection of sheep-testicle extract, later saying it didn't work as a muscle enhancer and only made him sick to his stomach.

By the 1960s and '70s, both amphetamines and steroids became widely used in baseball and were available in any clubhouse, according to many pro players. Hall of Fame third baseman, Mike Schmidt, has stated that "Amphetamine use in baseball is both far more common, and has been going on a lot longer, than steroid abuse." But comparing college or rookie photos of some players to their beefed-up appearance in professional days sure seems to suggest muscle enhancement that just working out can't provide.

In 1998, baseball fans were caught up in the race for the most home runs in a single season between Mark McGwire and Sammy Sosa, who were both on track to surpass the record of Roger Maris's 61 home runs hit in 1961. By the end of the season, McGuire hit 70 and Sosa 66. Seven years later, in 2005, both record breakers, as well as many other big-name stars, were called to testify before Congress on the use of performance-enhancing drugs in baseball. That was the same year that outfielder and power hitter, José Canseco, admitted to drug use in his book titled, *Juiced: Wild Times, Rampant 'Roids, Smash Hits, and How Baseball Got Big*. Canseco's claims, that the vast majority of other Major League Baseball players also used steroids, made him plenty of enemies, but he says he doesn't regret coming forward.

In 2006, baseball commissioner Bud Selig asked Congress to institute an investigation into the use of steroids and human growth hormones in his sport. Some thought Congress was overstepping its authority and wasting taxpayer money on an agency that should be policing itself, calling it a photo-op for congressmen. The investigation interviewed current and former players, most of whom denied any use of banned substances, as well as interviewing coaches; trainers; and even a former batboy, who proved to be a key witness by providing names, in exchange for a plea bargain. He had been accused of distributing of a controlled substance and money laundering and faced 30 years of sentence. The final report, released almost two years later, named 89 current and former players. The report also made suggestions to help eliminate the use of drugs in the sport. As a result, baseball enhanced its drug-testing policies, and penalties for users became more severe.

Now, it's bad enough for an athlete to enhance his or her own performance, but another thing entirely to get ahead by sabotaging a fellow competitor. In that regard, two promising U.S. female figure skaters started with a lot in common, but ended up as enemies. Nancy Kerrigan was born in Massachusetts in 1969, and Tonya Harding the following year in Oregon. Nancy learned to skate playing hockey with her older brothers. Tonya is said to have skated at a nearby shopping mall rink as an escape from a difficult family life; her mother was said to be abusive, and Tonya was the daughter of mom's fifth husband.

Neither girl came from money, even though figure skating is an expensive sport, but before they had reached teenage years, the girls' trainers had noticed clear, natural talent. Tonya was a strong, athletic performer, and Nancy was tall, elegant, and graceful. Both began entering and winning competitions, and each earned a spot at the 1992 winter Olympics, where Nancy won the third-place bronze medal, and Tonya came in fourth.

In January of 1994, the U.S. Figure Skating Championships were scheduled for Detroit. The day before the competition began, as Kerrigan was leaving the practice ice to head to her dressing room, a man rushed up behind her and smacked her just above the right knee with some type of club, knocking her to the ground. The suspect then ran off, and witnesses reported he got into a waiting car and sped away. The next day, Harding won first place, while Kerrigan nursed a painful and bruised right thigh.

By the time of the attack, a man named Shawn Eckardt had already confessed to his minister friend, that he was part of a conspiracy to attack Kerrigan. Eckardt was Tonya Harding's personal bodyguard, as well as the life-long friend of her ex-husband, Jeff Gillooly. At the time of the confession, the minister was a college classmate of Eckardt in a paralegal program, and he ended up reporting the incident to two of their professors, an ethics professor, who was also an attorney, and another professor, who was a private detective.

The friend told the professors that Eckardt, Harding's bodyguard, had played him an audiotape of the four conspirators discussing their plan. Eckardt told his friend he had made the tape in case he ended up needing to blackmail the other three guys. The friend said Eckardt even thought a side benefit of the Kerrigan attack would be that other skaters would want protection, and the bodyguard could then expand his business by hiring and managing friends to watch over frightened skaters.

Now, this attorney/professor agreed to represent Eckardt in any subsequent legal actions against him, if Eckardt's friend would take the allegation to the FBI. After the man agreed, the FBI asked Eckardt's friend to wear a wire and meet Eckardt at a local restaurant to try to get him to talk. Agents told Eckardt's friend, not to get in the car with the bodyguard, because they couldn't protect him there, and they knew Eckardt carried a loaded gun. But as soon as Eckardt got to the restaurant, he asked his friend to take a ride. When the man refused Eckardt's invitation, the FBI got nothing; they'd have to pick up the bodyguard up for questioning.

When they did, Eckardt confessed, claiming it was Gillooly's idea to get Nancy Kerrigan out of the picture, so Harding would have a better chance at the U.S. Championships in January, which would then, in turn, take her to the Olympics in Norway the following month. Eckardt, as Harding's bodyguard, and Gillooly, who was back living with her despite their divorce, were both too closely associated with Harding, so they had to come up with somebody else to assault Kerrigan. Eckardt had asked his friend, Derrick Smith, who refused. But Smith, instead, offered the plan to his nephew, Shane Stant, from Phoenix. Stant was originally asked to slit Kerrigan's Achilles tendon, but he refused to cut her, instead, he agreed to damage her right knee, her landing leg, for $6,500.

The conspirators were initially planning to attack Kerrigan at her home rink in Massachusetts, but couldn't figure out her exact practice schedule. The FBI later used phone records to verify that four phone calls were placed from the Gillooly-Harding home in Portland, Oregon to Kerrigan's practice rink.

But by the time Stant finally got to the Massachusetts location, Kerrigan was already on her way to the Figure Skating Championship, so the plan had to shift to an attack at the competition arena in Detroit.

Stant and his uncle, Derrick Smith, cased the arena in the days before the assault. The weapon Stant used was a collapsible police baton. And after hitting Kerrigan, Stant tried to run to the exit door he had scoped out the day before, but now it was chained shut. He literally rammed through the Plexiglas panel to make his escape, threw the baton underneath a car in the parking lot, and jumped into his uncle's car to make a getaway. Nobody really got a good look at Stant, since all eyes and cameras were on Nancy. The video of her on the ground immediately following the injury was broadcast all over the world by the next day.

After winning the championship, Harding, along with Gillooly, Eckardt, and Smith, returned to Portland, Oregon. But one week after the attack, Eckardt and getaway driver, Derrick Smith, were arrested and charged with conspiracy to commit assault; that was based on the confession Shawn Eckardt had made to his minister friend. FBI agents arrested Shane Stant the next morning in Phoenix. Rumors swirled that Gillooly and Harding were also connected to the case and would probably be arrested. They both denied involvement, but hired attorneys. During the FBI interviews, Harding changed her story, saying she didn't know about the plot until after she came home to Portland. She admitted to not coming forward earlier, but claimed she was only keeping quiet to protect Gillooly and the others. But one piece of evidence ended up solidly implicating Harding in the plot.

Now this was found in a dumpster outside a Portland-area bar. Strangely enough, the bar's owner routinely punished people who put their trash in the dumpster she paid for. She did this by opening the bags, figuring out who had capitalized on her trash pickup, and then returning the garbage to them. In late January of 1994, when the bar owner found unwanted bags of trash in her dumpster and opened them, she found paperwork belonging to Jeff Gillooly, whose name was already heavily in the news. Rather than calling Harding's ex-husband to get his trash, she instead phoned the FBI. Among the items in the bag was a check stub from the U.S. Figure Skating Association and handwriting that forensic investigators later determined belonged to Tonya Harding. What she had written was the name and phone number of Nancy Kerrigan's practice rink in Massachusetts, linking Harding to the scheme.

In February of 1994, Jeff Gillooly pleaded guilty to racketeering. After a plea agreement to testify against Harding, he ended up getting two years in prison and a $100,000 fine. The other three conspirators were indicted and all received jail sentences. Shawn Eckardt pled guilty to racketeering; Derrick Smith and Shane Stant pled guilty to conspiracy to commit second-degree assault.

Despite not skating in the U.S. Championships due to her injury, Nancy Kerrigan was selected for the 1994 winter Olympics in Norway. During the same month, Gillooly pled guilty to his involvement in the attack. At that point, the extent of Tonya Harding's participation wasn't known, so she was also given the green light to skate for the U.S. in Norway. And during her Olympic performance, Harding fell apart; she had trouble with her skates, and she delivered a very poor showing, after which she finished eighth. But Nancy was near perfect on the ice, and won the silver medal.

In the month following the Olympics, Harding pled guilty to conspiracy to hinder prosecution. That was just days before she was to leave for the World Championships in Japan. So when the judge sentenced Harding to three years' probation, 500 hours of community service, and a fine of $160,000, she was forced to resign from the U.S. Figure Skating Association and never made it to Japan.

Now 20 years later, Harding still alleges she had no prior knowledge of the plan to sabotage Kerrigan. Kerrigan has said she's tried to believe that for a long time. Harding's career was ruined. And if things weren't bad enough, Harding's ex-husband, Jeff Gillooly, released videotape to the tabloids of he and a very drunk Tonya Harding having sex. She's had no product endorsements and limited opportunities. Harding's exploits have amounted to nothing much more famous than winning a Celebrity Boxing match against Paula Jones, the woman who tried to sue President Bill Clinton for sexual harassment, which opened up the door to the whole Monica Lewinski sex scandal.

Now, let's finish this lecture by returning to two baseball scandals connected to my hometown favorites, the Cincinnati Reds. Breaking with tradition, the 1919 World Series was, like a few other years, a best-of-nine series, rather than the usual best of seven. It was to be a contest between the Reds and the Chicago White Sox, but, the Sox's first baseman, Arnold "Chick" Gandil, apparently convinced seven of his teammates to deliberately lose the series. Their incentive was payment promised from organized crime members, who would bet against them. Gandil apparently wanted to get back at White Sox owner, Charlie Comiskey, who had a reputation for paying really low salaries. And the players were really in a difficult place at that time, because American pro baseball had a rule that if a man refused the salary offered, he couldn't play for another Major League team.

The bad guys got a lucky break in the scam when one of the leading Sox pitchers, Red Faber, a guy who would not have allowed the scandal, got the flu and was benched. So, Eddie Cicotte pitched the first game of the series for the White Sox and lost. The Claude "Lefty" Williams pitched them another loser. When the series moved from Cincinnati to Chicago for game three, White Sox pitcher Dickie Kerr—who wasn't in on the scam—won for his team. But Cicotte threw the fourth game, losing for Chicago, after which, four of the dirty players got a payment of $20,000 to split, leaving Williams more than happy to lose the Sox's fifth game.

The teams then moved back to Cincinnati, where the Sox's coaches put Dickie Kerr back in for the sixth game, and he won. By then, rumors of the fix were swirling. Cicotte may have had second thoughts, or was mad about money he was owed, because he won game seven. At this point, the Sox had lost four games, but won three. On the night before the eighth game, Williams was visited in Chicago by some of the mobsters who threatened to kill him and his wife, so he lost the series for the Sox.

During the following season, the rumors continued, and in September of 1920, Eddie Cicotte and outfielder "Shoeless" Joe Jackson confessed to a grand jury. Despite being in the race for the American League pennant at the time, all seven of the bad guys still on the team were suspended. Ringleader Chick Gandil had already been sent down to the minor leagues. Some sources have doubted whether Shoeless Joe Jackson really was involved, saying he was naïve, illiterate, and had played quite well during the 1919 World Series. But in October of 1920, the Grand Jury handed down indictments, naming eight Chicago players. Shortly after the defendants were arraigned, though, the confessions [that] were signed by Cicotte and Jackson went missing, so that evidence wasn't available at trial in the summer of 1921. Despite eyewitness testimony from teammates, without those confessions, all eight players were acquitted. Still, the Commissioner of Baseball banned all of them from the game for life.

The infamous incident is now popularly called the Black Sox Scandal, but some say the team was already called The Black Sox before the series even started. That's because their tightwad owner, Comiskey, rarely had the team's uniforms washed, saying if players wanted them clean, they should do it themselves. Comiskey did get the uniforms laundered for the 1919 series, but deducted the cost from the players' salaries! Oh, and years later, the missing confessions were found in the possession of Comiskey's attorney.

Now, let's finish with Cincinnati's Mr. Baseball. Born and raised in what's now a poor, rundown, west-side Cincinnati neighborhood, just miles from where I grew up, Pete Rose eventually became baseball's all-time hit king. He was named Rookie of the Year in 1963, during his first season for the Cincinnati Reds, and went on to 17 All Star Games. Rose's speed earned him the nickname Charlie Hustle, plus, he was a switch hitter who played five different positions during his career. Now, don't tell anybody, but I remember occasionally skipping school with my girlfriends to take the bus down to the stadium, and man, could that guy play!

Rose's first 15 professional years were spent in Cincinnati, until 1979, when the Philadelphia Phillies offered him a four-year, 3.2-million-dollar contract, at that time, the highest salary in baseball. Following that, Rose played a year for the Montreal Expos before coming back to his hometown in 1984. In Cincinnati, Rose finished his career as a player/manager for two years, and then managed full time, until 1989.

Now, back in Cincinnati, Pete was known to have problems with gambling. He was often seen at the horse tracks, but suspicions also emerged that he had been betting on baseball. In early 1989, the Baseball Commissioner appointed a lawyer to investigate. By spring, a 225-page report was compiled showing evidence of Pete Rose betting on a total of 52 games.

A guy I actually knew, named Tommy Gioiosa, was the bouncer at a popular downtown nightclub and manager of a Cincinnati gym. He and another guy, named Paul Janszen, were friends of Rose's. The lawyer's investigative report detailed that Tommy and Paul had placed bets for Pete in 1985 and 1986 to the tune of up to $10,000 per game. There was clear forensic evidence in the form of cancelled checks. Two of Rose's bookies stated Pete's bets may have surpassed $1,000,000 during those two years. Although the report showed no evidence that Rose bet against the Reds or tried to do anything to throw a game, betting on baseball is strictly forbidden by anybody professionally associated with a team.

In the summer of 1989, Rose agreed to voluntary take a lifetime ban from the sport he loved and a settlement with Major League Baseball. Rose's compromise had three provisions. First, Major League Baseball would stop its investigation and make no finding of fact; second, Rose was neither admitting nor denying the charges; and third, he could apply for reinstatement after one year. Eight days later, Baseball Commissioner Bart Giamatti died of a heart attack.

In 2004, just before the release of his autobiography, *My Prison without Bars*, Rose finally admitted to betting on baseball on ABC's *Good Morning America*. Pete has said he picked the wrong vice, claiming, "I would have been better off taking drugs or beating my wife," because he'd probably have gotten a second chance. Rose has applied for reinstatement twice and been refused. In Cincinnati, we all wonder if he'll ever reach his dream of being inducted into the Baseball Hall of Fame.

Bad Boys of U.S. Politics
Lecture 10

You might be surprised to learn the name of the president whom many consider to be the worst in all of U.S. history. This man had extramarital affairs and paid large sums of money to keep them under wraps, took part in a major political scandal involving natural resources, and placed his friends in high government positions for which they had no experience. In this lecture, we'll discuss this president—Warren G. Harding—as well as some other well-known politicians who have been involved in their own scandals in more recent times.

Warren G. Harding

- Warren Harding was born in north-central Ohio in 1865, just months after the end of the Civil War and Lincoln's assassination. He worked on the family farm and graduated with a degree in journalism from Ohio Central College. Shortly after graduation, he and two friends bought the struggling *Marion Daily Star* in Marion, Ohio, and turned it into a success. Harding often wrote the paper's conservative editorials and, within a couple of years, became its editor and sole owner.

- Said to be charming, handsome, and eloquent, Harding became a local leader, with roles in various fraternal organizations and charitable groups. In 1891, he married a wealthy divorcee, Florence Kling DeWolfe. She was the daughter of a rival newspaperman and helped make Harding's paper a success. Florence also encouraged her husband to go into politics.

- Harding probably married Florence for her status in society—not for love—but he is said to have loved many other women during his life. He had a 15-year affair with a friend of his wife's, Carrie Phillips. Indeed, as events led up to Harding's presidential nomination, Phillips threatened to expose the affair to the public using love letters from Harding. In response, the Republican

National Committee paid her a monthly stipend to keep quiet and sent her on an extensive overseas trip. Harding was elected president in 1920.

- In 2009, Ohio attorney James Robenalt published an exposé of the affair based on the love letters. His book points out that Phillips was a German sympathizer and, perhaps, a German operative. She may have persuaded Harding not to run against the Democratic incumbent, Woodrow Wilson, in 1916.

- Robenalt notes that if Harding had run and beat Wilson in 1916, the United States may never have entered World War I—all due to Carrie Phillips's love of Germany and her influence over Harding. Minimally, Harding would have been the war president rather than Wilson, and things may have proceeded quite differently during the war.

Anthony Weiner

- Carrie Phillips had a paper trail of love letters from Harding, but she took the hush money from the Republican National Party and quietly went away. A modern politician, Anthony Weiner, was not as lucky in covering up his sexual exploits.

- In 2011, Meagan Broussard accepted payment to reveal the indiscretions of Weiner, who was at the time a Democratic congressman.
  - Today's digital technology has revolutionized the ability to nearly instantly trace communications to their source, and that's what happened when the married congressman sent a graphic image of himself—using his public Twitter account—to a 21-year-old college student named Gennette Cordova.

  - Although Weiner's post was almost immediately removed, it had already been forwarded to a political blogger. When that source made Weiner's naughty image public, the congressman said that his account had been hacked and that the images were not of him or that they had been digitally modified to appear sexual in nature.

  - Weiner claimed that the incident was a prank intended to damage his political career but said he would hire his own private investigators rather than have the FBI look into the case.

- Apparently, the incident with Cordova was just the tip of the iceberg. Weiner had been sending similar images to young ladies for several years. When Broussard learned about the photos Weiner sent to Cordova, she confided to several friends that she, too, had received similar images. One of her friends convinced Broussard to share her experience with a conservative blogger, who broke the lid off the story.

- Shortly after that, *ABC News* reportedly paid Broussard between $10,000 and $15,000 to supply images, emails, and cell phone logs that documented Weiner's shady behavior. As soon as Weiner learned that ABC was in possession of his messages to Broussard, he held a press conference and admitted he had lied when denying that he sent the images.

Paternity Testing

- Among Harding's longtime mistresses was another woman named Nan Britton, 30 years younger than her lover. In 1919, she gave birth to a daughter. Allegedly, Harding never met Britton's child and never claimed to be her father, but he provided Britton with financial support. Whether or not the child was Harding's has never been conclusively proven.

- ABO blood groups were discovered at the turn of the 20th century, and by the 1920s, it was established that blood types were inherited; this was the key to their use in determining paternity. But it wasn't until about 1935 that some states began to accept ABO testing in paternity suits.

- Since the mid-1980s, modern paternity testing has relied on nuclear DNA comparisons. Nuclear DNA is inherited from both parents and can be found in all body cells, except red blood cells, which lack a nucleus. Given the availability of this approach, it may seem surprising that it isn't always used to determine paternity in high-profile political scandals.

- For example, Arnold Schwarzenegger, the former governor of California, never even suggested DNA testing when the media learned that he had fathered a son with his longtime mistress, Mildred Patty Baena. Schwarzenegger said that he realized the boy was his because of the strong resemblance that began when the child was a toddler.

- Schwarzenegger's case contrasts sharply with that of John Edwards, John Kerry's vice presidential candidate in 2004. Edwards cooked up an elaborate scheme to cover up the fact that he had fathered a child with his lover, cinematographer Rielle Hunter.
  - Edwards convinced one of his political aides and closest friends, Andrew Young, to publically claim that he was the father of Hunter's child. When details about the scheme eventually began to emerge, Young revealed that Edwards had asked him to get one of the baby's diapers and have a paternity test done to determine if Hunter's child was really his.

  - As time went on and allegations continued to swirl that he was the baby's father, Edwards agreed in an interview to take a paternity test, but Hunter refused. Finally, when Young decided to come clean in early 2010, Edwards admitted that he was the father of Hunter's child.

Scandals in the Harding Administration

- During his rise to the presidency, Harding served as a U.S. senator and lieutenant governor of Ohio, a role through which he made many state connections. As a conservative Republican, Harding sought to return the country to "normalcy" after World War I and the progressive policies of his predecessor, President Woodrow Wilson.

- But Harding told close friends and advisors that he was "not prepared to be president." He delegated decisions to others and rewarded old friends by appointing them to political positions.

Members of his "Ohio Gang" included Harry Daugherty, who became attorney general; Jesse Smith, Daugherty's assistant; Albert Fall, secretary of the interior; and Charles Forbes, director of veterans affairs.

Harding and his "Ohio Gang" are said to have met weekly at what was known as the "Little Green House on K Street" in Washington, DC, for poker, drinking, and wild parties.

- Before Watergate, the Teapot Dome Scandal was largely regarded as America's greatest political scandal.
    - Before Harding's presidency, the government had acquired three large plots of land—two in California and one in Teapot Dome, Wyoming—that contained huge reserves of oil. This oil was set aside for use by the U.S. Navy, and Congress had given the naval secretary power to oversee the reserves.

    - But just three months after Harding was elected, Albert Fall convinced the naval secretary to turn responsibility for the reserves over to Harding, who then put Fall in charge. Fall leased the land to some of his cronies in the oil business, netting himself $360,000 in bribery money. After the story broke in 1922, a Senate investigation was undertaken, and Fall became the first former cabinet member ever to be sent to prison.

- Another scandal involved Charles Forbes, whom Harding had appointed director of veterans affairs. In less than two years at his post, Forbes embezzled almost half of the $5 million he had been allocated for the care of veterans. An investigation revealed that Forbes had allowed improper bids from contractors, resold hospital supplies at a profit, turned down legitimate disability claims, and pocketed money set aside to fill jobs in veterans' programs. He was convicted of defrauding the government and sentenced to two years in jail.

Harding's Final Illness

- In 1923, as the political scandals heated up, Harding took a damage-control trip out west. Before leaving, he allegedly told his attorney general, Daugherty, that he wanted Jesse Smith, Daugherty's assistant and, possibly, his lover, gone by the time the trip ended. Smith was later found with a gunshot wound to his head, and his death was ruled a suicide, but many believed that he was murdered because he knew too much about illegal operations in the Harding administration.

- While on the trip, Harding became ill. Although he was accompanied by his personal homeopathic doctor, Charles Sawyer, the president was treated by a U.S. Navy physician, who diagnosed the problem as an enlarged heart.
    - Sawyer, however, believed that Harding had food poisoning and treated him with purgatives to induce vomiting. Harding managed to deliver a speech in Seattle but was too weak to make a planned appearance in Oregon.

    - Fluids accumulated in Harding's lungs, and he developed pneumonia. From a clinical standpoint, an enlarged heart, body weakness, and fluid in the lungs are all classic symptoms of congestive heart failure.

    - When Harding was taken to a hotel in San Francisco, a number of doctors examined him, but on August 2, 1923—with only his wife in attendance—Harding died. On the advice of Dr. Sawyer, Mrs. Harding refused to permit an autopsy, and the president's body was quickly transferred to the train depot and on to Washington for a funeral.

    - Suspicions were immediately raised about Dr. Sawyer's medical care and whether Florence Harding may have poisoned her philandering husband, but the conferring physicians agreed that the cause of death was a stroke.

- Our look at both the life and death of Warren G. Harding illustrates the types of scandals and mysteries that have surrounded politicians since the dawn of government. In the past, the true stories of many such incidents may have been lost to history or personal discretion. But today, modern forensic technologies can be used to seek out the truth whenever dirty politics or dirty politicians are suspected.

Suggested Reading

Britton, *The President's Daughter.*

Farquhar, *A Treasury of Deception.*

———, *A Treasury of Great American Scandals.*

Mee, *The Ohio Gang.*

Primorac and Schanfield, eds., *Forensic DNA Applications.*

Robenault, *The Harding Affair.*

Questions to Consider

1. Were you surprised by the choice of history's worst U.S. president presented in the lecture?

2. Do you think men in politics are involved in more sex scandals than those in other professions? If so, what do you think is the cause of this phenomenon?

3. What lines of evidence do we have now to unravel political scandals (of any type) that investigators didn't have 50 years ago, 100 years ago, or 200 years ago?

Bad Boys of U.S. Politics

Lecture 10—Transcript

Any idea who quite a few authorities believe was the worst President in all of U.S. history? A man who had extramarital affairs and paid large sums of money to keep them under wraps? A president who took part in a corruption scandal involving natural resources? A president who placed his friends in high government positions for which they had no background, resulting in those cronies embezzling large sums of taxpayer money through their posts?

Sadly, those are common themes seen in political figures from around the world and at all ranks of government, but when you put these dishonorable acts together and examine the records of past U.S. Presidents, one name rises to the top—Warren G. Harding, the nation's 29th president. Harding even died what many consider a suspicious death. He provides a wealth of historical forensic issues for us to examine.

Warren Harding, the oldest of eight children, was born in north-central Ohio, in 1865, just months after the end of the U.S. Civil War and Lincoln's assassination. Surprising to me, given the times, both of Harding's parents were doctors. He grew up in a small town, worked on the family farm, and graduated with a degree in journalism from Ohio Central College at age 17. Shortly after that, he and two friends bought the struggling newspaper, the *Marion Daily Star* in Marion, Ohio, and turned it into a success. Harding often wrote the paper's conservative Republican editorials, and within a couple of years, became its editor and sole owner.

He was said to be charming, popular, handsome, and eloquent, and Harding became a local leader, with roles in various fraternal organizations and charitable groups. In 1891, he married a wealthy divorcee, Florence Kling DeWolfe, who was five years older than Harding and had a son from her previous marriage. She was also the daughter of a rival newspaperman. DeWolfe was a driven woman; she allegedly pursued Harding until he proposed to her. She was extremely business-oriented and helped make the paper a success. She was also the one who encouraged Harding to go into politics. Harding referred to Florence, or Flossie, as she was known to many, as The Duchess. And according to biographer Francis Russell, Harding said he married The Duchess for status and acceptance into society, not for love.

Throughout his life, however, Harding is said to have loved many other women. In fact, one was a close friend of his wife's, named Carrie Phillips. Phillips was also married to the owner of the local Marion department store, and although the Harding and Phillips couples frequently traveled together as a foursome, Warren Harding and Carrie Phillips carried on an affair behind their spouses' backs for 15 years. As events led up to Harding's presidential nomination, Carrie Phillips decided to leverage the clandestine relationship she had with him against the Republican National Committee. As a result, the organization paid her a monthly stipend to keep quiet and sent her on an extensive overseas trip.

By this time, both Mr. Phillips and Mrs. Harding were embarrassingly aware of the tryst between their respective spouses, but Carrie Phillips had threatened to expose the affair to the broader public by using the numerous "sweet and sappy love letters" she had from Harding. She didn't, and Harding was elected President in 1920, in what was called a landslide victory. As an aside, those love letters have their own controversial legal history. Phillips originally gave them to Harding biographer, Francis Russell, to be released upon her death. But after she died in 1960, Phillips's daughter took up court proceedings to bar Russell from including them in his writing. With the help of some nephews of Harding, the love letters were sealed in the National Archives, with copies going to the Ohio Historical Society. They were slated for release to the public in 2023 upon the 100th anniversary of Harding's death. However, on the basis that the letters were no longer covered by copyright 75 years after the death of their author, who was Harding, Ohio attorney James Robenalt obtained microfilm copies of the love letters and used them to write a 2009 exposé of the affair.

Not only does Robenalt's book reveal the intimate details of the long-standing Harding and Phillips relationship, it hints at other allegations. Both during and after her romance with Harding, Phillips spent many years in Germany, which may have changed Harding's life, as well as world history. Robenalt points out that Phillips was certainly a German sympathizer and could have been a German operative, and, may have persuaded Harding not to run against Democratic incumbent Woodrow Wilson for the presidency in 1916 four years before Harding did run.

Robenalt's book points out that had Harding beat Woodrow Wilson in the 1916 election, which was a likelihood, since Harding would have carried his home state of Ohio, rather than Wilson, the U.S. may never have entered into World War I, all due to Carrie Phillips' love of Germany and her influence over Harding. Minimally, Harding would have been the war president, rather than Wilson, and things may have proceeded quite differently in World War I.

Now, let's take a side journey to compare Carrie Phillip's extortion threat against Harding to a modern case, in which another politician's sexual indiscretions got him into hot water. As I like to say, technological advances change both the tools to commit crimes, as well as the methods to solve them. Plus, the evolution of society in general changes how we look at both criminal and civil matters. So, while Carrie Phillips had a paper trail of love letters from Harding, letters that forensic document examiners at the time likely could have easily validated as coming from Harding's own hand, Phillips took the hush money from the Republican National Party and quietly went away. Even the biographer to whom she entrusted the letters kept true to his word not to release them.

Anthony Weiner was definitely not so lucky. In 2011, Meagan Broussard went before *ABC News* to help expose, pardon the pun, the sexual photography exploits of Weiner, who was then a U.S. Democratic Congressman. In fact, rather than take hush money, Broussard accepted payment to reveal Weiner's indiscretions. Today's digital technology has revolutionized the ability to nearly instantly trace communications to their source, and that's what happened when the married congressman sent a graphic image of himself, unbelievably using his public Twitter account, to a 21-year-old college student named Gennette Cordova in 2011.

Although Weiner's post was nearly immediately removed, it had already been captured by another Twitter follower and then forwarded to a political blogger. When that source made Weiner's naughty image public, the congressman's reaction was to use his own modern technological excuses. He said his account had been hijacked, and those were either not images of him, or they'd been digitally modified to appear sexual in nature. Weiner claimed this was a prank intended to damage his political career, but said he would hire his own private investigators, rather than have the FBI look into the case. That's pretty curious.

Apparently, this sexting (or would we call it sex-twittering?) was just the tip of the iceberg. Not only had Weiner been sending similar and worse images to young ladies for several years, but suspicious political adversaries had been monitoring Weiner's electronic communications for months, even going so far as to use the forensic technique of posing as potential victims, in this case, young women, to try to bait the congressman. When Meagan Broussard learned about the photos Weiner sent to Gennette Cordova, Broussard confided to several friends that she, too, had gotten similar images from Weiner. One of those friends, with ties to the Republican Party, convinced Broussard to share her story with a conservative blogger who broke the lid off it.

Shortly after that, *ABC News* reportedly paid Broussard between $10,000 and $15,000 to supply them with a batch of images, emails, and cell phone logs that documented Weiner's shady behavior. Broussard claimed she exposed him because she was disgusted, not so much by Weiner's behavior, but by his adamant denials. She said she didn't do anything wrong, didn't have anything to hide, and was afraid if she didn't tell her story, the media would. And we all know that's true. In fact, just as soon as Weiner learned that ABC was in possession of his messages to Broussard, he held a press conference and admitted he had lied when denying he sent such images.

Okay, now let's return to Harding's infidelities. Among his lovers, and concurrent with his affair with Carrie Phillips, was another Marion, Ohio, woman named Nan Britton. She was 30 years younger than Harding. In fact, Harding and Britton's father were friends. Apparently Britton became infatuated with Harding when she was a teenager and he ran the local newspaper. In the 1927 book she wrote, entitled *The President's Daughter*, Britton claimed that in 1917, at the age of 20, she lost her virginity to 51-year-old Warren Harding in a New York City hotel room, and that moments later, the New York City Vice Squad broke down the door. But discovering the older man with this young girl was a U.S. senator, the authorities left the couple alone. Imagine anyone keeping that secret today!

As the affair continued, Harding set Britton up in a house in New Jersey near a casino where he gambled. In time, Britton became pregnant, and in 1919, gave birth to a daughter, at home, rather than in a hospital. Britton asserts the romance continued into Harding's presidency, and that she and Mr. President had sexual encounters in the White House, including in a closet off the Oval Office. Sound familiar? Allegedly, Harding never met Britton's child, let alone claimed to be the little girl's father, but nevertheless, he did provide Britton with financial support. Some sources suspect Harding may have been infertile and therefore couldn't have been the girl's father, since he had no children with wife, yet she had a son from previous marriage. Still, if Harding claimed to have married Flossie DeWolfe for status and money, rather than love, perhaps his sex life was exhausted with his mistresses, not his wife.

So, did Harding have a child with Nan Britton? That has never been conclusively proven—but what about the possibility of paternity testing in the case? You may know that blood tests were the original means of assessing paternity. ABO blood groups were discovered at the turn of the 20th century in Austria, but also, concurrently in Czechoslovakia. By the 1920s, it was established that blood types were inherited, and this was the key to their use in establishing paternity. Britton's daughter was born in 1919, and blood-group paternity testing began being used around 1925. But it wasn't until about 10 years later that some U.S. states started accepting ABO testing in paternity suits. The thing about blood testing, though, is it can only exclude a man as being the alleged father of a child; it can't prove he is. The four basic blood types are just too common in the population for a simple blood group match to be positive evidence of parentage.

Since the mid-1980s, modern paternity testing, which is a form of forensic analysis, has relied on nuclear DNA comparisons. Nuclear DNA is inherited from both parents and can be found in all body cells, except red blood cells, which lack a nucleus. Although the original DNA-based paternity tests did use blood, they extracted the DNA from white blood cells, which do have a nucleus. Mitochondrial DNA can't be used in paternity testing, because that form of DNA is inherited only through a person's mother. Still, given today's nuclear DNA capabilities, today's high-profile political paternity scandals surely always involve DNA testing, right? Well, surprisingly, that's not as common as you might think.

In 2011, the day after California Governor and action film star, Arnold Schwarzenegger, left office, he accompanied his wife, Maria Shriver—from the Kennedy family of U.S. politics—to a couple's counseling session that she had planned. The basis of the meeting was allegedly to discuss the family's adjustment to life after the governorship. But the first issue Mrs. Schwarzenegger raised was to confront Arnold with her long-standing suspicions that he was the father of their former housekeeper's son. Although Shriver had questioned her husband about this infidelity for years, and he had denied it, that day he finally admitted it. Shortly thereafter, Schwarzenegger made the affair public and expressed regrets for what he'd put his family through.

The "other woman" was the couple's recently-retired housekeeper and personal assistant, who was a 20-year veteran of the Schwarzenegger family's staff. Mildred Patty Baena had given birth to a son 14 years prior, but separated from her husband three weeks after the boy's birth. Not only did Baena raise her son in the household's extended family, Shriver and Baena were literally pregnant with their respective Schwarzenegger sons at the exact, same time. The two boys were born only five days apart, and they likely played together at times.

Early media allegations that Schwarzenegger took a paternity test were false; he never even suggested DNA testing. In his 2012 autobiography, Schwarzenegger admitted he realized the boy was his due to the strong resemblance that started when the child was only a toddler. At that point, much like Harding and his mistress, Nan Britton, Schwarzenegger started to give Baena extra money and kept her on staff to, in his words, "control the situation." He even bought her a new home in 2011 after her retirement.

Baena later said that Maria Shriver confronted her before the couple's counseling session, and that she told Shriver the truth even before Mrs. Schwarzenegger forced her husband's admission. It's amazing to me that this affair was kept quiet for over 14 years, especially considering it happened even prior to Schwarzenegger's first and successful run for governorship, in 2003.

Contrast this with the infidelity of John Edwards. In 2004, while still a North Carolina U.S. senator, and during the time his wife was battling breast cancer, Edwards was a Democratic presidential hopeful. He didn't get the nomination, but ultimately ran for vice president in 2004, alongside John Kerry. Although they were defeated, Edwards announced his intention to seek the presidential nomination four years later. But beginning as early as 2007, allegations began to surface about an extramarital affair with one of his campaign staff, cinematographer and sometimes actress, Rielle Hunter, resulting in considerable damage to Edwards' political career. Things got worse when it became obvious Hunter was pregnant.

Hunter gave birth to a daughter in early 2008, but didn't provide the father's name on the birth certificate. Instead, she went to the tabloids and said that Andrew Young, one of Edwards' top political aides and closest friends, was the father of her child. Unbelievably, Young not only went along with this, he and his wife took Hunter into their home, where they lived with their three children. But the whole thing was an elaborate scheme Edwards cooked up to cover the fact that Hunter was his mistress, not Young's. Among the more outrageous details was Young's eventual revelation that Edwards asked him to get one of the baby's diapers and have a paternity test done, to see if Hunter's child really was his. DNA can be obtained from nearly all bodily tissues and excretions these days, not just blood. Young also said Edwards wanted him to find a doctor that would exonerate him by falsifying any paternity test results that turned up positive.

As time went on and allegations continued to swirl that he was the baby's father, Edwards agreed in an interview to take a paternity test, but Rielle Hunter said she'd have no part in it. Finally, when Andrew Young decided to come clean on *ABC News* in early 2010, Edwards admitted he was the father of Hunter's child, who by this time was almost two years old. Edwards said he had been, and would continue to financially support the child. And that's what he was ultimately indicted with in 2011, since that support had come in the form of over a million dollars of misappropriated campaign funds.

Although he was not convicted, the situation did serious damage to any political future Edwards may have had.

One far-lesser-known political sex scandal did ultimately lead to a paternity test. You may not have heard of Minnesota Senator David Durenberger, but in 1993 a woman alleged he raped her back in 1963, causing her to become pregnant. Durenberger's denials were just as strong as Edwards's had been, but paternity blood testing by a Minnesota lab definitively excluded the senator as possibly being the father of the woman's nearly-30-year-old son.

Of course, paternity testing isn't the only reason we do DNA testing in sex scandals. We all remember the famous blue dress worn by White House Intern, Monica Lewinsky, while having sex with then-President Bill Clinton in 1997. As the tryst became public, despite Clinton's steadfast statement that he "did not have sex with that woman," the FBI ordered the President to submit to a blood sample in 1998, and it showed that the semen stain did, in fact, contain Presidential DNA.

Now, let's look beyond the sex scandals and examine the disastrous political life of Warren Harding. During his rise to the presidency, Harding served not only as a U.S. senator, but also as Lieutenant Governor of Ohio, a role through which he made many state connections. In fact, one of his main political advisors from Ohio, Harry Daugherty, promoted him toward the presidency because he thought Harding "looked like a president."

Harding made what have been called lofty speeches and was a Republican conservative. He wanted to return the country to what he called "normalcy" after World War I and the progressive policies of his predecessor, President Woodrow Wilson. But Harding had what some might call a mixed platform. He progressively backed women's voting rights, perhaps due to his strong mother and motivated wife, but conservatively supported prohibition. Harding lowered taxes and helped end strikes and race riots, and while he supported efforts to rebuild post-war Europe, he opposed joining the League of Nations.

Overall, though, Harding told close friends and advisors he was "overwhelmed with the job and not prepared to be president." Harding delegated both work and decisions to others, preferring to be what he called a more ceremonial president. As such, he rewarded old friends by appointing them to political positions, including making his Ohio advisor, Daugherty, the nation's Attorney General. Along with that came Mr. Jesse Smith as Daugherty's assistant, who had additional roles as "general gofer and bootlegger," despite Harding's lip service to prohibition.

Other members of what is now known as the Ohio Gang will figure into our story. Albert Fall became Secretary of the Interior, Charles Forbes became the first-ever Director of Veteran Affairs, and a private investigator named Gaston Means, who was said to be exceptionally adept at extortion, was made an FBI agent early in Harding's administration. Means was a consummate con artist who had an extensive history with the legal system—in a bad way. He was a forger, thief, and had been accused of murdering one of his private investigation clients. An entire lecture could be created about Gaston Means. In fact, after Harding's death, Means even attempted to extort ransom money during the kidnapping of Charles Lindbergh's son. Means wasn't actually involved in the kidnapping, but ended up dying in prison for his extortion stunt. But back to Harding. The guys from the Ohio Gang were involved in among the worst of the Harding scandals. They met weekly in what was known as the Little Green House on K Street in Washington, DC, where they played poker, drank, and had wild parties that some sources have called orgies, but that's not been substantiated.

I should also mention not all of Harding's appointees were bad boys. In fact, Herbert Hoover—who Harding made Secretary of Commerce and would later become the 31st President of the U.S.—seriously denounced the Ohio Gang. So let's dig into some of Harding's political scandals. If you studied U.S. history, you may recall the Teapot Dome Scandal. Prior to Watergate, it was largely regarded as America's greatest political scandal. Here's the gist of it. Before Harding's presidency, the government acquired three large plots of land—two in California and one in Teapot Dome, Wyoming. These contained huge reserves of oil. The oil was set aside for use by the U.S. Navy to help shift their fuel sources from coal to oil. Congress had given the Secretary of the Navy power to oversee the reserves.

But just three months after Harding was elected, his Secretary of the Interior, Ohio Gang member, Albert Fall, convinced the Secretary of the Navy to turn the responsibility over to President Harding. Harding, in turn, put Fall in charge of the oil reserves. Then Fall leased the land to some of his cronies in the oil business, netting himself $360,000 in bribery money. The Wall Street Journal had been digging into the story, which they broke on April 14, 1922. As a result, a senate investigation was undertaken, resulting in Harding's appointee, Albert Fall, becoming the first former cabinet member to ever be sent to prison.

Here's another one. The conclusion of World War I saw many injured soldiers returning home. So Harding created a position called the Director of Veterans Affairs, which many people applauded. But then Harding appointed Ohio Gang member, Charles Forbes, to the post. Forbes was allocated $500 million of government money to devote to the care of the veterans. He was also assigned to hire 30,000 people to build and staff VA hospitals throughout the country. But in less than two years, Forbes had embezzled 225 million of those dollars; that's almost half the allocated money. When the dust settled, the investigation revealed Forbes allowed improper bids from contractors that he bribed, and, he purchased and resold hospital supplies for huge profit. Forbes also turned down legitimate disability claims and didn't fill jobs at veterans' programs. He instead pocketed the money that was set aside for those obligations.

One of Harding's confidants was Charles Sawyer. Sawyer was Mr. and Mrs. Harding's personal homeopathic doctor, but some gave him little credibility as a physician. It's said he took no salary from the couple, but Harding did make Sawyer Brigadier General of the Army Medical Corps. In return, Sawyer was at least the voice of reason with regard to the Forbes case, that abuse of the new Veteran's Affairs system, since Sawyer was the one that persuaded Harding to call for Forbes's resignation in 1923. Forbes was convicted of defrauding the government, fined $10,000, and sentenced to two years in jail, but was released early. We'll talk more about Dr. Sawyer in just a few minutes.

Now earlier, I mentioned Jesse Smith, who had no official title, but was Attorney General Daugherty's aide. Smith was also Daugherty's roommate, in fact, there were rumors the two men were lovers. Among Smith's illegal activities for Harding and the Ohio Gang was obtaining liquor for both the K Street parties and the White House. Smith also sold licenses for alcohol sales, licenses that were only supposed to be given to registered pharmacies, and, he took money from various bootleggers in exchange for federal protection. There was no great forensic investigation necessary to figure all this out; Smith kept meticulous records on his alcohol-related crimes and transactions.

In 1923, as political scandals and suspicions heated up, Harding decided to take a damage-control campaign trip, hoping for a 1924 re-election. The trip was supposed to take him out west, and then on to the gold rush territories of Alaska. Before leaving, Harding supposedly told Attorney General Daugherty he wanted his alleged boyfriend, Jesse Smith, gone from Washington by the time the campaign trip ended. Smith was later found with a gunshot wound to his head in what was ruled a suicide. But people who knew Smith said he had an intense fear of guns, and many believed he was actually murdered because he knew too much about the illegal operations of the Harding administration. Although those murder allegations were even raised within the U.S. Senate, the suspicions related to Smith's death were never proven.

While in Alaska, accompanied by Dr. Sawyer, Harding took ill. A Vice Admiral U.S. Navy physician named Dr. Boone, diagnosed the problem as a heart condition, in particular, that Harding's heart was enlarged. But on his return voyage from Alaska, Harding got worse, and Dr. Sawyer believed Harding had food poisoning from some seafood he'd eaten. Sawyer treated the President with purgatives to induce vomiting. Now, anybody that knows much about physiology knows a body has to have the proper balance of ions, called electrolytes, in the blood, especially to maintain proper cardiac function, and vomiting can drastically change the body's ion balance.

The ship docked in Seattle, and Harding was able to deliver a planned speech, but by the time he traveled by train back to Oregon, Harding was so weak he had to cancel his address. Fluids had been accumulating in his lungs, and as a result he had developed pneumonia. From a clinical standpoint, an enlarged heart, body weakness, and fluid in the lungs are all classic symptoms of congestive heart failure.

When Harding was taken to a hotel in San Francisco, a number of doctors examined him; one even prescribed the heart medication digitalis. But on August 2, 1923, with only his wife, Florence, in attendance, Harding died in that San Francisco hotel room. On advice of Dr. Sawyer, Mrs. Harding refused to permit an autopsy, and the President's body was quickly transferred to the train depot and then on to Washington for a funeral. Can you imagine a sitting U.S. President that died under those circumstances not being autopsied today?

As you might assume, suspicions were immediately raised about Dr. Sawyer's medical care, and whether Florence Harding may have just finally had enough of her husband's philandering and poisoned him. But the conferring physicians agreed the cause of death was apoplexy, what we'd today call a stroke. Dr. Sawyer resigned from his position with the Army Medical Corps and returned to Marion. Mrs. Harding stayed briefly in Washington, DC, but on the advice of Dr. Sawyer, she also soon returned to Marion, where she died in Sawyer's care the following year. Florence's son, who had been raised by her parents, had died nearly a decade before when he was just 34 years old, from alcoholism and tuberculosis.

A look at both the life and the death of Warren G. Harding illustrates the types of scandals and mysteries that have surrounded politicians since the dawn of government. In the past, the true stories of many such incidents may have been lost to history or personal discretion. But today's media, along with modern forensic technologies, like DNA paternity testing, digital evidence of communications and financial transactions, as well as advanced autopsy and toxicology techniques, those can all be used to seek out truth whenever dirty politics or politicians are suspected.

Criminals of the Wild, Wild West

Lecture 11

The 19th-century westward expansion of the United States beyond the Mississippi River was a colorful time in American history, full of prospectors, cowboys, bad boys, and lawmen. Three such legendary characters were born in the span of just six years: Wyatt Earp in 1848, Jesse James in 1847, and Alferd Packer in 1842. In this lecture, we'll take a look at these notorious figures of the great frontier known as the Wild West and see what connections they have to forensic science, past and present.

Wyatt Earp

- In 1869, when he was 21 years old, Wyatt Earp took the position of constable in the town of Lamar, Missouri. He soon married, but within a year, his pregnant wife died from typhoid fever, and other parts of his life began to unravel.

- Earp was caught embezzling money from local schools and falsifying legal documents and was charged with horse theft. He was taken into custody in 1871 but escaped before trial, heading back to Illinois, where his brother Virgil was a saloonkeeper.

- Wyatt and Virgil eventually wound up in the silver-mining town of Tombstone, Arizona, where Virgil was appointed Deputy U.S. Marshall. Three of their other brothers—James, Morgan, and Warren—also came to Tombstone, as did Wyatt's friend Doc Holliday, a dentist and skilled gunman.

- Among the most notorious outlaws in Tombstone were two sets of brothers: Frank and Tom McLaury and Ike and Billy Clanton. The four were well-known for stagecoach robberies and horse theft and had threatened the Earp brothers after Virgil's attempts to bring them to justice. In September of 1881, friends of the McLaurys—Pete Spence and Frank Stilwell—robbed a stagecoach, and Virgil and Wyatt went out with the sheriff's posse to track the bandits.

- We may not think of the Wild West as a place where forensic science was practiced, but while examining the scene, Wyatt noticed an odd-shaped boot print. And just as investigators might do today—comparing footwear evidence and manufacturer's data—Wyatt consulted the local cobbler. When Wyatt described the unusual print, the shoemaker identified the boot as belonging to Stilwell, who was subsequently arrested, along with Spence. But the robbers had arranged alibis, and the charges against them were dropped.

- During the course of the proceedings, however, one of the McLaury brothers confronted Morgan Earp and threatened to kill the Earp brothers if they ever tried to arrest any of the gang again. These events seem to be among the last straws leading to the infamous gunfight at the O.K. Corral.

- On October 25, 1881, Ike Clanton got into a drunken argument with Doc Holliday and vowed that his gang would settle the score with Doc and the Earp brothers. The next day, October 26, the Clanton and McLaury brothers were seen near the O.K. Corral, brandishing guns, a violation of an ordinance against carrying guns in the city.

- Because they were lawmen, Virgil and Morgan were permitted to carry guns, and Virgil quickly deputized Wyatt and Doc. That afternoon, when the Earps and Doc found the outlaws, Virgil told them to turn over their weapons, and that's when the shooting began. The legendary gunfight lasted only about 30 seconds. Billy Clanton and both McLaury brothers were killed, and Ike Clanton and a companion got away. Virgil, Morgan, and Holliday were all injured, but Wyatt Earp was unharmed.

- Testimony at the hearing after the incident was conflicting and, in some cases, not credible. Had the incident taken place today, ammunition collected during autopsies or at the scene could be compared to bullets test-fired from any confiscated guns to identify specific shooters. Detailed autopsy methods could help show the direction of bullets fired, which could be compared to eyewitness accounts.

- Ultimately, the judge ruled not to bring the Earp brothers and Holliday to trial due to insufficient evidence. But the gunfight wasn't over.
  - Virgil was ambushed and shot in December of 1881—possibly by Ike Clanton—and lost the use of his right arm. Morgan died in March of 1882 after being shot through a window while shooting pool; witnesses said that they had seen Frank Stilwell running from the scene.

  - Wyatt vowed to kill the remaining outlaws involved. Wyatt, his brother Warren, Doc Holliday, and a few others formed a posse that may have killed Stilwell and claimed to have killed more than a dozen others suspected in Morgan's death or Virgil's injury.

Jesse James

- Jesse James was born in 1847 in Clay County, Missouri. Although he was never in the army, he began his lawless days, along with his older brother, Frank, engaged in guerrilla warfare against Union troops and their supporters in Missouri.

- Anti-slavery unionists ran many of the banks in Clay County; thus, the James brothers began robbing banks to show their contempt. The James gang joined forces with the Younger brothers, fellow Confederate sympathizers. In 1872, the group robbed the ticket office of the Kansas City exposition in front of more than 10,000 people, firing a shot that wounded a young girl. Despite the injury to the child, a cult status began to grow around the James gang.

- As the gang grew in size, their crimes became more notorious; they now held up stagecoaches and trains, as well as banks. The estimated tally of their crimes included 12 banks, 7 trains, and 4 stagecoaches. In response, two freight companies hired the Pinkerton Detective Agency to find and apprehend the outlaws.

- By 1882, Frank James decided to give up the life of robbery, as did several other gang members. As the group grew smaller, Jesse asked two brothers, Bob and Charlie Ford, to move in with him.

- On the morning of April 3, 1882, as Jesse and the Ford brothers were preparing for a robbery, Bob allegedly shot Jesse in the back of the head, and he died later that day.

- It has been claimed that the Ford brothers turned on Jesse because they had a deal with the Missouri governor. When they took credit for the killing and went to collect the reward, they were charged with murder, pled guilty, and were sentenced to hang—all in the same day. A few hours before the execution, however, the governor pardoned them.

- Almost immediately, rumors emerged that the Ford Brothers had staged Jesse's death to enable the outlaw to run away for good. There were claims that someone else was actually buried in James's tomb.
    - In 1948, a 101-year-old man named J. Frank Dalton claimed to be the living Jesse James, but his story didn't hold up under questioning.

    - In 1995, around the time mitochondrial DNA testing first came into forensic use, a team of researchers exhumed Jesse's grave. The samples they took from the remains matched two of Jesse's living relatives, but some people claimed that the body in the grave could be some other relative of the James family.

    - In 2007, an FBI facial recognition expert compared photographs of J. Frank Dalton and Jesse James and concluded that they were definitely not the same man; however, the expert also compared antemortem photos of Jesse to his autopsy photos and found discrepancies there, too.

    - In the end, it seems safe to say that while the legend lives on, the man himself probably did not.

Alferd Packer

- In November of 1873, 21 prospectors left Bingham Canyon, Utah, in search of gold. A couple of months later, they reached the camp of the Ute Native Americans, where Chief Ouray urged them to spend the rest of the winter before braving the mountains in the spring.

- But two weeks later, six members of the group struck out on their own. Among the six was 32-year-old Alferd Packer, who agreed to guide the other five about 75 miles away to an Indian agency near Gunnison, Colorado. They took provisions for 10 days.

- In April of 1874, 65 days after the group had started out, Packer walked into the Los Pinos Indian Agency alone. According to his story, the group had been snowbound for two months in the San Juan Mountains. After consuming their supplies, they tried to scavenge in the wilderness. Because Packer was weak and couldn't keep up, the others left him behind; he found his way to the Indian agency on his own.

- About a month later, when none of the other five men showed up, a search party was organized. This prompted Packer to tell a second story, which has come to us in slightly varying accounts. According to some reports, Packer said that three men had died along the way, and the remaining members of the party had been forced to cannibalize them. In other reports, one of the men had shot another, and Packer and the shooter ate the flesh of the dead man. Packer said that he was forced to kill the remaining man, Shannon Bell, in self-defense.

- In the summer of 1874, a traveling artist for *Harper's Weekly* was in the area. While at a pass near the Gunnison River, the artist came upon the partial remains of five men sprawled across the ground. He quickly drew what he saw—essentially, a crime scene sketch—and took his artwork into town.

- Based on the sketch, authorities believed that there had been a struggle and that the victims had been killed with an axe or hatchet. A coroner traveled to the scene and did an examination, but there are no surviving records of his findings. Packer was charged with first-degree murder, but before he could be tried, he escaped from jail and was on the run for nearly a decade.

- In 1883, one of the original 21 prospectors recognized Packer in a saloon in Wyoming, and he was rearrested. At his trial, Packer told yet another version of his story: While Packer was out searching for food, Shannon Bell went berserk and killed the other four prospectors with a hatchet while they slept. When Packer returned, Bell came at him with the hatchet, and Packer was forced to shoot Bell. Ultimately, Packer was convicted of five counts of voluntary manslaughter and, in 1886, was sentenced to 40 years of hard labor. In 1901, he received an early parole.

- In 1989, a group of investigators exhumed the remains of Packer's companions and examined their bones. From cut marks found on the bones, a toolmark expert concluded that an axe or hatchet-type implement had made many of the injuries. Cut marks around muscle attachment sites were thought to reflect defleshing. The study suggested murder and cannibalism, but it couldn't corroborate or dispute Packer's final account of events.

Suggested Reading

Gantt, *The Case of Alfred Packer, the Man-Eater.*

Guinn, *The Last Gunfight.*

Hodgson, *Lone Survivor.*

Primorac and Schanfield, eds., *Forensic DNA Applications.*

Sifakis, *Encyclopedia of Assassinations.*

Stiles, *Jesse James: Last Rebel of the Civil War.*

Tefertiller, *Wyatt Earp.*

Triplett, *Jesse James.*

Questions to Consider

1. Some criminals have become cult figures, especially those from the Wild West days. What is your theory on how or why that happens?

2. What realities of the Wild West do you think have not been accurately captured or presented by most Hollywood depictions of the era?

3. Are there any places in the world today that seem similar in any way to our impressions of how life was in the Wild West?

Criminals of the Wild, Wild West

Lecture 11—Transcript

The 19th-century westward expansion of the United States beyond the Mississippi River was a colorful time in American history, full of prospectors, cowboys, bad boys, and lawmen. Three such legendary characters were born in the span of just six years, Wyatt Earp in 1848, Jesse James in 1847, and Alferd Packer in 1842. Let's take a look at these notorious figures of the great frontier known as the Wild West and see what connections they have to forensic science, past and present.

Young Wyatt Earp lived in Illinois, where his father was the local constable. But after Constable Earp was caught bootlegging whiskey, the family relocated to Iowa. When the Civil War broke out, Wyatt was too young to enlist with his three older brothers Newton, Virgil, and James, but he ran away several times to join up anyway. Each time, Wyatt's father brought him back to stay home with younger brothers Morgan and Warren. After the older Earps returned from the war, the family settled in Missouri, where Wyatt's father became constable and then justice of the peace. At that point, Wyatt, who was 21, took the constable position over in 1869. He soon married, but within a year, Wyatt's pregnant wife died from typhoid fever, and other parts of his life began to unravel.

While still the Lamar, Missouri, constable, Wyatt Earp assumed his father's legacy as an unlawful lawman. He was caught embezzling money from local schools, falsifying legal documents, and also charged with horse theft. Wyatt was taken into custody in 1871 but escaped before trial, heading back to Illinois, where his brother, Virgil, was a saloonkeeper. Younger brother, Morgan, joined them there, too. Wyatt lived above a brothel, serving as its bouncer and pimp. After several arrest raids on the place, he moved to a floating whorehouse and gambling boat on the Illinois River. When the boat was raided, Wyatt earned a stiff fine for being what they called an old offender. Even in those days, the legal system more harshly penalized repeat criminals.

Wyatt went to Kansas then and spent time in cattle boomtowns before Virgil suggested they move to the silver-mining town of Tombstone, Arizona, where Virgil was appointed Deputy U.S. Marshall. Three of their other brothers, James, Morgan, and Warren, also came to Tombstone to dabble in water rights and silver mines. Wyatt's friend, Doc Holliday, a dentist and skilled gunman he met gambling, moved to Tombstone too, in 1880. Those days, Holliday didn't work as a dentist, because his patients didn't appreciate his tuberculosis coughing fits, plus, he could make more money gambling than in dentistry.

In Tombstone, Wyatt and another man were both vying to be sheriff, and, also involved in a love triangle with the same woman. Wyatt got the girl, but the other guy made sheriff. You can imagine the trouble this caused Wyatt when faced with the local law, something that happened often, since he and hot-headed Doc Holliday were fond of barroom brawls with cowboys. In those days the term "cowboy" was used for western outlaws and troublemakers who would come into town after cattle drives with money in hand for gambling, prostitutes, and whiskey. Only the ruffians were cowboys; legitimate cowmen were known as cattle herders and ranchers.

Among the most notorious Tombstone cowboys were two sets of brothers, Frank and Tom McLaury, and Ike and Billy Clanton. The four were well known for stagecoach robberies and horse theft, and had threatened the Earp brothers several times after Virgil's attempts to bring them to justice. In September of 1881, friends of the McLaurys, Pete Spence and Frank Stilwell, robbed a nearby stagecoach. The driver reported one of the masked gunmen use the phrase "Give me the sugar," a figure of speech Stilwell was known to use during holdups. Virgil and Wyatt Earp went out with the sheriff's posse to track those bandits.

Now, you may not think of the Wild West as a place where much went on in the way of forensics, but while examining the scene, Wyatt noticed an odd-shaped boot print. And just as investigators might do today, comparing footwear evidence and manufacturer's data, Wyatt consulted the local cobbler. When he described the unusual print, the shoemaker handed Wyatt a matching heel he had recently taken off a boot he repaired. He had fixed the boot for none other than Frank Stilwell. Armed with only the boot-print match and the linguistic evidence of using "sugar" to mean money, Stilwell was arrested, as was his buddy Spence.

Can you imagine that meager amount of evidence leading to an arrest today, especially based on a description of a boot print recorded only in memory? Minimal evidence wasn't what caused the judge to drop the charges, though. Stilwell and Spence had arranged alibis, which couldn't be easily refuted, since no one had really seen the faces of the masked bandits. While Virgil and Wyatt Earp were in Tucson for the hearing, Stilwell's good friend, one of the McLaury brothers, confronted Deputy U.S. Marshall Morgan Earp, who was left behind in Tombstone to mind the store. McLaury told Morgan he would kill the Earp brothers if they ever tried to arrest any of their gang again. These events seem to be among the last straws leading up to the infamous gunfight at the O.K. Corral.

Just as we often see in today's gang-related shootings, alcohol and guns do not mix. One night at the Occidental Saloon, one of the Clanton brothers, Ike, got in a drunken argument with Doc Holliday. He vowed that in the morning he and his brother, and the McLaury boys, would settle their scores with Doc and the Earp brothers. Now, Tombstone had an ordinance against carrying guns; cowboys were supposed to deposit their weapons at a corral or saloon at the city limits, something they never show in old westerns. But many of them carried concealed handguns, just as outlaws do today. The next day, October 26, 1881, the Clanton and McLaury brothers were seen near the O.K. Corral brandishing guns.

Now, Virgil and Morgan were permitted to carry, since they were lawmen, but Virgil quickly deputized Wyatt and Doc, giving sharpshooter Holliday a short-barrel shotgun to hide under his coat, in addition to the revolvers each man carried. That afternoon, when the group found the cowboys, Virgil told them to turn over their weapons, and that's when the shooting began. The legendary gunfight lasted only about 30 seconds, with only about 30 bullets fired. Billy Clanton and both McLaury brothers were killed, and Ike Clanton and a companion got away. Virgil, Morgan, and Holliday were all injured, but Wyatt Earp was unharmed.

A legal hearing was held to look into the incident, and according to the cowboys, the Earps fired first. The coroner said a shotgun had killed Frank McLaury, but the others died from handguns. The sheriff, the same guy whose girlfriend Wyatt stole, served as a witness for the prosecution. A far more impartial eyewitness, a railroad man who had come to town the night before the gunfight, testified for the defense, claiming he heard cowboys near the corral on the day of the battle talking about killing Virgil Earp on sight. Curious, the man asked one of the townspeople who this Virgil Earp guy was, and the person pointed out the U.S. Marshall. The eyewitness further stated he saw the man he heard talking about killing Virgil at the funeral for the dead cowboys. He pointed that man out as Ike Clanton.

When analyzing the historical accounts, many say the testimony of the sheriff and cowboys just isn't credible. Holliday would have to have been switching between his revolver and shotgun faster than humanly possible if their stories were true. But there's really no way to know who really shot first. Just imagine, though, how different this incident might have been if it happened now, especially with today's automatic weapons. A group of men could fire many more shots than just 30 in 30 seconds without changing clips, let alone guns.

And, in a modern gunfight, any ammunition collected from surgery, autopsy, or the scene could be compared to bullets test fired from the barrels of any confiscated guns to show who shot who. That type of gun-related science wasn't discovered until the late 1880s, when a French forensic professor connected ammunition to guns using lands and grooves; those are the hills and valleys produced on bullets by their journey down the length of a gun barrel. Had this gunfight happened today, detailed autopsy methods could help show the direction of bullets fired, which could also be compared to eyewitness accounts. And if the O.K. Corral were a place where horses were parked as cars are today, there might have been surveillance cameras to capture the whole thing.

Ultimately, the judge ruled not to bring the Earp brothers and Holliday to trial, due to insufficient evidence. But the gunfight wasn't over. Virgil was ambushed and shot in December of 1881, resulting in the loss of the use of his right arm. Ike Clanton's hat was found at the scene, which, while compelling, was circumstantial evidence.

Morgan died in March of 1882 after being shot through a window while he was shooting pool, and though the killer escaped, witnesses said they saw

Frank Stilwell running away from the scene. At a preliminary hearing, Pete Spence's wife said she heard the cowboys bragging about killing Morgan, but her testimony was ruled inadmissible because it was hearsay, and by law, she couldn't testify against her husband.

Wyatt vowed to kill the remaining cowboys involved, and of course, Frank Stilwell and Ike Clanton had more Earp brothers they wanted dead. So before Virgil was to accompany Morgan's body on a train to California for burial, he deputized his brothers Wyatt and Warren, as well as Doc Holliday and a few others. This was to ensure the train's safe passage and to apprehend anyone who interfered. There were rumors Frank Stilwell and Ike Clanton were planning an attack in Tucson, but by the time the train got there, Stilwell had already been shot dead. The justice of the peace issued warrants for Wyatt Earp and his posse, but they fled into the mountains. They later claimed to have hunted down and killed over a dozen outlaw cowboys suspected in Morgan's death or Virgil's injury.

Wyatt Earp's amazing life lasted a full 80 years. He was involved in mining, real estate, racehorses, and was a saloon owner in the Alaskan territories during the Klondike Gold Rush. But his brushes with forensics, on both sides of the law, were not yet over. In 1901, Earp moved to California and was hired by the Los Angeles Police Department at the age of 62 to help take care of things they considered, well, let's just say outside the law. But in 1911, his shady dealings got Earp arrested, yet again, by the very LAPD for which he worked.

Now, let's dig again into the Wild West forensic files to talk about Jesse James, but for different reasons than just his crimes. Let's look at how outlaws can grow so notorious over time, even popular, that their deeds become larger than their lives, and maybe even larger than their deaths. Jesse James was born in 1847 in Clay County, Missouri, a place they called "Little Dixie" due to its high number of Confederate sympathizers. Although Jesse was never in the army, he began his lawless days, along with his older brother, Frank, engaged in guerrilla warfare against Union troops and their supporters in the border state of Missouri.

Anti-slavery unionists ran many of the banks in Clay County, so as the local clashes heated up, some of the secessionists, including the James brothers, began robbing banks to show their contempt. During one robbery in 1869, Jesse deliberately shot and killed a clerk, not because the man resisted, but because Jesse thought he recognized him from his earlier Civil War skirmishes. The James brothers were brash; they staged Liberty, Missouri's first-ever bank robbery in broad daylight, and became prominently and publically known as part of an outlaw gang.

The James gang joined forces with the four Younger brothers, fellow Confederate sympathizers. And in 1872, in a particularly brazen attack, the group robbed the ticket office of the Kansas City exposition in front of over 10,000 people, firing a shot that wounded a young girl. They got less than a $1,000, though if they had been there half an hour earlier, they would have gotten away with the $12,000 that had already been taken away for safekeeping. Newspaper reports said that the outlaws actually put on antics for the public during the incident and amazingly escaped the large police presence that was there. Despite the injury to the child, a cult status started to grow around the James gang. In fact, Jesse James's friend edited the Kansas City Times, and accounts in that newspaper made the James brothers look more like Robin Hood heroes than dangerous and despicable men.

As the gang grew in size, their crimes became more notorious, [not only] holding up banks, but now [also] stagecoaches and trains. So the Pinkerton Detective agency, which dates all the way back to 1850, by the way, was hired by a couple of freight companies to find and apprehend the outlaws. Suspecting the James brothers were hiding at their mother's place, Pinkerton firebombed her house. Although a younger half brother was killed in the incident, and their mother lost her arm, the James gang rode on.

In most of the robberies, the outlaws consisted of 5 to 12 men on horseback. The estimated tally of their crimes included 12 banks, 7 trains, and 4 stagecoaches. Part of the misguided heroism directed their way was because the bandits typically didn't steal from the passengers; they only took money from safes. They also never deliberately killed anyone, except Jesse's vendetta mentioned earlier; although in one robbery, a train engineer was accidentally crushed by an overturned engine.

Further adding to his mystique, for all the group's holdups and antics, and even though his picture was in all the newspapers and wanted posters of the day, Jesse James was never captured. Many attributed this to his skillful horseback riding and early experience with guerrilla-style fighting during the Civil War days. He was also said to plan his crimes very carefully in advance; he was not a disorganized criminal who acted on impulse. Jesse didn't drink alcohol and discouraged his men from doing so. Ironically, he was even said to read the Bible before robberies.

But even a life of crime grows old, and by 1882, Frank James decided to give up the business, as did several other gang members, especially after a few of them had been captured or killed. As the group grew smaller, the few remaining members needed to stick close together, so Jesse asked two brothers, Bob and Charlie Ford, to move in with him. But on the morning of April 3, 1882, as Jesse and the Ford brothers were getting ready for another robbery, Bob allegedly shot Jesse in the back of the head, and he died later that day, at age 34. Now, why would the Ford brothers turn on Jesse? It's been claimed they had a deal with the Missouri governor. But when they took credit for the killing and went to collect their reward, they were charged with murder, pled guilty, and were sentenced to hang, all in the same day. A few hours before the execution, though, the governor pardoned them. Sure sounds like a deal to me.

But legends die hard. Almost immediately, rumors started to emerge that the Ford Brothers had staged Jesse's death, so the outlaw could simply get away for good. But there were postmortem photos, which you can find today on the Internet, and James was positively identified by two prior bullet wounds and a partial amputation of his middle finger. Nevertheless, there were claims that someone else, a guy named Charley Bigelow, who did resemble Jesse James, was actually buried in James's tomb.

In 1948, a 101-year-old man named J. Frank Dalton claimed to be the living Jesse James, but his story didn't really hold up well to questioning. In fact, no serious scholars entertained any mystery surrounding the bandit's death. But need I repeat [that] legends die hard? In 1995, a team of researchers exhumed Jesse's Missouri grave, around the same time mitochondrial DNA testing first came into forensic use. They sampled two bones and four teeth and were able to get a mitochondrial DNA profile from some of the dental samples. They found two descendants of Jesse's sister for comparison, since mitochondrial DNA is inherited only through the maternal line, and the DNA samples from the grave matched both of those living relatives. The testing satisfied many, but because mitochondrial DNA is shared by all descendants in a maternal line, some claimed the remains could simply be some other relative of the James family. After all, many people in a county in those days could likely be blood relations. So, the search for Jesse James continued.

In 2000, it was decided that the remains buried in 1948 as 101-year-old J. Frank Dalton, would be exhumed for testing. Why not? The man had himself buried beneath a tombstone that read, Jesse Woodson James, supposedly killed in 1882. But the forensic anthropologist involved, my good friend, David Glassman, immediately knew, without DNA testing, that the guy exhumed wasn't Jesse. In fact, David knew the remains didn't even belong to Dalton. The exhumed man had only one arm. Somehow the headstone of the real Dalton grave had been accidentally relocated at some point. Plans were made to continue digging, but when local authorities figured out tourism would take a real hit should DNA testing prove once and for all the Dalton remains were not those of the legendary Jesse James, they decided to stop digging.

In 2007, on a History Channel special, an FBI facial recognition expert compared photographs of J. Frank Dalton and Jesse James and concluded they were definitely not the same man. However, that same expert also compared antemortem photos of Jesse James to his autopsy pictures and said that there were discrepancies there too. And in 2009, another anthropologist from Wichita, Kansas, my friend Peer Moore-Jansen supervised the exhumation of a farmer named Jeremiah James to take DNA samples from him. This James was alleged to have died from a broken heart in 1935, but he's not Jesse. My personal forensic conclusion? While the legend lives on, the man probably did not.

Okay, our last Wild West forensic tale begins in November of 1873, when 21 prospectors left Bingham Canyon, Utah in search of gold. A couple of months later, they made it to the camp of the Ute Native Americans, where Chief Ouray urged them to spend the rest of the winter with the tribe and wait until spring to brave the mountains. But two weeks later, six members of the group decided to strike out on their own to beat the rest of the westward prospectors to the gold in the Colorado Mountains. Among the six was 32-year-old Alferd Packer, sometimes spelled Alfred, who agreed to guide the other five about 75 miles away to an Indian Agency near Gunnison, Colorado. They took provisions for 10 days, figuring that's how long it would take them.

Sixty-five days later, in April of 1874, a lone Alferd Packer walked into the Los Pinos Indian Agency. He didn't ask for food, but rather, a stiff shot of whiskey. He had pockets full of money, a rifle and skinning knife belonging to other members of his party, and he had a story to tell. Packer said the group had been snowbound for two months in the San Juan Mountains. After consuming their supplies, they tried to scavenge what meager resources they could find in the wilderness, and were even forced to eat their shoes. Packer said he was so weak he couldn't keep up, so the others left him behind, and he found his way to the Indian Agency all on his own.

About a month later, when none of the other five men showed up, a party was organized to search for them. This prompted Packer to spin a second version of his story; accounts vary slightly. Some report Packer saying three of the men died along the way from a combination of exposure and starvation, while another man died in an accident. Due to the extreme conditions, the living had been forced to cannibalize the dead. Other reports of Packer's second version say the oldest man died first; the others took some of his flesh with them and traveled onward. When a second man died, they cannibalized him too, but later, while Packer was away from camp looking for provisions, one of the two remaining men shot the other. When Packer returned to camp, he and the other survivor ate the flesh of the man who was shot. The end of both of these second accounts is the same though. Packer said he was forced to kill the other remaining man, Shannon Bell, in self-defense.

Now, that might have been the end of things, but in the summer of 1874, a traveling artist for *Harper's Weekly* was in the area capturing images of western landscapes to illustrate the New York-based magazine. While at a pass near the Gunnison River, just outside of Lake City, Colorado, the artist came upon the partial remains of five men, partly skeletonized, partly mummified, sprawled across the ground. He quickly, but painstakingly, drew what he saw, essentially a crime scene sketch, and took his artwork into town.

His image showed feet still wrapped in scraps of blanket fabric, pieces of flesh strewn around the area, limbs missing large areas of muscle, and signs of a definite struggle. One man's head was missing. There was little in the way of personal affects depicted, but it was immediately suspected these were the five missing prospectors. Based on what they saw, authorities believed the victims had been killed with an axe or a hatchet. The bodies were all in one place, though, not left here and there along the trail, as in Packer's latest version of the story.

A group traveled to the scene, where the coroner did an examination, but there are no surviving records of his findings. The bodies were buried on a ridge nearby in Lake City, Colorado; I've been to the spot. The authorities couldn't charge Packer with cannibalism; there was no law against that in the Colorado Territories, but they did charge him with first-degree murder. Before he could be tried, though, Packer escaped from jail; some say he was slipped a key and was on the run for nearly a decade.

In 1883, near Cheyenne, Wyoming, one of the original 21 prospectors recognized Alferd Packer in a saloon, and he was rearrested. At his trial, Packer told yet a third version of what happened, blaming it all on Shannon Bell, the man Packer claimed to have killed in self-defense. In his final story, which he did maintain the rest of his life, Packer was out searching for food, and while he was gone, Bell went berserk and killed the other four with a hatchet while they slept. When Packer returned, Bell came at him with the hatchet, and he was forced to shoot Bell in the gut, before taking the hatchet away and plunging it into Bell's forehead.

Probably due to his ever-changing story, Packer was convicted and sentenced to hang. The town actually had to send away for design plans for a gallows, since they had never made one before. But in 1876, while Packer was awaiting execution, Colorado was made a state, and as such, the statutes of the former Colorado Territory no longer applied. Packer had to be retried. After a change of venue to Gunnison, Colorado, he was convicted again, this time of five counts of voluntary manslaughter, and in 1886, Packer was sentenced to 40 years of hard labor.

Fifteen years later, in 1901, largely due to a crusade led by a local newspaper, Packer got an early parole from prison. Ironically, the lawyer hired by the newspaper got into a dispute with the paper's two owners for money they owed him, and shot them. The newspapermen survived and hired Packer to be their office security guard.

This incredible story doesn't even end there. In 1989, a group of scientific investigators exhumed the remains of Packer's companions and examined their bones, which were pretty well preserved. Three forensic anthropology colleagues of mine, Walt Birkby, Bruce Anderson, and Todd Fenton, were among the researchers. They found cut marks on many of the bones, and a tool-mark expert concluded an axe or hatchet-type implement had made many of the injuries.

Using the *Harper's Weekly* sketch for comparison, the anthropologists even figured out who was who in the image. Based on known ages and features, including an old Civil War injury one of the men reportedly suffered, they assigned probable names to the remains. The locations of the hatchet marks on the back and sides of the skulls, suggested the men were likely struck when unaware, so maybe sleeping. Some showed defensive wounds on their arms. Cut marks around muscle attachment sites could reflect defleshing, most of them were on the victim's back sides, probably implying the impending cannibal didn't want to look at their faces. Dozens of hack marks were found on the youngest, and perhaps strongest, member of the group, showing he put up a fierce struggle for his life.

What about any bullet wounds? One hipbone had a hole in it, Could this have been Bell's pelvic bone, shot clean through by Packer? Although it's been debated by others, the anthropology team believed it was from animal gnawing, not a gunshot. The toolmark expert agreed. So, though an interesting study, the research couldn't really corroborate or dispute Packer's final account of the events, but it did suggest murder and cannibalism by one or more of the men involved.

There are plenty of versions to the Alferd Packer story, not only his, but also differing published accounts, but one thing is clear; the Colorado Cannibal, as he is sometimes called, has become a cult figure. Particularly at the University of Colorado, where student protests over the food quality in 1968 led to one of the dining facilities actually being renamed the Alferd Packer Restaurant and Grill, famous for its ribs and steak tartare. T-shirts and buttons bear slogans like, "Alferd Packer, serving his fellow man," and, "Have a friend for dinner."

In 1985, Colorado citizens worked to get Packer a posthumous pardon, but the then-governor refused, saying, he'd be, "eaten alive" by the public if he did. Several theatrical groups have covered the Packer story, and there's even a campy movie called *Cannibal! The Musical*. In 2006, I was privileged to travel the country as part of a National Geographic film project, visiting historians and scientists involved with the Packer investigation.

So what became of Packer? After his parole, he lived just south of Denver for about six years, but died of stroke in 1907. His death certificate listed the cause as "senility, trouble, and worry." It's said after his release from prison, Packer never ate meat again.

Investigating Incredible Bank Heists

Lecture 12

The United California Bank was located in an affluent neighborhood and had numerous high-powered patrons. On March 24, 1972, when employees tried to open the bank's vault, the state-of-the-art door seemed to be jammed from inside. When the vault was finally opened, workers discovered a large hole blown in the vault's 18-inch-thick reinforced-concrete ceiling. About 500 of the vault's double-key, armor-plated safe deposit boxes were scattered around the room, empty, and an estimated $8 to $12 million in cash, stocks, bonds, and jewelry had been taken. In this lecture, we'll delve into this theft—the largest from a bank in U.S. history at the time. We'll also investigate a bank robbery in Australia that took eight years to solve.

Initial Leads

- After the discovery of the theft at the United California Bank in Laguna Niguel in 1972, local law enforcement immediately brought in the FBI. The bureau notified its offices around the country to learn whether the MO at the bank might match any other crimes.
  - Agents from Cleveland, Ohio, reported a string of bank burglaries around their jurisdiction that showed similarities

© whitetag/iStock/Thinkstock.

Inside the vault of the United California Bank in Laguna Niguel, police found a customized sledgehammer, specially designed with a homemade device on its end to help open safety deposit boxes.

to the California heist. Even more important, Cleveland had a suspect.

- This man was 36-year-old Amil Alfred Dinsio from Youngstown, Ohio, an area heavily associated with organized crime. Dinsio was linked to a series of smaller bank burglaries in northeast Ohio and the eastern United States.

- The FBI needed to find out whether Dinsio had been in California at the time of the robbery. One way law enforcement typically tracks people is by tracing financial transactions, especially purchases. In this case, investigators learned from ticket records that Dinsio and his six-man crew had come through Los Angeles International Airport a week before the heist.

- FBI agents got an identification from an airport taxi driver, who remembered Dinsio for the $100 tip the man had given him. Records showed that the taxi had driven the men to the home of Dinsio's sister in Orange County, California. Investigators then took photographs of Dinsio and the other suspects to nearby hotels, where one of the desk clerks recognized some of the men.

- Next, investigators examined phone records from the hotel on the night the crew came in and found that one of the numbers called belonged to a real estate agency. They learned that two gang members had leased a furnished condominium that sat in the direct line of sight of United California Bank.

- When FBI agents reached the condo, it had already been professionally cleaned. Agents thought they had reached a dead end until one of the men opened the condo's dishwasher and found a load of dirty dishes. Although months had passed, forensic analysts found the fingerprints of all seven members of Dinsio's gang on items in the dishwasher.

- Other agents also followed up on another phone number called from the gang's hotel. That number belonged to a man who had previously lived in Youngstown. When agents went to his house, they found the getaway vehicle in the garage with burglary tools in the trunk. Ultimately, fingerprints from five of the suspects were lifted from the car.

Details of the Job

- Once they caught up with the mobsters, authorities learned many more details about the burglary: Dinsio had first made an initial reconnaissance trip to prepare for the job. He had cased the bank and surveyed the typical activities of the local police department. Once he had a plan mapped out in his mind, Dinsio returned to Youngstown to assemble his crew.

- After Dinsio's men arrived in California, they purchased the tools they'd need for the heist, including a ladder, empty sandbags, a variety of hand tools, and a police-band radio.

- On the night of the burglary, Dinsio and four other men drove to the bank, while one of the remaining two monitored the police radio and another stood guard outside the bank. Two of the robbers filled the empty sandbags with soil from a nearby flowerbed. The others put the ladder up to the bank's roof in the vicinity of the outside alarm. Dinsio sprayed the alarm with foam insulation so that when it went off, the clapper couldn't hit the bell.

- The bank's exterior flat roof was made of 0.5-inch plywood, and the burglars easily cut an opening it in large enough to pass through. Once inside the attic crawl space, they patched into the bank's air conditioning unit for electric power.
    - Next, they drilled 15 holes in the 18-inch steel-reinforced ceiling over the vault and put a stick of dynamite in each hole. The bags of soil were then piled on top of the dynamite to force the explosion to travel downward and to act as a silencer for the blast.

    - After setting off the explosion, the men used acetylene torches to cut the reinforcing rebar, then dropped down into the vault.

They also had to bypass another alarm that was wired to go straight to the sheriff's office.

- Amazingly, the bank heist took place over the course of three nights. Dinsio and his crew initially broke in on Friday night and left before dawn, lightly sealing their entry hole in the bank's roof. The men rigged a mirror on top of the entry hatch, knowing that as long as they could see the mirror from the condo window, the break-in hadn't been discovered. They returned to the vault Saturday night and Sunday night.

- After his arrest, Dinsio was convicted of burglary and sentenced to 10 years in prison; his crew members were also prosecuted for their part in the theft. In late 2013, Dinsio published a book called *Inside the Vault: The True Story of a Master Bank Burglar*.

Heist in Melbourne

- In the mid-1990s, the Bank of Melbourne in Victoria, Australia, was a much simpler target for thieves. Easy access to the building could be made through an alley, and the beeps that signaled the opening of the vault could be heard from outside the bank's back door. The bank was also close to a train station, where bandits could easily make a getaway.

- For this heist, the thieves chose Saturday, September 14, 1996, the day of an Australian football semifinal match in Melbourne. Most of the police force was at the stadium for security. The thieves planned to attack after bank workers arrived but before the bank opened for business. They knew that the vault had to be opened then to stock the tellers' cash drawers.

- The night before the robbery, the men used a power saw to cut a large, square groove in the lower part of the bank's back door, cutting only about halfway through the door's depth. This makeshift door-within-a-door would allow them to get in and out of the bank without actually opening the back door and tripping its alarm. This would buy them a few seconds on the way in, which was important if they were to catch the vault while it was open.

- On the morning of the robbery, the two men waited by the back door for the beeps that would signal the opening of the vault. When they heard their cue, they rammed the weakened back door with steel fence posts, ducked through the opening they had created, and jumped in. While one robber used a gun to control the employees, the other emptied the vault and the tellers' boxes. The pair left with $137,000 in cash, plus a stash of blank traveler's checks and bank drafts.

Eight-Year Investigation

- During the initial investigation, police collected statements from the three tellers. All described both men as being slim; about 5 feet, 6 inches tall; and with native Australian accents. The pair wore black Janus masks—the traditional pair of theater masks depicting comedy and tragedy—over black, open-faced ski masks. They also wore military-green coveralls and black ski gloves with red stitching.

- A police dog tracked the robbers' scent to the parking lot of the nearby train station but then lost the trail, suggesting that the criminals had gotten into a car left there. On a second trip from the bank, the dog led officers to a dumpster in the alley behind the bank, where they found the theatrical masks and other clothing.

- Police followed a number of leads, checking local theatrical suppliers and stores that sold surplus army clothing, but were unable to learn more information about the robbers. Officers checked out local figures known to be involved in crime and looked into other bank robberies in Victoria but found no patterns. In addition, none of the bank checks surfaced. It seemed as if the case might go cold.

- Scientists in the Victoria police forensic lab decided to test the clothing left behind near the robbery scene for *trace DNA*—that is, background DNA from a possible second wearer. Behind the area of the mouth opening in each of the Janus masks, they found enough saliva and shed skin cells to develop a genetic fingerprint—assumed to be from the robbers. Unfortunately, to positively identify a suspect, a known sample of DNA is needed for comparison, and the police had no suspects.

- A break in the case seemed to come a few months later, when some of the stolen bank checks surfaced. Police traced the checks to two men, but they weren't the original bank robbers, and the trail from them led nowhere. The case went cold, although the two DNA profiles were entered in Victoria's new DNA database in 1999.

- In 2000, in a rural area outside Melbourne, police tried to pull a driver over because the numbers on his license plate were only partly visible. Rather than slowing down, the driver led the police on a high-speed chase that finally ended in the parking lot of a Melbourne shopping mall, where he was caught by security guards.

- Police arrested 35-year old Goran Stamenkovic, who was tried and convicted for a number of offenses and sentenced to four years in prison. During Stamenkovic's incarceration, he was required to submit a DNA sample for the forensic database, which was found to match the genetic profile from the bank heist eight years earlier. Investigators reanalyzed the original stored evidence and again found a match for Stamenkovic.

- Detectives tried to bargain with Stamenkovic to give up his partner in the bank heist, but he wouldn't cooperate. In a stroke of luck, just a couple of months later, 43-year-old Mikael Mann was arrested for the armed robbery of a woman in her Melbourne apartment. Mann's DNA sample was taken almost immediately and found to be a match for the second bandit in the Melbourne robbery. In 2007, both men were sentenced to seven years for the crime.

Suggested Reading

Dillow and Rehder, *Where the Money Is*.

Dinsio, *Inside the Vault*.

Primorac and Schanfield, eds., *Forensic DNA Applications*.

Questions to Consider

1. Why do you suppose many people enjoy crime stories, such as those of the great bank heists in this lecture, and some people even root for the "bad guys"?

2. How have bank robberies changed over time? What kinds of security measures have been developed to thwart them?

Investigating Incredible Bank Heists

Lecture 12—Transcript

The United California Bank in Laguna Niguel, California was part of a very affluent neighborhood. The bank sat adjacent to a retirement community for the rich and famous in an area known for lavish homes, upscale hotels, beautiful beaches, and exclusive golf courses. During his presidency, Richard Nixon was rumored to be one of the bank's high-powered patrons, along with other local elites. The bank was recognized as a secure place to keep cash, especially among people who didn't want the IRS or anyone else to know about it.

But on March 24 of 1972, when employees tried to open the bank's vault, they realized something was wrong. The vault's state-of-the-art door wouldn't open. The mechanism seemed to be jammed from the inside. They called in technical support, and when the vault was finally opened, workers found a complete mess. There was a large hole in the ceiling and a thick layer of gray dust covered everything inside. The dust was from the hole blown in the vault's 18-inch-thick reinforced concrete ceiling, but curiously, there was also soil on the vault floor. Passing down through the ceiling was an extension cord and one of those hanging work lamps you might use in your garage.

About 500 of the vault's double-key, armor-plated safe deposit boxes were scattered around the room, empty. The thieves hadn't gotten into all of the boxes, though. Police also found burglary tools in the vault, too, including a customized sledgehammer, specially designed with a homemade key-type device on its end to help open the deposit boxes. Investigators even found milk cartons filled with water, which they figured were used to cool down power drills. A single, cotton work glove was also found in the vault.

An estimated $8 to $12 million in cash, stocks, bonds, and jewelry had been taken, at that time, the largest theft from a bank in U.S. history. It would equal about $100 million today. Local law enforcement immediately brought in the Federal Bureau of Investigation, and eight agents were assigned to the case. But over the next few weeks, when the FBI came up with almost no physical evidence, including no fingerprints, the number of agents rose to 125.

One of the advantages the FBI has is its extensive network of agencies throughout the U.S. So the Bureau notified its other offices to see if the MO, modus operandi, or, in English, the method of operations, well, if that might match any of the other crimes. Agents from Cleveland, Ohio reported a string of bank burglaries around their jurisdiction that showed similarities to the California heist, including one in their area not long after the Laguna Niguel job. Even more importantly, Cleveland had a suspect.

The man was 36-year-old, Amil Alfred Dinsio, from Youngstown, in my home state of Ohio. Now, you might never have heard of Youngstown, but it sits about midway between Cleveland, Ohio and Pittsburgh, Pennsylvania, home of the United Steelworkers. Although once prosperous because of its steel mills, Youngstown, like so many larger cities around it, eventually suffered the so-called rust-belt depression that hit during the second half of the 20th century. In both its heyday and hard times, this area was heavily associated with organized crime. Syndicates from both Cleveland and Pittsburgh battled over Youngstown as part of their turf, leading to a string of car bombings attributed to the mob in the late 1950s and early 1960s. Things were so bad that in 1963, the Saturday Evening Post dubbed Youngstown "Crimetown, USA."

Though Dinsio wasn't known for violence, he had been part of organized crime since his teenage years. Dinsio was linked to a series of smaller bank burglaries in northeast Ohio and in the eastern U.S., and stretching as far south as Florida. In some cases, Dinsio was just considered a person of interest by police, but he did have convictions, just nothing serious enough to put him in prison for any significant amount of time. Dinsio man was no dummy; he was good with his hands and had a fair amount of engineering and mechanical know-how. But in order to consider him a serious suspect for the United California job, the FBI would have to start by figuring out if Dinsio was in California at the time.

One way law enforcement typically tracks people is using the trails left by financial transactions, especially purchases. But back in 1972, that type of investigation was much more labor intensive than it is today. I mean, the Internet wasn't around, and far more people used cash back then, than today's easily-tracked debit and credit card transactions. There were still plenty of traceable purchases in those days, though, especially to try and figure out if someone traveled between Ohio and California.

As you might have guessed, investigators headed straight to the California airport closest to the bank, Los Angeles International. There, the ticket archives showed Dinsio and his six-man crew had come through the airport a week before the heist. FBI agents questioned airport taxi drivers, using photographs of Dinsio and his men, to see if anyone looked familiar. One of the cabbies remembered Dinsio for the 100-dollar tip the man had given him. Records showed that the taxi had driven the men to the home of Dinsio's sister, who just happened to live in Orange County, the same county as the bank. Police figured the five men wouldn't all have stayed at his sister's modest home, so they took the photographs of Dinsio and the other suspects to nearby hotels, where one of the desk clerks recognized some of the men.

Next, investigators examined phone records from the hotel on the night the crew came in, and they found one of the numbers called belonged to a real estate agency. Then they learned that Dinsio's two nephews, who were part of his gang, had leased a furnished golf-resort condo that sat in the direct line of sight of the United California Bank.

As soon as FBI agents walked into the condo, though, they had a sinking feeling they were too late. The place had been completely and professionally cleaned—the carpet, walls, everything. Even the furniture had been wiped down with a thick layer of polish. The agents thought they had reached a dead end, until one of them opened the condo's dishwasher, and miraculously found a load of dirty dishes. Although months had passed, forensic analysts found the fingerprints of all seven members of Dinsio's Youngstown gang on items in that dishwasher.

Now, at the same time agents were investigating the condominium lead, others had been following up on another California phone number that was called from the gang's first-night hotel. That number belonged to a man who had previously lived in Youngstown. And when agents went to his house, the guy said not only did he know some of the suspects, but they had left their car in his garage. FBI agents found an Oldsmobile Super 88 in the garage; it was purchased in California, and it was, no doubt, the getaway vehicle. How did they know? Well, the seats were covered with a layer of concrete dust, there were burglary tools in the trunk, and they found the mate to the cotton glove that had been discovered in the bank vault. Ultimately, they found fingerprints from five of the suspects and lifted them from the car.

Once they caught up with the mobsters, authorities learned much more detail about the burglary. Dinsio had first made an initial reconnaissance trip to prepare for the job. He had cased the bank and surveyed the typical activities of the local police department, which was a surprisingly small agency. Once he had things all mapped out in his mind, Dinsio returned to Youngstown to assemble his crew, based on the expertise he felt the job needed. When the gang arrived in California for the heist, after renting the condo and buying the car, Dinsio's men attended to other matters. They took their time assembling the tools they'd need; they sent different guys to different stores in order to not raise suspicion. Among the things they bought were a ladder; empty sand bags; a variety of hand tools, including an acetylene torch; as well as a police-band radio.

On the night of the burglary, one man stayed at the condo and monitored the radio, while another one walked to the bank to stand guard with his own radio, and Dinsio and four other men got in the car and drove over to the bank. Two of them got out and began filling the empty sand bags with soil from a nearby flowerbed. The others put a ladder up to the bank's roof in the vicinity of the outside alarm. Now, while two of the men held Dinsio's ankles, he sprayed the alarm with foam insulation, so that when it went off, the clapper couldn't hit the bell.

The bank's exterior flat roof was made of a half-inch plywood, and the burglars easily cut an opening it in large enough to pass through. Once they were inside the attic crawl space, they patched into the bank's air conditioning unit to get electric power. Then, they drilled 15 holes in the 18-inch steel-reinforced ceiling over the vault and put a stick of dynamite into each hole. The bags of soil were then piled on top of the dynamite to force the explosion to travel downward, and also act like a silencer for the blast. Once inside, they could use the empty to carry their loot. After setting off the explosion, the men used the acetylene torches to cut the reinforcing rebar, and then dropped down into the vault. Next, they had to bypass another alarm in the vault that was wired to go straight to sheriff's office.

Now, one of the most amazing things to me was that this bank heist lasted three nights. Dinsio and his crew broke in on Friday night, took everything they could before dawn, and then lightly sealed their entry hole in the bank's roof. Then the men rigged a mirror on a stand and put it on top of the entry hatch, knowing as long as they could see the mirror from their condo window, the break-in hadn't been discovered. So then they came back Saturday night and Sunday night, draining the vault until literally just hours before the bank was set to open on Monday morning.

After his arrest, investigators found some of the stolen property in Dinsio's house, including rare coins that their California owner could identify. Dinsio was convicted of burglary and sentenced to 10 years in prison; his crew was also prosecuted for their part in the theft. In late 2013, Dinsio published a book called *Inside the Vault: The True Story of a Master Bank Burglar.*

Now, let's travel to Australia and look at another interesting heist. In the mid-1990s, the Bank of Melbourne in Victoria was a much simpler prime target for thieves. The building was situated with a back door that led to a walled-in courtyard adjacent to an alley. Now, by scaling the wall from the alley-side, a person could easily get onto the roof of a small outbuilding that was just inside the bank's courtyard, and from there it was an easy drop down into the bank's yard. Now, if a person listened carefully near the back door, they could hear the beeps that marked the opening of the vault from outside. The bank was also close to a train station, so bandits could easily disappear into the crowd and be far away in no time.

The thieves even chose the perfect day to strike—September 14, 1996. That was the Saturday morning of the Australian football semi-finals, being played in Melbourne. They knew the bulk of the police force would be at the stadium for security and that most people would either be at the game or watching it on TV, not out and about in town. The men planned to attack as the bank was opening for the day, between the time workers arrived and when they opened the bank for business. This was because the thieves knew the vault had to be opened then in order to stock the teller's cash drawers for the day. Their timing would also allow fewer witnesses, and there wouldn't be any customers who might panic and complicate things.

The night before the robbery, the men used a power saw to cut a large, square groove in the lower end of the bank's back door; they only went about halfway through the door's depth. They already knew the relative thickness of the door because they'd cased the bank and they had watched employees go in and out of it. The crooks knew once they entered, the tellers would probably hit the bank's silent panic buttons, but the makeshift door-within-a-door they had created would let them in and get out quickly, without actually opening the back door and tripping its alarm, and this would at least buy them a couple of seconds on the way in, which was really important if they were going to catch the vault while the door was still open.

On the morning of the robbery, the two men waited out back for the beeps that would signal the vault was opening to start the bank's day, and when they heard their cue, the men rammed that weakened back door with steel fence posts, they ducked through the opening they created, and jumped in. While one robber used a gun to control the employees, the other emptied the vault and the tellers' boxes. The whole heist took just a bit over two minutes, and the pair left with $137,000 in cash, plus a stash of blank traveler's checks and bank drafts. That wasn't a huge haul, but these guys were well prepared, and they were careful not to leave clues.

During the initial investigation, police collected statements from the three tellers. All three described both men as being slim, about 5′ 6″, and with native Australian accents. The most remarkable aspect was that the pair wore these black-colored Janus masks; those are the traditional pair of theater faces, depicting comedy and tragedy. One man wore the smiling mask, and his partner wore the sad face—both of those over black, open-faced ski masks. So together, the ski hood combined with the theater mask completely hid the perpetrators' skin, hair color, eye color, and any detail of their ears, mouths, or their teeth. Not only did their disguises hide the men from the witnesses, but also from the video surveillance cameras as well.

As far as the rest of their clothing, the man with the gun wore a military green style coverall suit that had a small German flag sewn onto one arm; the other guy had a similar coverall that had an elastic waist, and it had pockets, but not a German flag. One of the tellers said she saw an oblong shape in the back pocket of that man's outfit that looked to be a cell phone, which, in 1996, were not nearly as common, and much larger than today's cell phones. The employee also noticed a light blinking on the object, which in those days usually meant a phone message had been left. The robbers wore black ski gloves with red stitching, and when one man had taken his off, the teller saw a surgical-type glove underneath it. The gun involved was a small, black handgun, and the guy handling the money had a backpack he stuffed with the cash.

You might wonder how the employees gathered so much detail, especially when staring down the barrel of a gun. After all, one of the problems with eyewitness testimony is a phenomenon we call weapon focus; in other words, a witness to a crime that involves a gun or knife is understandably preoccupied with the weapon. Well, bank employees, just like security guards and others who work in occupations that might be criminal targets, well those types of people are specifically trained to study every detail of a robber's appearance, if they possibly can, for situations just like this. In fact, after herding the three bank employees under a table, one robber noticed one of the women studying him carefully, and he commanded her, do not look. Luckily, the robbers also told the workers they would not be hurt, and they weren't.

Because the bad guys were last seen on foot, investigators brought in a police dog named Boss from their canine unit. Boss was able to track the robbers' scent to the nearby train station, where the handlers worried the trail would be lost on the platform, but to their surprise, the dog crossed the tracks, instead, suggesting the criminals didn't get on the train. Once on the other side, though, Boss got all confused in the station's parking lot; either the football game's extensive crowd that morning caused the dog to lose the scent, or more likely, the men had gotten into a getaway car parked in the lot.

As the officers walked the dog around the parking area, they evaluated other possible options. Even in those days, investigators could check pings off cell phone towers, so, they did consider that angle, since a getaway car might have called the robber's cell phone, or the bad guys could have also called the getaway driver for a pickup. Police could also investigate stores where high-tech power saws were sold, because they figured the tool had to be battery-operated, not electric, since there was no outlet in the bank courtyard, and a gas-powered saw would have made a lot of noise to attract attention. Detectives also knew they would probably catch anybody who tried to pass any of the bank or traveler's checks, since the numbers of those were already on record.

But they decided instead to take the dog back to the bank for another try, even if just to find additional evidence—and good-old Boss led them straight to a dumpster in the alley behind the bank. Inside was a pair of coveralls, a pair of outer gloves with surgical gloves layered inside them, a black ski mask, and the pair of theatrical masks. Good dog! In the yard adjacent to the bank, police located another pair of gloves, a second ski mask, and a baseball cap. They also found a couple of shoe impressions in the soil at the bottom of the wall just behind the small shed in the bank's courtyard. Well, the direction and proximity of the footprints suggested they could have been left by someone trying to scale the wall and reach the bank.

So, even though investigators didn't have the bad guys, they had a lot of leads to explore. Some officers were sent to local theatrical stores to see if they could get anywhere with those Janus masks. They couldn't, even though these were fancy masks with a fabric liner. The serial numbers on the military clothing indicated that the coveralls were the type that were routinely used in training, but investigators couldn't track them beyond that point, not even by checking local Army surplus stores. Police couldn't find a nearby source for any type of hand-held battery-operated saw, either. Even the cell phone tower analysis led nowhere.

So, the next step was to look up any questionable characters to see if some of the local bad guys had suddenly struck it rich. Nothing. None of the bank checks surfaced, either. There had already been 28 bank robberies in Victoria between July of 1995 and February of 1996, and 11 of those were at Bank of Melbourne branches. So they started making comparisons between the September heist and those other robberies, to look for any possible patterns. But again, nothing. Ironically, things had gotten so bad for the Bank of Melbourne that the branch robbed in September was already set for relocation to a more secure facility. The scheduled date for that move turned out to be two weeks too late. Despite all the leads they had, the case looked like it might go cold.

Now, this was back in 1996, and DNA technology was certainly being used at the time, but up to this point, a significant amount of biological material was typically needed, or at least used, to generate a DNA profile. A sufficient amount of cells for analysis could be found in something like a semen stain in a rape victim's panties, or maybe on a drop of a wounded killer's blood on a homicide victim's shirt. But one of the things that DNA analysts were often finding around this time was, when they got criminal DNA profiles from stains found on clothing, there was often a small amount of DNA from a second person. It turned out almost invariably that the other DNA profile came from the person who had been wearing the clothes, rather than from the perpetrator. The source was a background-level of skin cells from the wearer of the clothing, which analysts had been previously overlooking. In fact, it was actually a pair of forensic scientists from the Victoria Australia police lab, not too far from the bank we've been talking about, who, by the time of the robbery had already submitted an article to the prestigious journal *Nature* about this phenomenon they were calling *trace DNA*.

So the forensic lab decided to test the clothing left behind near the robbery scene for trace DNA. They got nothing out of the ski masks, no DNA from any of the gloves, or the baseball cap, but what about the cloth lining of the two Janus theatrical masks? Behind the area of the mouth opening in each mask there turned out to be enough saliva and shed skin cells to develop a genetic fingerprint. Just six days after the crime, the testing confirmed a different male profile from each mask; they were assumed to be from the robbers. Great news, right? But, to nail a suspect, you need a known sample of his DNA for comparison, and the police didn't have that; they didn't even have any firm suspects. In addition, investigators didn't yet have the type of computerized DNA databases we have today, in which genetic profiles can be compiled and then compared at local, state, and even national levels. At that point, the best analysts could do was manually compare the recovered

DNA profiles to any possibilities they found, such as maybe guys who had committed other bank robberies.

The next big break came a few months later, when some of the stolen bank checks started surfacing. Two guys claiming to be from a catering company went into a butcher shop and used one of the checks to buy—are you ready for this?—over $22,000 worth of filet mignon steak. Turns out they sold the steaks to another butcher to get cash. And a couple of days later, the pair went into a wine store and tried to buy $86,000 of wine; that's when police arrested them. It turned out these weren't our bad guys. They had gotten the checks from somebody else who said he got them from another guy, and so on, and that trail led nowhere.

Police had yet another lead when they learned from the wine shop owner that somebody had called that morning to make sure the big order was all set for pickup. One of the numbers pulled from the store's phone logs turned out to be from a woman known to hang with a guy who had four prior convictions for armed robbery. Pay dirt, right? Nope, the Melbourne bank heist couldn't be traced to him, either. The case just went completely cold; although the two DNA profiles were entered into Victoria's brand new DNA database in 1999, they sat there waiting.

Then, in 2000, in a rural area outside Melbourne, police tried to pull a driver over because the numbers on his license plate were only partly visible. But rather than slowing down, the guy sped off in what was to become a one-hour, high-speed police chase along a country road leading up to the city. Approaching Melbourne, the driver started weaving in and out of traffic, and eventually wrecked his car. He then took off on foot, ran to another vehicle, and used a screwdriver to carjack its owner. From there, the maniac drove the stolen car to a shopping mall, where he got out and grabbed the keys from a woman who was trying to put her baby in her car. But as the bad guy ran around to the driver's side, security guards raced up and tackled him.

Police arrested 35-year old Goran Stamenkovic, who was soon tried and convicted on a long string of offenses, not just the violent assaults and the driving violations, but also for drug possession. He was sentenced to prison for a minimum of four years. Now, during Stamenkovic's incarceration, a law was passed requiring DNA to be taken from certain classes of convicts and then put into the Victoria forensic database. Stamenkovic's sample was submitted before his release, which had been scheduled for 2004. Wouldn't you know, eight years after the Melbourne bank heist, Stamenkovic's genetic profile matched the trace DNA that had come from one of the theater masks.

And just to make sure there had been no mistake, investigators went back and found the 1996 police file and the stored evidence. And by 2004, DNA technology was significantly improved over its capabilities at the time of the first testing. When forensic scientists reanalyzed the surgical gloves that had been found at the bank heist, they were able to pull out two complete DNA profiles, as well as a partial third. One of the full profiles, again, proved to be a match to the DNA sample taken from Stamenkovic in prison. But that loser that fled police on that country road in 2000 sure didn't seem to match the sophisticated criminal style that was demonstrated back during the 1996 robbery. So as a final check, they even had Stamenkovic retested. It was him, all right. Either he hadn't been the mastermind, or drugs had turned the man's brain to a desperate mush.

Detectives tried to bargain with Stamenkovic to give up his partner in the bank heist, but he wouldn't cooperate. But in an amazing stroke of luck, just a couple of months later, a man was arrested for the armed robbery of a woman in her Melbourne apartment. Due to the nature of the offense and the perpetrator's extensive criminal history, they took a DNA sample from him almost immediately. As soon as the man's profile went into the Victoria database, it came up as a match to the surgical gloves worn by the second bandit in the Melbourne robbery. How 43-year-old Mikael Mann was walking free to rob that woman, I will never know, since he already had a record that included 63 crimes dating all the way back to his teenage years. In fact, he had recently been rearrested and was out on bail when he robbed the woman's apartment.

Finally, after eight long years, investigators knew the identity of the two men who had broken through the back door of the Bank of Melbourne, frightened the staff, and stolen cash and checks. But the pair absolutely refused to talk, so investigators still have no idea about a possible getaway driver or anything else related to the case. Based on the DNA recovered from the Janus theatrical masks, Mann had worn the frown of tragedy during the holdup, while Stamenkovic wore the comedy mask, but neither of them was smiling in 2007 when they were sentenced to seven years each.

How Reliable Is Eyewitness Testimony?

Lecture 13

Jennifer Thompson and Ronald Cotton are two national experts who work on exposing the dangers of eyewitness testimony in police and judicial processes. Their work helps to demonstrate how misidentification can lead to miscarriages of justice in the U.S. court system. These two authorities acquired their expertise on this topic in a case in which they were both involved—on opposite sides. Jennifer Thompson was raped in 1984 and, later, picked Ronald Cotton out of a police lineup as her attacker. Thompson's testimony against Cotton sent him to prison for more than 10 years, even though he was not her rapist. It's almost impossible to imagine that today, the two are close friends and colleagues.

The Attack and Examination

- In July 1984, in Burlington, North Carolina, 22-year-old college student Jennifer Thompson was raped in her apartment by a man who had broken in while she was asleep. During the attack, Thompson had the presence of mind to study her rapist so that she would be able to identify him later. Afterwards, she managed to run from her apartment to a stranger's home nearby and call the police.

- Thompson was taken to the hospital so that she could be examined and a rape kit could be taken. The rape kit procedure involves collecting physical evidence that may be present on a victim's body. Investigators—often, forensic nurse examiners—swab body parts for saliva or semen and collect hairs or fibers that might be from the perpetrator.

- While Thompson was in the emergency room, she heard that another rape victim was also in the hospital—a woman named Elizabeth—who might have been attacked by the same man after he left Thompson's apartment. Elizabeth lived only a half mile from Thompson.

- The next stop was the police station. There, Thompson worked with a police identification kit, selecting facial and other features to be put into a composite drawing. Thompson was also questioned about the rape and the physical examination at the hospital. Because officers believed that the initial examination had not been performed correctly, they sent Thompson to another hospital to have the rape kit redone.

Thompson's Identification

- The sketch made from Thompson's description was released to the public through the media. Based on an anonymous tip, police rounded up Ronald Cotton, a 22-year-old man who resembled the description and worked near the area where the two young women had been raped.
    - Cotton had a breaking-and-entering charge on his record, as well as an attempted sexual assault charge from his teenage years, for which he served 18 months.

    - Cotton's alibi for the night of the rapes was quickly shown to be inaccurate. In his defense, Cotton claimed that he had confused a couple of days of the past week.

- When police brought Thompson six photographs of possible suspects, they told her that her rapist might or might not be among them. That's an important part of police lineup procedure because it doesn't force the witness to make a choice.
    - One of the problems with the police photo presentation, however, was that Thompson was given all six images at the same time. That method allows the witness to compare each face against the others and use a process of elimination to exclude individuals, rather than simply positively identifying one perpetrator.

    - In fact, Thompson herself later said that she treated the photo array like a standardized test question: She compared the features of each man against the others, excluding those she was sure were not right, and settled on Cotton as the "best fit" for her rapist.

- Cotton was later one of seven similar-aged African American males brought into the police station for Thompson to observe. Each man was asked to step forward and repeat a few sentences that Thompson remembered from the rape. Although she later admitted that she almost picked another man in the lineup, Thompson ultimately selected Cotton. When police told Thompson that she had picked the same man as in the photo array, she was convinced that she had identified the man who raped her.

Trial and Retrial

- In 1985, Cotton was brought to trial, charged with rape and burglary. The prosecution presented police testimony about his false alibi and physical evidence, including a flashlight found at Cotton's house that was alleged to be similar to one used in the rape and a small piece of foam rubber found at the scene that was consistent with the inside of a pair of sneakers owned by Cotton. After Thompson pointed to Cotton in the courtroom and stated that he was her rapist, it took a jury just 40 minutes to find him guilty of both rape and burglary, for which he was sentenced to life in prison.

- Eyewitness testimony is considered direct evidence because it involves a person testifying to what he or she directly saw, heard, felt, and so on. Perhaps surprisingly, tangible physical evidence—such as hair, fibers, fingerprints, and even DNA—is considered indirect evidence because the judge or a jury must make inferences based on that evidence. Eyewitness testimony, if true, is a statement of fact with no interpretation necessary. Of course, as we all know, people can make mistakes or even lie on the witness stand.

- Once Cotton was incarcerated, he began writing to attorneys and the media, trying to convince them of his innocence. About a year after he entered prison, a new inmate arrived, Bobby Poole. Cotton later heard from another prisoner that Poole had admitted to the rape of Thompson and another woman in July 1984, as well as other rapes since that time. Poole even joked to fellow inmates that Cotton was doing some of his jail time for him.

- Cotton soon earned a new trial, although not because of the information about Poole. It had been revealed that the second rape victim on the same night as Thompson had picked a different man in the police lineup, but that evidence had been withheld during Cotton's original trial. By this time, however, the second victim, Elizabeth, had concluded that Cotton was her rapist, too. Thus, at his second trial, Cotton was charged with two rapes.

- Elizabeth testified—although less convincingly than Thompson—that Cotton was the rapist. Poole was brought into the courtroom, but Thompson stuck to her identification of Cotton. At the end of the trial, Cotton was convicted of two counts of rape and two counts of burglary, for which he was given two life sentences.

Release and Forgiveness

- In 1994, after seven years in prison, Cotton learned about DNA testing in forensic cases. Richard Rosen, an attorney and law professor, agreed to help him pursue DNA testing but warned him that if the DNA evidence identified Cotton as the rapist, there would be no other legal means to get him out of prison. Cotton insisted that the test be done.

- Much of the evidence in the two 1984 rape kits had degraded over the course of 10 years, but in Elizabeth's rape kit, a fragment of a single sperm cell was found. That tiny bit of evidence yielded a DNA profile that proved Cotton was not the rapist and Poole was. Cotton was exonerated and released in June of 1995. Poole died in 2001 while still incarcerated.

- When she heard the news of Cotton's innocence, Thompson felt tremendous guilt and shame. She wondered if people thought she had reacted too quickly after the crime and had just tried to identify any man to pay for what happened to her. She also speculated that people might think of her as a racist, given that she is white and Cotton is black.

- About two years after Cotton's release, Thompson asked if he would meet with her. Thompson burst into tears and apologized to Cotton, who told her that he just wanted them both to be happy and to move on with their lives. The pair hugged and promised to stay in touch.

Lessons from the Case

- It has long been known that eyewitness testimony is problematic. Many scientists, probably most famously forensic psychologist Elizabeth Loftus, have extensively studied the phenomenon.
    - One interesting facet of the issue is that people tend to differentiate the faces of people from their own race better than they can distinguish among the faces of people from other races.

Eyewitness testimony is valuable in criminal cases, but law enforcement officers, investigators, attorneys, juries, and judges should also be aware that it can be faulty.

    - Loftus thinks that when an individual sees a person of a different race, the primary things noticed may be the stereotypical characteristics associated with that race. For instance, a Caucasian might note eye shape and hair form in Asians rather than other details. During a lineup, when multiple people from that race are presented together, we can't differentiate because we never focused on the finer details in the first place.

- It's also known that human memory is malleable. As we ponder something that happened or try to recall someone we saw, our minds can easily distort the details, especially if new information becomes available. We may begin to confuse or blend the original story or image with the new data. Experts have also noted that a witness who at first may be uncertain can often become more certain over time. It's as if we want so badly to be sure that we subconsciously force ourselves to become more confident in our assertions.

- Since the time Cotton gained his freedom, several hundred people have been exonerated through DNA testing; about three-fourths of them were convicted partly by faulty eyewitness identification, making it the most common source of wrongful convictions in the legal system.

- To atone for the error that cost Cotton more than 10 years of freedom, Thompson has devoted her life to talking about the mistake she made. She regularly speaks at conferences to educate those involved in the legal system of the pitfalls of convictions based solely or too heavily on eyewitness testimony. She also opposes the death penalty because of the possibility of wrongful executions.

- Through their story and activism, Thompson and Cotton have lobbied for reform in police procedures, especially those involving eyewitness identifications. Their efforts and their book, *Picking Cotton*, have helped prompt procedural reform in some parts of the country, but many agencies still resist change.

- After being released from prison, Cotton got a job with the DNA testing company that performed the analysis that earned his exoneration, and the state of North Carolina awarded him $110,000 in restitution. Cotton married, had a daughter, and to this day, remains friends with Thompson and her family.

Suggested Reading

Borchart, *Convicting the Innocent.*

Innocence Project, http://www.innocenceproject.org/.

Loftus, *Eyewitness Testimony.*

Loftus and Ketcham, *Witness for the Defense.*

Münsterberg and Hatala, *On the Witness Stand.*

Murray, *Overturning Wrongful Convictions.*

Petro and Petro, *False Justice.*

Primorac and Schanfield, eds., *Forensic DNA Applications.*

Scheck, Neufeld, and Dwyer, *Actual Innocence.*

Thompson-Cannino and Cotton, *Picking Cotton.*

Questions to Consider

1. Do you have any experience being an eyewitness to a crime or some other important incident? If so, how confident did you feel regarding your account?

2. What kinds of things contribute to faulty memories when trying to recall events or people?

3. Can you think of ways by which a person can be led to "recall" something that he or she didn't actually experience or previously remember?

How Reliable Is Eyewitness Testimony?

Lecture 13—Transcript

In July of 1984, on a hot night in Burlington, North Carolina, 22-year-old college student Jennifer Thompson was sleeping in her apartment after coming home from a date. Unbeknownst to her, a man had broken the light bulb outside of her back door, cut her telephone line, and pried open her door. He was crawling along the floor in her bedroom when Jennifer suddenly woke up. She started to scream, but then stopped when she felt a knife blade against her throat. She whispered that she would give the man her money, her credit cards, even her car, if he would just leave her alone. When he replied he didn't want any of those things, she knew she was about to be raped.

During the attack, one solitary thought continued to cycle through Thompson's mind, keeping her focused—she was going to live; she was determined to live; and because of that, she needed to study her rapist's face, his voice, look for scars or tattoos on his body, memorize his clothing, try to estimate his height by comparing it to her own body, in short, take in every detail she possibly could. That way, if police were somehow able to find this man, she could definitely identify him. Because she could smell he had been drinking, Jennifer thought she might have a mental advantage over her attacker, since she clearly didn't have any physical advantage. She decided to try to escape by using her wits.

During the assault, Jennifer thought maybe the man knew her from somewhere, but then realized more likely he must have been tracking her, or at least looking around her home for a while before she woke up. He knew her name, knew she was from Winston-Salem, and thought she had a boyfriend in Germany, since there were postcards in her apartment from Germany, signed by a man, though the mail was actually from her brother. She didn't know it at the time, but the alcohol on his breath was beer taken from her own refrigerator and liquor from her cabinet.

When Jennifer heard him call her "baby," even though it sickened her, perhaps subconsciously she thought he wanted a cooperative victim. Plus, she had heard that one of the worst things a woman can do during a rape is to fight, since that just exaggerates the sense of control the rapist is trying to exert. Jennifer told her attacker she would feel much more relaxed if he got rid of the knife and asked him nicely to put it outside of her front door. She thought, maybe, that if he complied, she could rush after him, close the door, lock it, and then call for help. She had no idea her phone line had already been cut. The man actually agreed to put the knife outside, but insisted that she go into the bathroom while he did so. She realized she'd have to think of another plan.

Next, Jennifer said she was really thirsty and needed to go to the kitchen for something to drink, and she said she could get him a drink too. She wrapped a blanket around herself, and headed for the kitchen. Jennifer flipped on the light, actively considering that doing so might keep him from following her, since she hadn't yet seen him in full light. When she saw the back door into her kitchen was open, she knew that's how he had gotten in. Jennifer fussed around for a minute or two with glasses and ice cubes, just to make some noise in the kitchen, before bolting for the back door and rushing outside. She ran to a stranger's home nearby, wrapped only in a blanket, and called the police.

Jennifer Thompson was taken to the hospital early that morning for an examination and so that a rape kit could be taken, that's when they try to collect any physical evidence that may be present on the victim's body. When Jennifer was in the emergency room, she heard there was another rape victim in the hospital that same morning, a woman named Elizabeth, who police thought was probably attacked by the same man, after he left Jennifer's apartment. Elizabeth lived only a half mile away from Jennifer, and maybe the rapist attacked her because he was frustrated by not being able to consummate his crime with Jennifer.

Now, today, many of the larger hospital emergency departments have what are called SANE nurses; that stands for sexual assault nurse examiners. These special investigators swab body parts for saliva or semen, and collect other physical evidence. That might include hairs from the perpetrator or maybe fibers from his clothing. But remember, this case occurred in 1984, before DNA technology was used in forensics, so at that point, while hairs and fibers gathered in the emergency room might be helpful, the most they could get from any semen found would be the perpetrator's blood type and maybe other blood-protein markers.

Jennifer's next stop was the police station. She was asked to use one of those police sketch artist kits from which the witness selects facial features, like the shape of the suspect's face, the type of eyes, nose, lips, hairstyle, eyebrows, or any other things the witness might have noticed. Then the selected components are put into a composite drawing. Jennifer was also questioned, not only about the rape, but also about how the physical exam had gone at the hospital.

When the officer found out she had not received any penicillin to counter a sexually-transmitted disease the rapist might have carried, and also had not been given the contraceptive commonly called "the morning after pill," both of which are supposed to be standard parts of the treatment of any rape victim, well, they sent Jennifer to another hospital to have the entire humiliating rape kit procedure redone. It was the next day before a detective took her back to her apartment. That's when Jennifer realized for certain that her attacker had been in her house for quite a while before the rape, drinking, smoking cigarettes, and going through her personal belongings.

The eyewitness details that Jennifer had given law enforcement, and the sketch made from them, were released to the public through the media. Based on an anonymous tip they received, police rounded up Ronald Cotton, a 22-year-old man who resembled the description and worked at a restaurant near the area where the two young women had been raped. Cotton did have a breaking and entering charge on his record, as well as an attempted sexual assault charge from his teenage years, for which he served 18 months. Making suspicions even stronger, Cotton's alibi for the night of the rapes didn't pan out. The story he told police was quickly shown to be inaccurate. But in Cotton's defense, he claimed he simply made a mistake and confused a couple of days of his past week.

When police brought Jennifer Thompson six photographs of possible suspects, they followed best practices; they told her that her rapist may or may not be among them. That's an important part of a police lineup procedure, because it doesn't force the witness to choose anyone from the group. But one of the problematic things done during the police photo presentation was that Jennifer was given all six images at the same time. Now, you might wonder, what's wrong with that? Well, it allows the witness to keep comparing each face against the others and use the process of elimination to exclude individuals and narrow down to one, rather than having the witness simply pick one photo from a whole group and say, "that's the guy."

Giving all the photos at once could be compared to a multiple-choice test question, where we tend to compare the possibilities before we choose, "the right answer." In fact, Jennifer Thompson herself later said that she treated that photo array like a standardized test question. She compared the features of each man against the others, excluding those she was sure were not right, and settled on Ronald Cotton as the best fit to her rapist. Jennifer also said that after she chose Cotton, the police told her, "Well, we thought that might be the guy, since he has a prior conviction."

Ronald Cotton was later one of seven similar-aged African-American males brought into the police station for Jennifer to observe. I read an interview with her in which she said she actually thought the lineup would be like what we see on TV; the men would be on the other side of a one-way piece of glass and wouldn't be able to see her. She was shocked when she had to stand in the same room and face the seven men, as each one of them held up a piece of paper with a number on it.

Ronald Cotton later mentioned he was so frightened at the lineup he was visibly shaking as he stood there holding the number-five sign. He had already been told that Jennifer had picked his image out of a photo array, and he was terrified. Each man had to step forward and then repeat a few of the lines that Jennifer had remembered from the rape, so she could hear each guy's voice.

Although she later admitted she very nearly picked the fourth man in the lineup, Jennifer ultimately did select Cotton. As you may have seen in legal dramas on television, police are not allowed to conclude a lineup by saying, "good job," or, "you picked the right guy," and they certainly didn't say that to Jennifer Thompson. But when police told her she had picked the same man whose image she'd identified during the photo array, Thompson was convinced she had chosen the man who raped her, and that he would end up rotting in jail for what he had done to her and the other woman he also attacked that same night.

In 1985, there was a week-long trial in which Ronald Cotton was charged with both the rape of Jennifer Thompson and the burglary of her home. Police gave testimony about Ronald Cotton's false alibi. Physical evidence was presented, including a flashlight found at Cotton's house that was alleged to be similar to one used in the rape. A small piece of foam rubber found at the scene was said to be consistent with the inside of a pair of sneakers Cotton owned. But after Jennifer Thompson actually pointed to Ronald Cotton in the courtroom and stated that he was her rapist, it took a jury just 40 minutes to find him guilty of both the rape and burglary, for which he was sentenced to life in prison.

Now, some of you may already know that things like the flashlight and foam rubber are called circumstantial, indirect evidence in the courtroom; they show the suspect may have been at the scene, but not that he definitely did the action accused. But eyewitness testimony is considered direct evidence. That's because a person testifies to what he or she directly saw, heard, felt, etc. Surprisingly to some people, tangible physical evidence, like hair, fibers, fingerprints, and even DNA, is indirect evidence, because the judge and/or a jury must make an inference based on that evidence. In other words, they have to use it to draw their own conclusion. Whereas eyewitness testimony, if true, is a statement of fact with no interpretation necessary.

I guess it's the "if true" part, where we really get into trouble. Not only can people lie on the witness stand, but they can also be mistaken, especially in times of extreme stress. And we all know that our senses can play tricks on us, leaving us to either doubt what we saw, or to extrapolate beyond what we saw. We may simply confuse things enough to make a mistake in our recall of events.

Based on everything I've read, heard, and seen, I truly believe that Jennifer Thompson acted in good faith when she chose Ronald Cotton from that photo display. And perhaps, when she saw him again in the physical lineup, she more strongly recognized him because she had already seen his photograph.

Either way, Thompson absolutely believed that Ronald Cotton was the man who had broken into her home and raped her, and she was glad to see him headed to prison.

Once Ronald Cotton was incarcerated, he started writing to attorneys and the media, just trying to get someone, anyone, to listen to his pleas of innocence. And then one day, about a year after he entered prison, things took an unbelievable twist; Cotton saw a new inmate being brought in, and he thought the man looked kind of familiar. So Cotton walked right up to the guy and said, "Hey, man, where are you from?" The man, who was named Bobby Poole, replied that he was from Burlington, North Carolina. And Cotton said, "Me, too, and you look just like the police artist's sketch of the suspect in the Burlington rape I'm sitting here in prison for, but didn't do."

The new inmate brushed Cotton off and denied the accusation. Later, though, Cotton heard from another prisoner that Bobby Poole admitted he was the man who had raped both Jennifer and Elizabeth on that night in July of 1984—not only that, but also other women since then, too. In fact, Poole was in prison doing two consecutive life sentences for other rapes, but joked to fellow inmates that Cotton was actually doing some of his jail time for him.

Ronald Cotton soon earned a new trial, although it wasn't because of the information about Bobby Poole. In fact, although there was testimony presented at the second trial in 1987 about Poole admitting in jail to raping Jennifer Thompson, the judge removed the jury from the courtroom during that testimony. The judge ultimately did not allow them to hear or consider that as evidence. The reason Cotton was actually awarded a new trial was it had finally been revealed that Elizabeth, the second victim raped the same night as Jennifer Thompson, had picked a different man out of the police lineup, not Ronald Cotton, but that evidence had been withheld during Cotton's original trial.

To make matters worse, though, by the time of the second trial, Elizabeth had come to the conclusion that Cotton had actually been her rapist on that fateful night, too. So in his 1987 trial, Ronald Cotton stood faced with not just the rape of Jennifer Thompson, as was the case at his first trial, but was now charged with the rape of the second victim, Elizabeth, as well. This time

Elizabeth was also present in court, and, although less convincingly, she, too, testified that Cotton was her rapist.

Now, that's not to say that Bobby Poole had nothing to do with the second trial. Due to the jailhouse talk, Poole was actually brought to the courtroom, and Jennifer Thompson was asked if he could have possibly been her rapist. She replied, "Absolutely not." She stuck to her guns and insisted that Ronald Cotton was the man who had raped her. In fact, Jennifer later said how angry she was that the defense would bring in some other man who vaguely resembled Cotton to try to confuse the jury. How dare they do that, and how dare they doubt her word?

So, in 1987, at his second trial, Ronald Cotton was convicted of not one, but two counts of rape and two counts of burglary, for which he was given two life sentences. When asked if he had anything to say to Jennifer and Elizabeth, he said he was sorry for what had happened to them, and although he had done some bad things in his life, he absolutely did not rape either of them.

Back in prison, Cotton said it was not uncommon for other inmates and guards to call him Poole, and also confuse Bobby Poole for being Ronald Cotton; that's how much the two men looked alike. And as the years went on, Cotton did what he could just to adjust to prison life, while Jennifer Thompson married, had triplets, and tried to put her terrible memories behind her, as much as that was possible for either of them. Then, in 1994, when most of the population of the United States and others around the world were captivated by the famous O. J. Simpson murder trial in California, Ronald Cotton learned about something that would ultimately be his salvation; he heard about the power of DNA testing in forensic cases.

Richard Rosen, an attorney and law professor, agreed to help Cotton pursue DNA testing. Rosen warned him, though, that if the DNA evidence fingered Cotton as the rapist, that was the end of the line as far as any legal means to get him out of prison. Cotton insisted the test be done. When authorities went back to the rape kits from the two 1984 cases, there was a lot of degradation to the evidence after the 10 years. But within the rape kit from Elizabeth, they found a single fragment of a single sperm cell. That tiny bit of evidence yielded a DNA profile, which not only proved, once and for all, that Ronald Cotton was not the rapist, but showed that Bobby Poole was. Cotton was exonerated and released in June of 1995 after 10-and-a-half long years in prison, over a decade of his life lost. By the way, Poole later died of cancer in 2001 while still incarcerated.

When she heard the news of Cotton's innocence, Jennifer Thompson felt a tremendous amount of guilt and shame. She felt like she'd betrayed herself and the whole judicial process. She wondered if people thought she reacted too quickly after the crime and just tried to find a man, any man, to pay for what happened to her. She also speculated people thought she was a racist, since she is white and Cotton is black. At the same time, Jennifer also thought it ironic, and even a little cruel, that Ronald Cotton was now getting all the attention as a victim, as well as restitution from the state. She would never be free from what was done to her, and there was no restitution she could ever be given to compensate her for what was taken away. Those were the kinds of things Jennifer Thompson said to friends and reporters just after Ronald Cotton's release.

But time has a way of changing things, and about two years later, Thompson did something most people could not fathom: She asked if Ronald Cotton would meet her in a neutral place for a private discussion. As she describes it, she was so nervous when she saw him that she couldn't even stand up. She started to cry and said, and I'm quoting her here, "Ron, if I spent every second of every minute of every hour, for the rest of my life, telling you how sorry I am, it wouldn't come close to how my heart feels. I am so sorry." And she described how Ronald Cotton took her hands and said, now I'm quoting from an interview with him, "Jennifer, I forgive you; I don't want you to look over your shoulder; I just want us to be happy and move on in life."

Can you imagine? Jennifer said before she understood the power of forgiveness and mercy, she used to actually pray that Ronald Cotton would be raped and killed in prison. And now here he was, sitting across from her, telling her that he forgave her for the over 10 years of his life her mistake had cost him. Their conversation lasted two hours, while their families waited outside. The pair hugged and promised to stay in touch when they parted. Clearly, Jennifer Thompson and Ronald Cotton learned a lot about themselves and each other from this life-altering experience. But what about the rest of us, and the justice system in general? What have we learned from situations like this?

Well, it's long been known that eyewitness testimony is problematic. Many scientists, particularly most famously, forensic psychologist Elizabeth Loftus, have extensively studied the phenomenon. They've even pinpointed specific facets of the issue, such as that people tend to differentiate the faces of people from their own race better than they can make distinctions among the faces of people from other races. Maybe that's simply because we tend to spend more time around people who look more like us, often beginning with our own family. Loftus thinks that when an individual sees a person of a different race, the primary things noticed are the main, or what we might even say, stereotypical characteristics associated with that race.

For instance, a Caucasian might note eye shape and hair form in Asians; we don't focus on the other details. During a lineup, when multiple people from that race are presented together, we can't differentiate, because we never focused on the finer details in the first place. Could that have played into Jennifer Thompson's original misidentification, even though she said she studied all the details of her attacker's face?

It's also known that human memory is pretty malleable, I'm sure you know this yourself. As we sit and ponder something that happened or try to recall someone we saw, our mind can easily distort the details, especially if new information is coming in. We begin to confuse or blend the original story or image with that new data. When she reviewed the photo lineup, Jennifer Thompson quickly eliminated four of the six images, but then spent close to five minutes comparing the mental image she had of her rapist to the last two of the photographs. That shows a pretty significant degree of uncertainty. Jennifer explained she just wanted to be right about her conclusion.

But experts have noted that a witness who at first may be uncertain can often become more certain over time. It's as if we want so badly to be sure, that we subconsciously force ourselves to become more confident in our assertions. And later, when Ronald Cotton was the only one of those six guys in the photos who she again saw at the physical lineup, well she became even more convinced she was right. And by the time she got to the steps of the courthouse, she was 100 percent sure, even though she originally spent almost five minutes comparing two men depicted the photo array.

But when you think about it, and this was even pointed out to Jennifer by Mike Gauldin, who was the detective assigned to her case, Jennifer was definitely not the only one who made a big mistake. The detective himself also believed Cotton was her rapist, as did other police officers. And the two judges and two juries, who heard the two legal cases against Cotton, well, they were also wrong in their guilty verdicts. That's why everyone—law enforcement officers, investigators, attorneys, juries, and judges—should take very seriously what we all intuitively know: Human memory is not infallible. We definitely need eyewitness testimony in criminal cases, but we must all realize how faulty it can be.

Since the time Ronald Cotton gained his freedom, several hundred people have been exonerated through DNA testing. About three-fourths of those people were convicted partly by faulty eyewitness identification, making it the most common source of wrongful convictions in the legal system. Even more shocking still, in almost 40 percent of those cases, more than one eyewitness incorrectly identified the same innocent person as being the perpetrator. Sadly, studies by organizations like as the Innocence Project in New York estimate that only 5 to 10 percent of criminal cases have any type of physical evidence that could yield DNA. And much of that evidence may now already be degraded or have been destroyed.

So armed with this kind of knowledge, and to atone in a very personal way for her error that cost Ronald Cotton over 10 years of freedom, Jennifer Thompson has devoted her life to talking to anyone who will listen about the mistake she made. She regularly speaks at conferences to make those involved in the legal system aware of the pitfalls of convictions based solely, or too heavily, on eyewitness testimony. She also opposes the death penalty, since we have no way to know how many innocent people may have already been executed without the truth ever coming out.

Through their story and activism, both Thompson and Cotton have lobbied for reform in police procedures, especially those involving eyewitness identification. Among the suggestions they and other advocates have made are that victim line-up photos need to be presented one at a time, what's called a sequential lineup, as opposed to a simultaneous lineups, where all images or all people are presented together. Also, officers presenting lineups should be blind to who the real suspect in the group is to avoid any subconscious cues they might be sending to the witness.

Some agencies have adopted a new software system for conducting photo lineups that was partly designed by Mike Gauldin, the detective in the Jennifer Thompson case. It involves no human cues. It presents lineup images on a computer screen one at a time, and there's just a "yes," "no," and "not sure" button the witness selects regarding recognition of each face shown.

The work that Thompson and Cotton have done together, including their 2009 book, titled *Picking Cotton*, has helped prompt reform of police procedures in some parts of the country. But many agencies still seem to resist suggested changes. Cotton's attorney, Richard Rosen, says it's important to all of society that mistaken identity be averted, and not just for the wrongfully accused, the victim, and the judicial system. The danger to all of us, and the reason we should all support suggested reforms in police and judicial procedures, is because when police settle on the wrong person, it allows the real perpetrator to harm others. That's what let Bobby Poole go on to rape again, and again. In fact, the Innocence Project estimates that when misidentification causes an innocent person to be incarcerated, almost half of the guilty perpetrators who are later convicted by DNA evidence, went on to engage in additional violent crimes, while the innocent person sat in prison.

After being released from prison, Ronald Cotton got a job with the DNA testing company that performed the analysis that earned his exoneration. The State of North Carolina awarded him $110,000 in restitution, but that was only after Jennifer Thompson lobbied for an increase to the original $5,000 he was given. Cotton married, had a daughter, and to this day, his and Thompson's families are friends.

To me, the biggest surprise of this story is Cotton's seeming lack of bitterness about this tragedy that was caused by a woman's flawed memory, and how together, he and Jennifer have advocated for changes in relevant police and legal procedures. As the Innocence Project has wisely concluded, "memory is evidence, and must be handled as carefully as the crime scene itself to avoid forever altering it."

The Truth behind False Confessions

Lecture 14

Marty Tankleff served 17 years behind bars for the murders of his parents based on a false confession—one that Tankleff never wrote or signed. It may seem nearly impossible, but Tankleff's story is just one of far too many, especially in cases involving juvenile suspects or people who are mentally incapacitated. Corrupt or simply overzealous law enforcement officers, misleading interrogation methods, and false technology have been used to dupe naïve people into confessing to crimes they didn't commit. In this lecture, we'll look at several cases of false confessions and the personal tragedies that result.

Murder in Suffolk County

- In early September 1988, at 6:15 a.m., 17-year-old Marty Tankleff called 911 in Suffolk County, New York. He had found his 62-year-old father, Seymour, on the floor in the den of the family's home, bludgeoned and bleeding from a slash in his throat. Marty discovered his 54-year-old mother, Arlene, dead in the master bedroom, also beaten and with her throat cut. The teenager tried to keep his unconscious father alive while he waited for an ambulance.

- When a police officer arrived, Marty said that he had gone to bed about 11:00 the night before, while his parents were hosting friends for a poker game. Marty, the couple's adopted son, was the only other person who lived in the house. Mr. Tankleff had recently sold a successful insurance business and invested the proceeds in several bagel stores and a gym.

- The crime scene investigation produced little evidence. Police believed that Mr. Tankleff had been hit over the head while sitting at his desk, then attacked with a kitchen knife, although no weapon was found. The lead detective, Jim McCready, took Marty to the police station to gather more information.

Marty's "Confession"

- After four hours of intense questioning, officers arrested Marty Tankleff for the murder of his mother and the critical injuries to his father. According to McCready, Marty had admitted his guilt. But the confession was in the detective's handwriting, and the young man steadfastly denied admitting the crime.

- In fact, Marty told police that he believed a man named Jerry Steuerman was responsible. Steuerman was a disgruntled business associate who owed Seymour Tankleff $500,000. He had been at the poker game the night before and was the last guest in the house.

- According to Marty, McCready told him that Mr. Tankleff had regained consciousness at the hospital and identified his son as his attacker. The detective also told Marty that some of his hair had been discovered in his mother's hand and that tests had shown Marty had washed off blood in the shower. McCready explained to Marty that people sometimes block out horrible incidents from their minds. Marty began to question his own sanity, wondering if he could have attacked his parent in some altered state of mind.

- At that point, McCready read Marty his Miranda rights and wrote out the confession. Although Marty was confused, he refused to sign. Seymour Tankleff never regained consciousness and died from his injuries about a month later. McCready had fabricated the story about his identification of Marty.

- Within a week of the attack, Steuerman took $15,000 out of a business bank account he had shared with Tankleff. He then assumed a false identity and left New York for California. Nevertheless, Suffolk County Police never seriously investigated Steuerman, perhaps because of alleged payoffs related to earlier illegal activity. McCready later located Steuerman and brought him back to New York but refused to connect him with the deaths of Seymour and Arlene Tankleff.

Trial and Conviction

- At trial in 1990, McCready indicated that he first became suspicious of Marty because the teenager seemed devoid of emotion at the crime scene. But Marty said that he had been taught by his father to put emotions aside in tense situations in order to deal effectively with the problem at hand. Since the original questioning, Marty had passed two polygraph tests, and he believed that the jury would realize he had not committed the murders.

- There was no forensic evidence linking Marty to this crime, and forensic testing found no blood evidence on either the barbell or the kitchen knife that were listed as weapons in the fabricated confession. Further, Arlene Tankleff had been so fearful of Steuerman that she had written down specific threats he had made. Nevertheless, Marty was found guilty of murdering both his parents and sentenced to 50 years to life in prison.

- During this same time, the New York State Commission of Investigation issued a report about rampant corruption in the Suffolk County Police, and it was discovered that McCready had perjured himself in an earlier murder trial.

New Evidence and Appeal

- In prison, Marty contacted organizations that help exonerate those who are falsely accused. By 2001, a retired New York homicide detective, Jay Salpeter, agreed to help Marty. Salpeter found new witnesses and evidence that pointed to Steuerman as having arranged the murders. Salpeter even located eyewitnesses who could testify to seeing Steuerman and McCready together on several occasions before the murders, despite McCready's claims that he had never met Steuerman before the Tankleffs were killed.

- All this new evidence and dozens of witnesses who were willing to testify against Steuerman were presented at Suffolk County evidentiary hearings in 2006. Presiding at those hearings was District Attorney Thomas Spota, who had represented McCreary on two other legal issues. Needless to say, Marty's attorneys did not win the motion for a new trial.

- Marty's appeals finally reached the New York Supreme Court; in December of 2007, the 36-year-old Tankleff's conviction was unanimously overturned, and he was awarded a new trial in Suffolk County. However, based on allegations of corruption in Suffolk County and lack of evidence in the case, Attorney General Andrew Cuomo announced in June of 2008 that the state would not retry Tankleff.

The Confession of Michael Crowe

- In 1998, 12-year-old Stephanie Crowe was stabbed to death in her Escondido, California, bedroom, with no signs of forced entry into the house. After 11 hours of coercive interrogation, police convinced her 14-year-old brother, Michael, that he had killed his younger sister, though he had not.
    - Investigators used a piece of bogus technology—a *computer voice stress analyzer*—to trick Michael by saying that the machine showed he was lying.

    - This instrument was developed specifically for interrogation purposes and is not based on any scientific evidence.

- According to Richard Leo, an expert in coerced confession, a video recording reveals Michael's interrogation to be nothing more than "psychological torture." Police told Michael's parents that their son had been taken into grief counseling.

- Luckily, the truth about Stephanie's murder ultimately came out: A 28-year-old mentally ill homeless man had been seen wandering the Crowes' neighborhood on the day of Stephanie's murder. On the morning after the murder, police picked the man up for questioning and found blood on his clothes. DNA testing later showed that the blood was from Stephanie Crowe.

- Amazingly, the Escondido police and prosecutor's office didn't pursue the case against the homeless man because they didn't want to face public scrutiny for their error. The California Department of Justice prosecuted the case in 2004. In 2011, the Crowe family was awarded more than $7 million in damages.

Christopher Ochoa and Richard Danziger

- In 1988, 22-year-old Christopher Ochoa was brought in by the police for questioning in the rape and brutal killing of the 20-year-old female manager of a pizza restaurant in Austin, Texas. Ochoa and his friend and roommate, Richard Danziger, worked at another franchise of the same pizza chain in Austin. When the pair visited the restaurant where the murder had taken place and talked to the security guard posted there, police began to suspect that they had something to do with the crime.

- Over the course of two days, Ochoa was subjected to a brutal interrogation. Officers screamed at him, threatened him with the death penalty, and warned that he would be raped in prison. Ochoa claims that he asked for an attorney early on, but that request was denied. Although much of the interrogation was audio recorded, there were repeated stops and starts in the recording, as well as three hours of missing tape.

- Police called Ochoa's mother and told her that she should convince her son to plead guilty or he would die. Ochoa's mother was ill and, in fact, suffered a stroke that her son believes was related to the pressure surrounding the situation. Ultimately, Ochoa signed a confession.

- In 1989, Ochoa was sentenced to life in prison for the crime, as was Richard Danziger. Eight years later, another prisoner contacted numerous authorities, claiming that he had been the killer in the pizza restaurant. Eventually, Austin investigators questioned Ochoa further about the crime, but Ochoa believed that recanting his confession at that point would stall his chances of parole.

- By 1999, Ochoa decided to contact the University of Wisconsin's Innocence Project. Once again, DNA evidence from the victim was still in storage, and testing proved that the semen from the case belonged to neither Ochoa nor Danziger.

- Exonerated in 2001, Ochoa has since graduated from the University of Wisconsin law school. Unfortunately, while in prison, Danziger received a vicious beating from a fellow inmate that resulted in permanent brain damage. Although Danziger is also a free man, he still requires full-time care.

Eddie Joe Lloyd and the Innocence Project

- Problems also arise with confessions when authorities interrogate persons of diminished mental capacity. In 1984, while being held in the Detroit Psychiatric Institute, a man named Eddie Joe Lloyd wrote to police that he had supernatural abilities that would help solve crimes. Police questioned Lloyd without an attorney present, and he ultimately confessed to killing a 16-year-old girl. His confession clearly shows that police fed him details about the crime that he could not have known.

- In addition to police injustices, Lloyd's case was also affected by legal malpractice. In 1985, his first court-appointed attorney withdrew from the case about a week before trial, leaving Lloyd's second attorney little time to prepare. The second attorney called no witnesses in defense of his client—not

Some people who sign confessions to end an interrogation do so with the belief that science and the law will sort things out later, never dreaming that may not happen for decades.

even someone to attest to Lloyd's mental health history—and did not cross-examine critical witnesses. After less than an hour of jury deliberation, Lloyd was convicted of felony murder.

- After the Innocence Project in New York took on the case, DNA evidence exonerated Lloyd, who was set free after 17 years in prison. Unfortunately, he died 2 years later. His estate was awarded $4 million from Wayne County, Michigan.

Suggested Reading

Borchart, *Convicting the Innocent*.

Firstman and Salpeter, *A Criminal Injustice*.

Innocence Project, http://www.innocenceproject.org/.

Münsterberg and Hatala, *On the Witness Stand*.

Murray, *Overturning Wrongful Convictions*.

Petro and Petro, *False Justice*.

Primorac and Schanfield, eds., *Forensic DNA Applications*.

Scheck, Neufeld, and Dwyer, *Actual Innocence*.

Questions to Consider

1. At what age do you think young people should be permitted to be questioned without their parents present?

2. Were you surprised to learn how common false confessions are in criminal cases? Why or why not?

3. Do you think you could or should refuse a police officer's request to search your vehicle or personal possessions if you had done nothing wrong?

The Truth behind False Confessions

Lecture 14—Transcript

In early September of 1988 at 6:15 in the morning on the first day of his senior year of high school, 17-year-old Marty Tankleff called Suffolk County New York 911 dispatch. He had found his 62-year-old father, Seymour, on the floor in the den of the family's home, bludgeoned and bleeding profusely from a slash in his throat, just clinging to life. Marty discovered his 53-year-old mother, Arlene, lying dead in the master bedroom, also beaten, and with her throat cut. The teenager did his best to keep his unconscious father alive, while he waited for an ambulance.

When a police officer arrived, Marty said he didn't know what happened. He had gone to bed about 11:00 the night before, while his parents were hosting friends for a poker game. Marty was the only other person who lived in the house; he was the couple's adopted son. Mr. Tankleff had just wrapped up a successful insurance career, having recently sold the business due to some health issues, and he invested the proceeds into several bagel stores and a fitness gym. Marty's father was also the village constable in the exclusive neighborhood where the family lived, on a cliff overlooking New York's Long Island Sound.

The crime scene investigation produced very little evidence. Police believed that Mr. Tankleff had been hit over the head while sitting at his desk and then been attacked with a blade of some type. And although they didn't find a weapon, they suspected that a kitchen knife had been used to cut both Seymour and Arlene Tankleff's throats. The lead detective, a man named Jim McCready, took Marty down to the police station to gather more information.

After four hours of intense questioning, officers arrested Marty Tankleff for the murder of his mother and the critical injuries to his father. How did they come to this conclusion? Marty had admitted it, or at least that's what Detective McCready said, after interrogating the teenager. But the resulting confession was in the detective's handwriting, not Marty's. Marty steadfastly denied admitting to the crime, and he never signed any type of confession paperwork.

In fact, Marty said that when police first questioned him that morning, he told them who he thought was responsible, a man named Jerry Steuerman, who was a disgruntled business associate of his father's in the bagel stores the men owned together. Steuerman owed Seymour Tankleff a half-million dollars. Marty told police that Steuerman had been at the poker game at the Tankleff's the night before and was the last guest in the house. The teenager was adamant that if Steuerman had left their home while his father was still capable, his dad would have locked up the house, set the home alarm, and turned off all the lights before going to bed, as was his usual custom. When Marty found his parents that morning, the alarm was not set, the doors were not locked, and the lights were still on.

So what led to the teenager's admission of guilt? According to Marty, Detective McCready told him during the interrogation that Mr. Tankleff had briefly regained consciousness at the hospital and stated that his own son, Marty, was the person who attacked him. The detective also told Marty that some of his hair was discovered in his mother's dead hand, and that tests had shown Marty's shower was used that morning to wash off blood from the incident. McCready explained to Marty that people sometimes block out horrible incidents from their mind, especially when they are distraught or in shock. Ultimately, Marty had begun questioning his own sanity, wondering if somehow, in his sleep, or in some altered state of mind, he could actually have been the one who attacked his parents. Was he going crazy? Had he done this? Marty could hardly bear the thought of it.

Only after that, did Detective McCready read 17-year-old Marty Tankleff his Miranda rights. Next, McCready wrote out the confession, based on his assumptions. But even though Marty was confused, he absolutely refused to sign that statement of guilt. The teenager even asked to be given a polygraph lie-detector test to try to get at the truth, but police refused. Unfortunately, Seymour Tankleff died from his injuries about a month later. In fact, he never once regained consciousness and was never able to tell his son that McCready had completely fabricated the story about a phone call from the hospital that implicated Marty. Surprisingly, Detective McCready has never denied using those misleading tactic and out-and-out lies during the young man's questioning.

Within a week after the attack on the Tankleff couple, their bagel-store partner, Jerry Steuerman, took $15,000 out of a joint bank account the business associates shared. In a plan to skip town, and stage his own death, Steuerman named his girlfriend the beneficiary of his life insurance policy and left New York for California. He assumed a fake identity, got a new hairstyle, and shaved his beard. But despite all that, Suffolk County Police never went after Steuerman in any serious way.

Perhaps this had something to do with alleged connections between Steuerman and Suffolk County Police, including stories of police payoffs for cocaine deals conducted out of the bagel store by Steurman's son, in the late 1970s and early 1980s. Detective McCready did locate Steuerman and did bring him back to New York, but McCready refused to connect him with the deaths of Seymour and Arlene Tankleff. After all, why would he, once he had gotten their son to confess to the crimes?

At trial in 1990, McCready indicated he first became suspicious of Marty because he appeared somewhat devoid of emotions at the crime scene. Police, and some others, saw Marty as a spoiled rich kid who resented his parents' rules and restrictions. People who were present at Marty's trial also said he showed little emotion. But Marty said he was taught by his dad to put emotions aside in any tense situation, because if you don't, you can't effectively deal with the problem you're facing. Besides, Marty still maintained his innocence. Since the original questioning, he had passed two polygraph tests, although those aren't admissible in court, and Marty believed the jury would realize the truth. After all, the prosecution had no evidence, other than the confession he refused to sign that was written in Detective McCready's handwriting.

There was simply no forensic evidence linking Marty to this crime, no blood or skin under his fingernails, no cuts or scratches to his body on the morning of the crime. Forensic testing found no blood evidence on either the weight-lifting barbell, or the kitchen knife, which the fabricated confession alleged were the weapons used. Besides, Marty knew his mother had been so fearful of her husband's business partner, Jerry Steuerman, that she had actually written down the specific threats he had made to Mr. Tankleff and put that piece of paper in the family's home safe.

Marty felt sure that all these facts would prove to the jury he was not responsible for the death of his mother and father. So, the young man did show little emotion in the courtroom, that is, until he was found guilty of murdering both his parents, at which point he completely broke down. While still a teenager, Marty Tankleff was sentenced to 50 years to life in prison. Some had suspected that Marty probably killed his parents out of greed, hoping for a huge inheritance. But Marty knew his parents' will was written to exclude him from inheriting anything until he turned 25, so he couldn't understand why people thought he could have done such a terrible thing for financial gain.

After Marty's conviction, his half-sister and her husband, who, by the way, happened to be Marty's only relatives who did not stand up for him, they inherited Marty's portion of his parents' $3 million. You won't believe what they did with the money. They opened a restaurant and bar near the courthouse, with a new business partner, none other than Detective McCready. You just can't make these kinds of stories up!

During this same time, the State Investigation Commission for New York issued a report about rampant corruption they had discovered within Suffolk County police. It was discovered that Detective McCready had perjured himself in an earlier murder trial. McCready was also later charged with assault for his part in a barroom brawl. In both his perjury and assault charges, McCready's attorney was Thomas Spota, whose law partner, by the way, had been the District Attorney who prosecuted Marty's case. Oh, and Spota's firm had also represented Jerry Steuerman's son when he finally did get charged with dealing cocaine out of the bagel store back in the early '80s. We'll come back to Spota in a moment.

While in prison, Marty kept insisting he was innocent; he contacted organizations whose missions are to help exonerate those who are falsely accused. By 2001, a decorated, retired, New York homicide detective, Jay Salpeter, agreed to work for free to aid Marty's cause. Salpeter found new witnesses and clues that all pointed to Mr. Tankleff's business partner, Jerry Steuerman, as having masterminded the murders and gotten some of his goons to carry them out.

Salpeter even located a man who admitted to having driven the getaway car that night. This man described a pipe used in the attack having been thrown out of the car as the killers were fleeing the scene. Salpeter discovered a pipe fitting description in the spot where the driver said it would be. The getaway man passed a polygraph test and later signed a legal affidavit as to what he had seen the night he drove two men to the Tankleff residence. The driver said he thought they were going for a burglary, only to see the guys get back into his vehicle with blood on their clothes. Now, this guy also had previously confessed to his getaway role in that crime to both a Catholic priest and a nun, who were also willing to testify, so long as they had the confessor's permission.

Salpeter had even located eyewitnesses who could testify to seeing Steuerman and Detective McCready together on several occasions before the murders, despite McCready's claims that he never met Steuerman before the Tankleff couple was killed. They also found out that McCready was not the next officer in line on the police duty roster on the morning of Marty Tankleff's 911 call; he didn't even live close to the Tankleff residence. Yet he's the one that showed up to take the investigation, 19 minutes after the call went out.

All this new-found evidence, and dozens of witnesses who were willing to testify as to what they'd seen or heard Jerry Steuerman or his thugs do and say, was ultimately presented at Suffolk County evidentiary hearings. Presiding at those hearings was District Attorney Thomas Spota, the same guy who had represented Detective McCreary in two legal issues, and whose law firm had served as legal counsel for Steuerman's son's cocaine charges. Needless to say, at this 2006 hearing, Marty Tankleff's attorneys did not win their motion for a new trial.

The appeals that happened went all the way up to the New York Supreme Court. Finally, in December of 2007, after 6,338 days behind bars—that's 17 long years—36-year-old Marty Tankleff's 1990 conviction was unanimously overturned, and he was awarded a new trial in Suffolk County. Although he was released from jail, the case wasn't completely over. There was still the possibility of retrial. However, based on the allegations about corruption in Suffolk County, Attorney General Andrew Cuomo, who would later become

Governor of the State of New York, he announced in June of 2008 that the State would not retry Marty Tankleff for the murder of his parents, due to insufficient evidence.

That's just one case where a false confession was presented, false because the accused never even wrote the alleged admission of guilt, yet it got him convicted and sentenced to 50 years in prison. It may seem like a near impossibility, but Marty Tankleff's story is just one of way too many, especially in cases involving juvenile suspects or people who are mentally incapacitated in some way. Corrupt or simply overzealous law enforcement officers, misleading interrogation methods, and false technology have been used to dupe naïve people, for far too often and too long.

In 1998, 12-year-old Stephanie Crowe was stabbed to death in her Escondido California bedroom with no signs of forced entry into the house. After 11 hours of coercive interrogation, police literally convinced her 14-year-old brother, Michael, that he had killed his younger sister, though he had not. Investigators used a piece of bogus technology called a *computer voice stress analyzer* to trick Michael by saying the machine showed he was lying. That instrument, which was developed specifically for interrogation purposes, has not a shred of scientific evidence to back it up.

In fact, there is no such thing as a lie-detection machine. Even if the machine could measure stress, that's in no way the same thing as measuring truth. Who wouldn't be stressed in an interrogation situation, especially when being questioned about a murder? The voice stress analyzer is simply used to intimidate or dupe people into confessions. Still, police told Michael Crowe that objective computer science was on their side. In addition, the officers lied to the boy, saying they had found his younger sister's blood in his room.

The whole interrogation was video recorded, and by the time they broke Michael down, he was sobbing and saying, "I don't even know what I did. I can't even believe myself anymore. I've been told I did this awful thing, but I don't remember it." Keep in mind, this is a 14-year-old boy faced with two adult police officers playing the stereotypic good cop and bad cop. At one point, when Michael was left alone for a few minutes in the interrogation room, he just sat with his head in his hands and cried like a baby. It breaks your heart to listen to the interview. Just imagine how they scrambled that young kid's mind.

In fact, coerced-confession expert, Richard Leo, says the interrogation of Michael Crowe amounted to nothing more than, "psychological torture." And where were Michael's parents all this time? Police had told them their son had been taken into grief counseling, since he had just lost his little sister. And that's not all: After questioning two of Michael's friends, police got one of them to confess too, and amassed enough faulty information to also arrest the other one, too.

Luckily, the truth eventually came out before the three boys faced a murder trial. A 28-year-old, mentally-ill, homeless man had been seen wandering the Crowe's neighborhood on the day of Stephanie's murder. Several people reported the guy's strange behavior, even stating they saw him in the Crowe's driveway, looking up at the house with outstretched arms. On the morning after the murder, police did pick the guy up for questioning, and they actually found blood on his clothes, which they kept for DNA testing. But, despite that, on the very same day, two officers coerced a confession out of 14-year-old Michael Crowe; that is something I just cannot understand.

The eventual results of DNA testing on the blood from the homeless man's clothing, it showed the blood was from Stephanie Crowe. But get this; the Escondido Police and Prosecutor's Office didn't pursue the case against the homeless man because they didn't want to face the public scrutiny for their error in settling on Michael as guilty. The California Department of Justice had to take over the case and prosecute it, which they did in 2004—over five years after the truth about Stephanie's death had finally come out. In 2011, the Crowe family was awarded over $7 million in damages.

Surprisingly, using methods like lying to a suspect during interrogation and talking about evidence that doesn't exist is completely legal, just so long as the officers don't use it to either promise leniency or a worse punishment, although that happens, too, of course, in some of these cases, like when a suspect is told that a confession is the only way to avoid the death penalty. And it's not uncommon for officers to look for ways to keep parents out of the room when interrogating juveniles, which I'm sure makes those of us who are parents just shudder.

I'll never forget an experience I had while in the car of a public defender. He was giving me a ride from a hotel to a Tennessee courtroom for a second-opinion case in which I was to serve as an expert witness. We passed a traffic stop along the side of the road where a couple of teenage girls were standing behind the open trunk of their car while two uniformed officers were digging through the contents of the trunk.

The attorney said out loud, "The answer is no!" And I said to him, "What do you mean?" And he said something like, I tell everyone I know, especially kids, that if a police officer asks if he can search your car, or your purse, or something like that, the answer should always be a polite no, not until my parents or attorney are present. Those kids have no idea who could have put something in their trunk that might implicate them for something they had nothing to do with, especially if it's a used car or a rental. He said there's no way they should allow a spontaneous search like that. Still, I wondered how many of us would have the nerve to say no to the police, especially those of us who've done nothing wrong and who respect people in authority.

Someone who had never been in trouble in his life, and was raised to trust and respect authority, was Christopher Ochoa, a hard-working 22-year-old man, who had been an honor student in his high school and editor of his school's literary magazine. Ochoa was brought in for questioning in the brutal 1988 killing of a 20-year-old woman at a local pizza restaurant in Austin, Texas. The manager, who was alone at the time, had been taken into the back, tied up with her bra, raped, and then shot in the head. Ochoa, and his friend/roommate, Richard Danziger, worked at another franchise of the same pizza chain in Austin. And when the pair visited the pizza store where the murder had taken place, they talked to the security guard posted there about what had happened. That's when police began to suspect maybe they had something to do with the crime.

The things police did to intimidate Ochoa during 24-hours of interrogation over the course of two days are shocking. Ochoa was screamed at; he had furniture thrown at him; police threatened him with the death penalty; they even tapped the veins in his arm to show him where the lethal injection would be placed. They told him how fellow prisoners would repeatedly rape him while he awaited his death. What evidence did they have? None. But they lied to Ochoa, and they told him that his best friend, Danziger, had turned him in. And they told Danziger pretty much the same thing, but in reverse, putting both men in the same position and pitting the two friends against each other.

Danziger also denied having anything to do with the crime. I'm sure each of those guys was shocked at the thought the other might have done it, but was trying to implicate him. Important, too, was that Ochoa claims he had asked for an attorney early on, but that request was denied. Although much of the interrogation was audio-recorded, there were repeated stops and starts in the recording, as well as three hours of missing tape.

Ultimately, Ochoa signed a confession. He said he did so, because the police just wouldn't let up. He decided to plead guilty after police began calling his mother and telling her the horrible things that would happen to him in prison. The police worked on Mrs. Ochoa to convince her son to plead guilty, or he was going to die. Ochoa's mother was ill, and in fact, she suffered a stroke that her son believes was from all the pressure surrounding the whole situation. He says he knows people find it inconceivable that someone would sign a confession about something they absolutely did not do, but said when you're the suspect, it's different. Ochoa wanted his mother to be spared the torment. He just wanted the whole thing to end.

In 1989, Ochoa got life in prison for the crime, so did Richard Danziger. But eight years later, a different Texas prisoner wrote to numerous authorities, claiming that he had been the one that killed the young woman. Most of those letters were ignored. But eventually, Austin investigators went back to Ochoa to question him further, and believe it or not, he still said he committed the crime. That's because he thought recanting his confession at that point would stall his chances of parole. By 1999, though, after 10 years in prison, Ochoa decided to contact the University of Wisconsin's Innocence Project. DNA evidence from the victim was, luckily, still in storage, and proved that the semen from the case belonged to neither Ochoa nor Danziger.

He was exonerated in 2001 after nearly 12 years in prison, and Chris Ochoa was chosen by *ABC News* as their Person of the Week in May of 2006. That was also the day he graduated from the same University of Wisconsin law school that took on his exoneration case. Unfortunately, while in prison, his friend Richard Danziger received a vicious beating from a fellow inmate and wound up with permanent brain damage; he had already been transferred to a secured mental facility. Although Danziger is now a free man, he still requires full-time care.

The reasons for false confessions vary, ranging from intimidation or actual violence against the suspect, as we've seen, to interrogations that happen while the person being questioned is under the influence of drugs or alcohol. There are also problems when authorities interrogate persons of diminished mental capacity, which most police officers are never trained to do. That's what happened to Eddie Joe Lloyd, who was convicted of the 1984 rape and killing of a 16-year-old girl in Detroit Michigan. While Lloyd was being held in the Detroit Psychiatric Institute, he wrote to police that he had supernatural abilities that would let him help them solve local crimes. Police questioned Lloyd at the mental health facility, without an attorney present. Lloyd ultimately confessed to killing the young girl, but his confession now clearly shows police fed him details about the crime there's no way he could have known.

In addition to police injustices, Lloyd's case also suffered from legal malpractice. In 1985, Lloyd's first court-appointed attorney, who did little to investigate into the case, withdrew himself about a week before the trial, leaving Lloyd's second attorney just one week to prepare the case. The trial wasn't postponed. This second attorney was no better, though. He called not a single witnesses in defense of his client, not even someone to attest to Lloyd's mental-health history, nor did his attorney cross-examine critical witnesses, like the interrogating officer. After his attorney's five-minute closing argument, and less than an hour of jury deliberation, Eddie Joe Lloyd was convicted of felony murder. In open court, the judge actually regretted Michigan not having the death penalty, saying it was a shame that all he could do was put Lloyd in prison for the rest of his life.

Lloyd's court-appointed appeals attorney never once came to meet him in prison. When Lloyd wrote to the judge to complain about the bad lawyering he had received, the appeals attorney responded by telling the judge his client was, "guilty and should die." After the Innocence Project in New York took on the case, DNA evidence exonerated Mr. Lloyd, and he was set free after 17 years in prison. Unfortunately he died only two years later at 54 years old. Lloyd's estate was awarded $4 million from Wayne County, Michigan.

The personal tragedies and decades lost by these kinds of miscarriages of justice are incomprehensible, not to mention how costly these situations are to all of society. Innocent people are locked up at taxpayers' expense; the real perpetrators are left to strike again; and huge sums of money are paid out in damages to those who are exonerated—and rightfully so, in my opinion. But that money can never make up for the years lost or the stigma and memories these exonerees will carry the rest of their lives. Fortunately, a number of organizations have begun to tackle these cases, using today's forensic science technology and challenges to police and legal procedures.

The Innocence Project is just one such groups. It was founded in 1992 at the Cardozo School of Law at Yeshiva University in New York. According to their website, over 300 convicted individuals in the U.S. have now been exonerated, and that's by DNA testing alone. The average prison time served was 13 years, and 18 of those people were on death row. Innocence Project studies document either confessions of guilt, or guilty pleas, in 25 percent of the cases in which DNA evidence proved the convicted person did not commit the crime. Many of those involve what the Project calls confession contamination, where details that were never released to the public were present in the alleged confession. The only way the interrogated person could possibly have known those specific facts is if they were fed to them by law enforcement during the interview.

Some people who sign a confession just do so to get out of the interrogation; they do that with the belief that good science and law will sort things out later, never dreaming it may be decades later. But even when suspects are not fed details of the crime by police and get all or most of the details wrong in a false confession, all the discrepancies are often thrown out in court as attempts on the part of the suspect to minimize their deeds. The single thing that is often clung to in the courtroom is that the person said they did it, even if they didn't get the details of the crime right in their confession.

Now, I would never want to end this lecture with you thinking I don't have the utmost respect for law enforcement or our legal system. I do believe that the vast majority of those incarcerated are guilty of their crimes, and that nearly all police officers and court-appointed officials do extremely good work. But like any other profession, there is malpractice that irrevocably damages the system and those caught up in it. A study by the Chicago Tribune examined 400 overturned homicide cases throughout the United States, and noted that not a single prosecutor got reprimanded, no attorneys lost their jobs, and no one representing the law went to prison. We certainly can't say that for those who were in the defendant's seat in those cases, and my heart goes out to those who are innocent, but found guilty.

Crooked Cops and Bad Convictions

Lecture 15

On June 22, 1970, a New Zealand man named Len Demler reported that his daughter, Jeanette Crewe, and her husband, Harvey Crewe, were missing from the couple's home. Demler had found the house covered in blood spatter but discovered his 18-month-old granddaughter, Rochelle, crying in her crib. The child had clearly been taken care of, for at least several days, at a crime scene that was estimated to be about five days old. In this lecture, we'll analyze the investigation of this puzzling crime in New Zealand and look at two other tragic cases involving wrongdoing on the part of police.

The Crewe Murders

- After Len Demler's discovery that his daughter and son-in-law were missing in June of 1970, neighbors of the couple reported hearing several gunshots five nights earlier. The witnesses also described other strange occurrences in the five days preceding the discovery: They had seen a woman at the Crewes' farm on several occasions and had witnessed Rochelle, the couple's 18-month-old daughter, running in the yard just two days before she was found alone in the house by her grandfather.

- At the house, there was no sign of forced entry or struggle, but evidence suggested that two people had lost extensive amounts of blood in the living room, one in a chair and the other on the floor in front of the fireplace. A pathologist noted that the blood loss was enough to have resulted in the death of both victims.

- Two months later, in a nearby river, Jeannette's body was discovered wrapped in a bedspread and bound by copper wire. Harvey's body was found upstream about a month after that, also trussed in wire and having apparently been weighted down by a car axle. Autopsies determined that both victims had been shot with a .22-caliber rifle, and Jeanette's facial bones were fractured, as were six of her teeth.

- So much of Harvey's blood was found on his easy chair that investigators assumed he had been shot first. Jeannette was probably hit in the face with the butt of the rifle when she tried to defend herself or come to Harvey's aid. Then, she was shot as she lay incapacitated on the floor in front of the fireplace.

- Although they did not initially find any ballistic evidence at the scene, police collected more than 60 rifles from homes in the area for comparison with the type of gun they suspected had killed the couple. About four months after the crime, a .22 cartridge case was discovered in the yard, and investigators determined that it had been shot from a rifle previously collected from one of the Crewes' neighbors, Arthur Thomas.

- Later, a detective named Johnston received an anonymous tip to search Thomas's farm, where he was told he might find evidence related to the axle used to weight down Harvey Crewe's body. Johnston searched the property and found a part that matched the axle type. Thomas was arrested in 1970 and convicted in 1971, based on the cartridge casing found in the Crewes' garden and the axle piece found on his property.

- Despite the evidence, Thomas's conviction led to an outcry by friends, relatives, and the media; Thomas had no motive and was known in the community as a good man. Mrs. Thomas said that her husband was home with her on the evening of the killings, but she was suspected of being the woman who took care of the baby after the Crewe couple was murdered. Thomas was granted a retrial, but in 1973, he was convicted for a second time.

- A number of public and private individuals became further involved in the effort to get Thomas exonerated. Ultimately, forensic scientist Dr. Jim Sprott was enlisted to reexamine the cartridge casing found at the scene. He determined that the casing had not been exposed to the elements for more than four months and believed that it had been planted at the scene.

- Based on Dr. Sprott's conclusions and other statements that had come to light, a Royal Commission of Inquiry reversed Thomas's conviction in 1980 and agreed that two specific investigators, Len Johnston and Bruce Hutton, had likely been involved in planting the cartridge casing. But the officers were never tried because the solicitor-general claimed to have insufficient evidence. Thomas was released from jail in December of 1979.

© tilzit/iStock/Thinkstock.

Supporting the theory that the shell casing had been planted at the Crewe crime scene was the fact that the area had been extensively searched two times before the casing was found.

Unanswered Questions

- Ian Wishart, author of a 2010 book about the Crewe case, paints Johnston as both the killer and the one who planted the evidence. Wishart's theory includes the fact that Johnston had investigated a burglary at the Crewe house in 1967 and, thus, would have known the couple, the layout of the house, and possibly, the hiding place for a spare house key.

- Wishart and other sources note that in the three years between the 1967 burglary and the 1970 killings, there were two arson attempts at the Crewe farm. It has also been alleged that the brake line in Jeannette's car was cut in 1969, but Jeannette did not report the incident to police. One suspicion arising from these incidents is that Johnston was stalking Jeannette. A second theory is that he was blackmailing the couple because he suspected the 1967 burglary was a scheme to defraud the Crewes' insurance company.

- A book released in 2012 by reporter Chris Birt exposes more possible evidence related to the framing of Arthur Thomas and its subsequent cover-up. Birt claims that Johnston was undoubtedly the killer, but in an earlier book, he had identified Jeannette's father, Len, as the murderer.

- Of course, another major unanswered question is: Who cared for Rochelle after her parents had been killed? As mentioned earlier, Thomas's wife was initially suspected of being the caregiver. Some have suggested that it was Jeannette Crewe, who had murdered her husband, then killed herself by the river five days later. Other sources claim that Len Demler was the killer and that other members of Jeannette's family took care of the baby. To this day, the mysterious caregiver remains unidentified.

Abner Louima

- In 1997, a Haitian immigrant named Abner Louima was tortured while in police custody in Brooklyn, New York. The perpetrator of the assault, an officer named Justin Volpe, ultimately admitted guilt and, in 1999, was sentenced to 30 years in prison without parole. Three other officers were originally sentenced to 5 years each for trying to cover up the incident, but their convictions were later reversed for lack of evidence. Those reversals have led many to argue that justice has not been served.

- The incident began when Louima was taken into custody during a brawl outside a New York nightclub. At the police station, he was tortured and sodomized by Officer Volpe. His internal injuries were so severe that he had to be taken to a local hospital, where Volpe claimed that his prisoner's injuries were from the nightclub fight. Later, a nurse called the authorities to report her suspicions that police were responsible for Louima's injuries.

- Volpe later admitted torturing Louima at the precinct. In July of 2001, Louima was awarded $8.75 million in damages by the New York Police Department; Volpe won't be eligible for release until 2025.

The Cerro Maravilla Massacre

- On the night of July 25, 1978, in Puerto Rico, three young men planned to sabotage some of the broadcast towers that are planted along Cerro Maravilla, the island's fourth highest summit. Their act was to be in retaliation for the U.S. government's continued imprisonment of two groups of Puerto Rican militants. The three men were Carlos Soto Arriví, Arnaldo Darío Rosado, and Alejandro Gonzalez Malavé.

- Arriví and Rosado were longtime members of a separatist group called the Armed Revolutionary Movement, known in Spanish known as MIRA. On the evening of July 25, they, along with Malavé, hijacked a taxi and ordered the driver to take them to the Maravilla broadcast towers. But when they reached the peak, they were ambushed by waiting police. The police reported that they ordered the men to surrender, but the revolutionaries fired first, and Arriví and Rosado were killed. The third man, Malavé, was only slightly wounded.

- During an interview with reporters two days later, the taxi driver said that the three men had tried to surrender to the officers and that Malavé had called out, "Don't shoot; I'm an agent." In fact, Malavé was an undercover police officer who had been assigned to infiltrate MIRA. Malavé was the one who had told the authorities about the revolutionaries' plan.

- When the news was first released, the governor of Puerto Rico, Carlos Romero Barceló, commended the police for preventing a terrorist attack by revolutionaries, and no charges were filed against any officers. But as the media began to circulate the taxi driver's conflicting stories, Barceló was pressured to investigate further. The driver's expanded story was that he had seen Arriví and Rosado being beaten by police in front of the cab.

- The opposition parties of Puerto Rico and members of the public continued to press for further investigation, claiming that the events on the mountain had been a setup. Between 1978 and 1980, both the

FBI and the U.S. Department of Justice looked into the matter and concurred with the official conclusions about the case: The incident was not a deliberate massacre of the freedom fighters.

- But in 1981, the Judiciary Committee of the senate of Puerto Rico reopened the case, granting immunity to certain police officers who had been at the scene in exchange for their testimony. Ultimately, testimony revealed that Arriví and Rosado had been dragged from the cab and beaten after an initial round of gunfire, then executed by the police.

- This evidence led to the conviction of 10 police officers on charges of perjury, obstruction of justice, and destruction of evidence. Four of the officers were also convicted of second-degree murder. The second round of investigations did not, however, confirm a cover-up by the Puerto Rican or U.S. governments, although such allegations were certainly made.

- Malavé was dismissed from the force because of public pressure, then later given immunity to testify against the officers who shot Arriví and Rosado. On April 29, 1986, he was assassinated in front of his mother's house; a group called the Volunteer Organization for the Revolution claimed responsibility.

- Fourteen years after the incident, the head of the Justice Department's Civil Rights Division during the investigation issued a public apology for what he thought had been a cover-up by the FBI in the case. Twenty-five years after the killings, former Governor Barceló stated that he acted prematurely in praising the officers involved in the incident, but he still admits nothing with regard to a cover-up. Pro-independence followers still convene on the mountain every year on July 25 to mark the anniversary and celebrate the Independentista movement.

Suggested Reading

Borchart, *Convicting the Innocent*.

Münsterberg and Hatala, *On the Witness Stand*.

Murray, *Overturning Wrongful Convictions*.

Nelson, *Murder under Two Flags*.

Petro and Petro, *False Justice*.

Scheck, Neufeld, and Dwyer, *Actual Innocence*.

Questions to Consider

1. What kinds of mistakes can police officers make, unintentionally (not maliciously), when honing in on alleged suspects?

2. What are some of the issues you believe contribute to police misconduct? How rampant—or not—do you believe the problem is?

Crooked Cops and Bad Convictions

Lecture 15—Transcript

On June 22, 1970, a New Zealand man, by the name of Len Demler, reported his daughter, Jeanette Crewe, and her husband, Harvey Crewe, missing from the couple's home. Demler had found the house filled with blood spatter, but discovered his 18-month-old granddaughter, Rochelle, crying in her crib. The child had clearly been taken care of, for at least several days, at a crime scene that was estimated to be about five days old. The baby showed no signs of having gone without food, and there was a partial fresh bottle of milk in the refrigerator. The baby's diapers had apparently been regularly changed; recently-dirty diapers were found in the house, and the clothes dryer was still running.

Neighbors reported hearing several gunshots five nights before the Crewes were reported missing and the baby found alive. The witnesses also described other strange occurrences in the five days preceding the discovery, like seeing a woman at the scene on several occasions, noticing different lights on or off around the house and farm buildings, and the curtains at the home being open and then closed. They also saw the 18-month-old child running in the yard, just two days before being found alone in the house by her grandfather. Witnesses also claimed to have seen a woman that was not Jeanette in the Crewe car, several times, traveling on nearby roads. At other times, people noticed the car parked near the road, in front of the couple's house, but when authorities arrived at the scene, the family car was in the Crewe garage.

With regard to the physical evidence, although there was no sign of forced entry or a struggle, crime scene investigation revealed two different blood types in the house, as well as blood spatter on the porch matching one of the two blood types. The scene suggested that two people had lost extensive amounts of blood in the living room, one in a chair, and the other on the floor in front of the fireplace. A pathologist noted the blood loss was enough to have resulted in the death of both victims. The seat cushion of the blood-stained chair and a rug that was normally lying in front of the hearth had been burned in the fireplace.

Two months later, in a nearby river, Jeanette's body was discovered wrapped in a bedspread, bound by copper wire; Harvey's body was found upstream about a month after that, also trussed in wire and having apparently been weighted down by a car axle. Autopsies determined both victims had been shot with a .22-caliber rifle, and Jeanette's facial bones were fractured, as were six of her teeth. Fragments from a lead slug were recovered from her head; the largest piece had the number eight embossed on it. So much of Harvey's blood was on his easy chair that investigators assumed he had been shot first, as he sat in the living room, at a distance of just a couple of meters from the direction of the dining room. They figured Jeanette was likely hit in the face with the butt of the rifle when she tried to defend herself or come to Harvey's aid. Then she was shot as she lay incapacitated on the floor in front of the fireplace.

This was a rural farming area of New Zealand, so while police were collecting statements, they also decided to also collect rifles from all the homes in the area so they can compare them with a type of gun they suspected had killed the couple. In all, they initially collected just over 60 similar weapons, and then, based on a handful of test firings, police narrowed the possible rifles in the group down to one belonging to the Crewes' neighbor, Arthur Thomas. He became their prime suspect. The other initial suspect had actually been Len Demler, Jeanette Crewe's father, but he was ruled out based on the results of the rifle comparisons.

Now, to back up, investigators initially did not find any ballistic evidence during several intensive searches at the crime scene, even though they had been told to look for shell casings. It wasn't until about four months after the crime, during their third search of the yard outside the Crewe home, that investigators found a .22 cartridge case. They determined it had been shot from the rifle they had previously collected from neighbor Arthur Thomas.

Later, a detective named Johnston received an anonymous tip to search Thomas' farm, where he was told he might find evidence related to the axle used to weight down Harvey Crew's body. Johnston searched the Thomas property and found a part that matched the axle type. Arthur Thomas was then arrested in 1970 and tried and convicted in 1971 based on the cartridge casing from his gun found in the Crewe's garden and the suspicious axle piece found on his property.

Despite the evidence, his conviction lead to an outcry by friends, relatives, and the media, since Thomas had no motive and was known in the community as a good man. His wife was the one suspected of being the woman who took care of the baby after the Crewe couple was murdered. Mrs. Thomas said her husband was home with her the entire night of the killings, and neither of them had anything to do with the case. Arthur Thomas was granted a retrial, but in 1973 he was convicted for a second time.

Especially following his second conviction, a variety of public and private individuals got involved in trying to get Arthur Thomas exonerated. Ultimately, the assistant editor of the Auckland Star Newspaper enlisted forensic scientist, Dr. Jim Sprott, to reexamine the cartridge casing found at the scene. Sprott determined the case was just too uncorroded to have been outside and exposed to the elements for over four months. Sprott believed the evidence had been planted at the crime scene. Also supporting this theory was the fact that the area had already been exhaustibly searched two times by police, before that shell casing was found. Remember, too, police already had Thomas's rifle in their possession at the time the spent shell casing was discovered, and had already test-fired it at least once.

So based on Dr. Sprott's conclusions and other statements that had come to light, a Royal Commission of Inquiry reversed Thomas's conviction in 1980 and agreed that two specific policemen had likely been directly involved in planting the cartridge casing. The inquiry's report stated that, "Mr. Hutton and Mr. Johnston planted the shell case … to manufacture evidence that Mr. Thomas's rifle had been used for the killings." But the police officers were never tried, because the Solicitor-General claimed to have insufficient evidence. There's been much public criticism since then, suggesting that in any other country, those police would have been arrested and tried. Arthur Thomas was released from jail in December of 1979, after more than nine years in prison; he was compensated almost a million New Zealand dollars.

So, what motive would Hutton and Johnston have to falsify evidence in the Crewe murders? You might speculate that sheer pressure to solve the crime could be enough, since you can imagine how much publicity this case got, particularly due to the baby found at the scene. But years of investigative reporting have suggested the motives of one of those officers may have run much deeper.

Hutton was the police officer in charge of investigating the killings, and Officer Johnston worked under Hutton. Johnston was the guy who allegedly got the anonymous tip to search Thomas's property for the axle part. Suspicions have continued to center on Johnston for a number of reasons and by a number of sources. Ian Wishart, author of a 2010 book about the Crewe case, paints Johnston as both the killer and the planter of the evidence, saying Johnston was actually known as "The Fitter" by his police colleagues, because of his ability to make evidence match his version of a crime. Wishart calls Johnston a "dirty cop" and says he had a vindictive temper and a tendency for physical violence. Wishart's theory includes the fact that Johnston had investigated a burglary at the Crewe house in 1967, so he would have known the couple, the layout of the house, and probably even knew where they hid their spare house key.

Wishart and other sources point out that in the three years between the 1967 burglary and the 1970 killings, there were two arson attempts at the Crewe farm, the first one while Jeanette was in the hospital after the birth of Rochelle. That fire was started by igniting the baby's unworn clothing. The second arson occurred precisely a year to the day before the killings. It's also been alleged that the brake line in Jeanette's car had been cut in 1969, shortly after the birth of the baby, but that she only told a few friends and did not involve police. There's been nothing concrete found to prove the brake incident, except the word of a few of Jeanette's friends, but in hindsight, those friends seriously question why Jeanette would not go to the police about what had happened, unless someone from the police, specifically Officer Johnston, was involved.

One suspicion is that Johnston may have been stalking Jeanette Crewe after meeting her during the burglary investigation. Jeanette was, at least in photographs, quite a beautiful woman. A second theory is that Johnston may have been trying to blackmail the couple since the 1967 burglary he had investigated was suspected as possibly being insurance fraud. Johnston himself admitted in court he thought the burglary was an inside job.

In a book released in 2012 by Chris Birt, a reporter who has researched the case for almost four decades, it exposes more possible evidence related to the framing of Arthur Thomas and its subsequent cover up. Birt claims that the release of previously unavailable documents has shed additional light on the truth. He insists that his are not wild theories and that he has paperwork to support all claims of injustice in his book, including the suppression of eyewitness statements and malpractice by judges, prosecuting attorneys, and other officials involved in the case. Birt purports that Officer Johnston was, without a doubt, the killer. However, an earlier book released by Birt in 2001, fingered Jeanette's father, Len, as the couple's murderer. So which do we believe?

There's one more major part of the mystery that has likely crossed your mind. Who cared for baby Rochelle after her parents had been killed? There are lots of theories here, too. Initially, as I mentioned, Arthur Thomas' wife was suspected of being the woman who cared for the child after her husband supposedly killed their neighbors. Early on, some even suggested that the mystery caregiver was Jeanette Crewe herself and that the case was a murder/suicide. This theory sets forth Harvey Crewe was killed by his wife after a domestic argument in which he broke her jaw. She then took his body to the river and weighed it down, returned to the house to care for the baby, but a few days later, went back to the river and shot herself. The evidence just doesn't support that at all, especially given Jeanette's body was bound in a bedspread. Whoever was spreading that rumor could not have had the full story, which, as you know, is not often released to the public in the early stages of an investigation.

Other sources continue to suggest that Jeanette's father, Len Demler, was the killer, and that her own family, either Jeanette's younger sister; or her father, perhaps dressed as a woman; or Demler's new wife, one of them took care of the baby. Jeanette's mother, the first Mrs. Demler, died several months before the murders. Before dying, though, she changed her will to give half the family farm to her daughter, Jeanette, and cut Jeanette's younger sister (who was daddy's favorite) and Mr. Demler out of the estate. But, there's also lots of evidence against that theory, too, including eyewitnesses who stated the woman they saw at the house after the shootings was clearly neither Jeanette's younger sister nor her new stepmother.

But if Officer Johnston was responsible for the killings, maybe he simply had somebody watching over the Crewe farm and directed that woman to care for the child until the couple was reported missing and the baby discovered. People argue, though, this act of kindness doesn't just doesn't match with a man who would kill the baby's parents in cold blood. Johnston died in 1978, less than 10 years after the crime and before many newer forensic techniques were developed that might have shed more light on the case, or found better evidence. So for now, no one knows who took care of the baby, not even Rochelle Crew herself, who is still alive and well, unless that mysterious caregiver is still somewhere hiding the secret.

In 2013, it was revealed that police have lost the fingerprint evidence from the 1970 crime scene. There were said to be 10 unidentified prints recovered from the home, which can now never be re-evaluated. I do understand how that can happen over time. I've encountered similar situations in some of my own casework; files get lost, they get purged, or irrevocably damaged. But it sure seems convenient in this case, where one or more dirty cops seem to have gotten away with murder.

Now, this lecture topic is not intended, in any way, to suggest that most law enforcement officers aren't trustworthy beyond measure, and are not only law-upholding, but also law-abiding citizens; I truly believe the vast majority are. I personally know many officers who are simultaneously incredible crime fighters and persons of great integrity. Many people say, though, there's a so-called Blue Wall of Silence among police officers to back each other up. And sometimes wrongdoing isn't even exposed by internal affairs investigations, but it's not true that police abuse and corruption are never prosecuted.

Consider the 1997 case, in which Haitian immigrant, Abner Louima, was tortured while in custody in Brooklyn, New York. The perpetrator of the assault, a second-generation former NYPD officer named Justin Volpe, ultimately admitted guilt and was sentenced to 30 years behind bars without parole in 1999, and you can just imagine the treatment cops may get in prison from other inmates. Another officer was sentenced to five years as an accomplice, even though Volpe denied that man's involvement. That alleged accomplice and two other officers were originally also sentenced to five years each for trying to cover up the incident, but those convictions were later reversed on appeal due to a lack of evidence. That led many to argue justice was still not served in this case.

The incident began during a brawl outside a New York nightclub during which Louima was taken into police custody. At the police station, he was brutally tortured and even sodomized by Officer Volpe. Louima's teeth were broken; he suffered such serious internal injuries, he had to be taken from the precinct to a local hospital. There, Officer Volpe alleged some of the injuries were from the fight in front of the nightclub, during which the officer claimed Louima sucker-punched him, although the blow was actually delivered by some other man.

Later, one of Louima's hospital nurses called the authorities to report her suspicions that police were responsible for her patient's injuries. Louima ended up needing three major surgeries over two months of hospitalization. Volpe later recanted his allegation that Louima struck him, admitted torturing the man at the precinct, and pled guilty. This incident became more or less a national symbol of police brutality in the United States. In July of 2001, Abner Louima was awarded $8.75 million in damages by the NYPD. Volpe won't be eligible for release until 2025.

For our final case in this lecture, we'll visit Cerro Maravilla, which is Puerto Rico's fourth highest summit. It's known as a place where you can see both the southern and northern coasts of the island on a clear day. But two things have really marred the area's beauty. One, the many television, radio, and cell towers that are planted along the peak, and secondly, related to those towers, the tragic events that took place on that mountain on July 25, 1978. It was on that night that three young men between the ages of 18 and 24 planned to set fire to, or otherwise sabotage, some of the Maravilla summit's broadcast towers. This was supposed to be in retaliation for the U.S. government's continued imprisonment of two groups of Puerto Rican militants. One of those groups of prisoners had attempted to assassinate

President Harry Truman in 1950, and the other was a group of jailed Puerto Rican nationalists, who in 1954 shot five members of the U.S. Congress from a balcony in the U.S. Capitol Building. The three young men who were to carry out the tower sabotage at the top of the mountain were Carlos Soto Arriví, Arnaldo Darío Rosado, and Alejandro Gonzalez Malavé.

Arriví and Rosado were longtime members of a separatist group, called The Armed Revolutionary Movement, in Spanish known as MIRA. Arriví was from a well-to-do family; his father was famous Puerto Rican novelist, Pedro Juan Soto. Rosado was from a more humble background. But both men became interested in the cause of Puerto Rican independence at young ages, probably in high school. As an organization, MIRA was founded in 1967 by Filiberto Ojeda Rios, who was credited with several acts of terroristic bombing in and around the New York area, including General Electric's corporate headquarters, a bank, and a police station. Ojeda Rios was captured in the 1970s, and as a group, MIRA is no longer active, though there are still organizations striving for Puerto Rican independence from the United States.

But back to the mountain and 1978. Arriví and Rosado met up with the third member of their sabotage group, Malavé, on the evening of July 25. The trio hijacked a taxi and ordered the driver to take them to the Maravilla broadcast towers. But when they got there, they were ambushed by waiting police; 18-year-old Arriví and 24-year-old Rosado died that night at the top of the mountain. The police reported that they ordered the men to surrender, but the revolutionaries fired first, and two were killed in self-defense. The third man, Malavé, was only slightly wounded.

The taxi driver said he didn't know what had happened, since he hid under the dashboard when the shooting started. But two days later, the cabbie changed his story during an interview by news reporters; he then said that the three men had actually tried to surrender to the officers, and that one of the men, Malavé, the lone survivor of the trio, had called out to police, "Don't shoot, I'm an agent." As it turns out, Malavé was a young undercover police officer, recruited while still in high school. He had been assigned to infiltrate the MIRA independence group, to which Arriví and Rosado belonged. Malavé was the one who had told the authorities about the plan that night, and he was the one that hijacked the cab.

When the news was first released, the Governor of Puerto Rico, Carlos Romero Barceló, commended the police for preventing a terrorist attack by revolutionaries. No charges were filed against any of the police. But as the media began to circulate the taxi driver's conflicting stories, Barceló was pressured to investigate further. The cabbie's revised and expanded version was that when he last saw Arriví and Rosado, they were being beaten by police in front of the cab. The taxi driver said the police had extracted him from the vehicle, and one of them kicked him before taking him away. Still, however, no criminal charges were filed against any of the police officers present when the two young men were killed.

The opposition parties of Puerto Rico, and other members of the public, continued to press for justice, or at least a further investigation, saying it was a setup. Public perception was that the police executed Arriví and Rosado to set an example for other members of MIRA. Between the years 1978 and 1980, Governor Barceló even asked the U.S. Federal Bureau of Investigation and the U.S. Department of Justice's Civil Rights Division to look into the matter. Both of those groups concurred with the official Puerto Rican conclusions about the case; they found that the incident was not a deliberate massacre of the freedom fighters.

But in 1981, the Judiciary Committee of the Senate of Puerto Rico reopened the case, deciding to grant immunity to certain police officers who had been at the scene the night of the incident in exchange for their testimony. Although it took a couple of years, things about the events of that night on Maravilla finally began to become clearer. In 1983, Officer Miguel Cartegena Flores stated, "When I arrived on the scene, I saw four police officers aiming their guns at the two activists, who were kneeling before them. I turned my eyes away and heard five gunshots." Flores also told the committee that several hours before the incident his commanding officer said to him and other officers, "These terrorists should not come down alive," meaning not survive the trip to the mountain. A second man who was also granted immunity, Officer Carmelo Cruz, corroborated these statements.

Additional reports, from other witnesses and the taxi driver confirmed there were two different rounds of gunfire. The cab driver said he was told by the police not to mention hearing the second set of gunshots. He also stated, when he was removed from his taxi the two young men appeared to have been disarmed and were being beaten by the police. Because of these statements, it appeared to investigators that both Arriví and Rosado were definitely still alive after the first round of gunshots and were summarily executed after being dragged from the taxi and beaten.

This evidence led to the conviction of 10 police officers in 1984 on charges of perjury, obstruction of justice, and destruction of evidence. Four of the officers were also convicted of second-degree murder. All received sentences ranging from 6 to 30 years. Barceló claimed that the police involved had lied to him, and that he was not involved in any type of a cover-up for their actions.

While this second round of investigation between 1981 and 1984 uncovered a police plot to assassinate the victims, it could not confirm a definite cover up by the Puerto Rican or U.S. governments, though those allegations were certainly being made. For instance, the two major Puerto Rican political parties who opposed Barceló definitely thought he and others, all the way up to United States President Jimmy Carter, were involved in a cover up of the Cerro Maravilla Massacre, as it became known.

For one thing, Barceló had always aligned himself with the Republican Party, but curiously, in late 1979, while Carter was in his first, and ultimately only term as President, Barceló began a Puerto Rican campaign to name President Carter as the Democratic Party nominee for the 1980 election. Carter won the nomination by a single vote; he would not have won without Puerto Rico's support. Carter lost his 1980 presidential bid to a Republican, but the involvement of Barceló in Carter's campaign, caused Puerto Rican opposition party members to really question whether a suspected cover up could have even gone all the way up to the President of the U.S. and the FBI.

As I said, the original United States Department of Justice's 1978–79 investigation of the Maravilla incident concurred with Barceló's conclusions, stating there was lack of evidence to charge anyone with anything in the killings of Arriva and Rosado. Just a month-and-a-half after Carter won the Democratic nomination for the presidential election, the Department of Justice proclaimed the Maravilla case closed, even while investigators were still exploring, "unexplained contusions," in other words, bruises, seen in photos of one of the dead victims, including on Rosado's forehead just above a black eye.

Bruising doesn't happen after death, but if police had beaten the young men before executing them, bruises could have been the result. In fact, the autopsy report was apparently sanitized of any mention of surface wounds, contrary to the photographs. Autopsy results further documented gunpowder residue on the chest wounds of both men that would not have occurred had they been shot at a distance. And consider the likelihood of both men being shot right in the chest, if they had died in some kind of chaotic shootout with police. Much later, in 1993, a U.S. District Court did consider a motion against a Puerto Rican forensic pathologist. It was alleged he altered autopsy results to make them more consistent with the police accounts of the Maravilla killings, but the allegations couldn't be proven.

So even though the second round of investigations convicted 10 dirty cops and led to several suspensions, demotions, and reassignments, in both police and government offices, no one was ever charged with a cover up. The suspicions of the opposition were never satisfied, and that's why some have called the entire incident "The Puerto Rican Watergate."

As for Malavé, he was dismissed from the force due to public pressure, and then later also given immunity to testify against the police officers who shot Arriví and Rosado. The police countered that even though Malavé was an undercover agent, he was the one that took the cab driver hostage at gunpoint, stating Malavé was the driver of the cab all the way up the mountain, and that he refused to let the cabbie out on the road, even though Arriví and Rosado told him that would be best. In hindsight, though, the taxi driver proved to be such a crucial witness to what really happened on the mountain that night, it's a good thing he was there and wasn't harmed.

As for Malavé, he was killed on April 29, 1986, just two months after his acquittal in the matter. He was assassinated in front of his mother's house, shot three times, and his mother was wounded in the attack. A group calling themselves Volunteer Organization for the Revolution claimed responsibility and vowed to target all the police involved in the Cerro Maravilla Massacre. Fourteen years after the incident, the man who had headed the United States' Justice Department's Civil Rights Division during the investigation issued a public apology for what he thought had been a cover up by the FBI in the case. Twenty-five years after the killings, former Governor Barceló told the press it was premature of him to praise the officers involved in the incident, as he initially had done, but he still admits nothing with regard to a cover up. Pro-independence followers still convene on the mountain every year on July 25 to mark the anniversary and celebrate the Independentista movement.

Guilty until Proven Innocent
Lecture 16

Joyce Gilchrist worked for the Oklahoma City Police Department for about 20 years. During that time, she became well known for her uncanny ability to make connections between evidence and perpetrators. Her contributions to cases and trials sent thousands of people to prison, with 23 of those individuals ultimately sentenced to death. Tragically, 11 of those people had already been executed by the time an FBI review of Gilchrist's work demonstrated her incompetence and deception. In this lecture, we'll look at how human error, bias, or wrongdoing can subvert even the most sophisticated forensic tools.

Misuse of Science: Joyce Gilchrist

- Forensic scientist Joyce Gilchrist worked in several capacities in the Oklahoma City forensic lab, including as an analyst of trace evidence and in forensic chemistry. She performed hair analysis, fiber examinations, and chemical and serological testing of body fluids, among other procedures.

- Sometime in 2001, a secret FBI review of Gilchrist's work was conducted. Investigators found that in five of eight cases reviewed, she had either made serious mistakes or gone beyond the "acceptable limits of forensic science."
    - One of the problems with analysis of such evidence as hair, fiber, and blood is that it can narrow down identification only to a relatively large category, not to a particular individual. For example, although there are different hair forms and colors, a tremendous number of people share the same patterns.

    - Gilchrist used hairs, fibers, and blood to place suspects at crime scenes with a confidence that was scientifically unwarranted. Other accusations of Gilchrist's malpractice included not performing tests that might demonstrate a suspect's innocence and withholding evidence from the defense that would help counter the prosecution's case.

The Case of Robert Lee Miller Jr.

- In 1986, two elderly women were raped and murdered in the same neighborhood in Oklahoma City. A man named Robert Lee Miller Jr. was convicted of the two murders, largely owing to Gilchrist's testimony, and sentenced to death. Ten years later, DNA evidence proved conclusively that Miller was not guilty. At the time of Miller's trial, DNA technology was not commonly used in forensics and was not accepted in Oklahoma courts, although serological testing was already providing some genetic information.

- Evidence collected from the crime scenes included semen, blood, saliva, and hair. The perpetrator was determined to be blood type A and a *secretor*, that is, an individual who expresses blood type markers in other body fluids, such as semen and saliva. The blood also contained a genetic marker that at the time was said to be more common in African Americans than people from other ancestries.

- Ultimately, police focused on Robert Miller as the prime suspect. Miller endured a 12-hour interview—despite the suspicion that he was high on drugs at the time—and gave a false confession. Miller's trial hinged largely on that coerced confession, combined with extensive testimony from Gilchrist about blood, saliva, and hair.

- Gilchrist correctly testified that she could not exclude Miller as being the murderer because he had type A blood. Although crime scene body fluid testing revealed four genetic markers that were not present in Miller's blood, Gilchrist claimed that those discrepancies did not exclude Miller because the crime scene samples could have been tainted by the victim's blood—even though the victim was not type A.

- Gilchrist's analyses were also said to exclude another suspect, Ronald Lott. DNA testing later proved that Lott was the rapist and murderer, and he was convicted in 2002. Lott is now on death row, while Miller finally walks free after 10 years in prison.

- Gilchrist was fired, and thousands of cases in which she was involved have been reevaluated. Former inmates are now suing her for her part in their wrongful convictions. Sadly, some of them—executed after Gilchrist's analysis and testimony—never lived to see justice served.

The Madrid Train Bombings

- On March 11, 2004, in and around Madrid, Spain, 10 bombs detonated almost simultaneously on four trains during morning rush hour. In one of the worst acts of terror in Europe since World War II, 191 people were killed and nearly 2,000 were injured. Investigators fairly quickly discovered that the bombs had been packed in backpacks and detonated by mobile phones.

- Because the incident occurred three days before Spain's general election, the attack was originally thought to be the work of a Basque separatist organization, but suspicions quickly turned toward al-Qaeda.

- A few weeks after the bombings, the Spanish National Police closed in on an apartment in west Madrid to make an arrest. But suspects in the building detonated a suicide bomb, killing 7 of themselves and 1 policeman and wounding another 11 officers. Some of the suspects fled the scene. Spanish authorities determined that the explosives used at the apartment were of the same type as the devices used in the train bombings.

- Although authorities had intelligence about the group responsible, they initially didn't have much physical evidence to definitively link the crimes to specific individuals. But while investigating around the scene of the train bombings, Spanish police had recovered a blue plastic bag. It contained seven copper detonating devices and had a single usable fingerprint on its surface.

Suspect: Brandon Mayfield

- More than 5,000 miles away, just outside Portland, Oregon, a 37-year-old attorney named Brandon Mayfield and his wife, Mona, began to notice some strange happenings around their home. When they'd return to the home after being gone for a time, things seemed slightly out of place. They suspected a possible burglary, but nothing was missing.

- Brandon Mayfield was born in Oregon in 1966. At age 20, he met his future wife, Mona. She was an Egyptian national and a Muslim, and as their relationship proceeded, Brandon converted to Islam. When he met Mona, Mayfield was already in the U.S. Army Reserve. In 1992, he joined the U.S. Army and was an officer until 1994; he earned a law degree in 1999.

- In addition to his family law practice, Mayfield worked for an Oregon State Bar Association organization dedicated to helping clients who couldn't afford legal representation. In 2003, he represented a man named Jeffrey Battle in a child custody matter. Battle was one of a group of American Muslims known as the Portland Seven who had been convicted of conspiring against the United States with the Afghani Taliban Islamic fundamentalist movement.

Surveillance and Arrest

- The Spanish National Police requested that the FBI run the fingerprint found on the bag of detonating devices through its Integrated Automated Fingerprint Identification System (IAFIS)—a database of digitized fingerprint records. On March 19, 2004, eight days after the Madrid train bombing, the FBI's Latent Print Unit identified Brandon Mayfield as the source of the fingerprint. The print was labeled latent fingerprint (LFP) 17.

- In actuality, the IAFIS search had generated a list of 15 to 20 most likely candidates, of which Mayfield was one. The results generated by IAFIS are not definitive; any system-suggested matches are always followed up with a comparison by a qualified fingerprint examiner. In the Mayfield case, three FBI examiners concluded that Mayfield was the source of the print.

- As its legal authority to begin surveillance of Mayfield and conduct clandestine searches of his residence and law office, the FBI used the Foreign Intelligence Surveillance Act (FISA) of 1978. Further justification came from the USA Patriot Act of 2001. The FBI obtained a "sneak and peek" warrant for Mayfield's home and office, which allowed agents to copy computer hard drives, install sound-monitoring devices, and gather bank records.

- Within about three weeks of identifying the print as Mayfield's, the FBI was contacted by the Spanish National Police, who had determined that the print was "conclusively negative" as a match to Mayfield. Nevertheless, an arrest warrant was issued for Mayfield. In addition to the fingerprint, the attorney's regular attendance at a local mosque and other personal and professional activities may have been considered as evidence by the FBI. Mayfield was arrested on May 6, 2004.

- In the affidavit used as justification for the search and arrest, the FBI stated that the identification of the fingerprint was 100 percent positive and that the Spanish National Police was satisfied with the FBI's conclusion. At his hearing, Mayfield produced his expired passport to show that he hadn't been out of the country, as well as witnesses who accounted for his whereabouts around the time of the bombing in Spain.

- After further investigation of the fingerprint, on May 19, Spanish authorities arrested an Algerian named Daoud, who not only was a definitive match to the print but was also linked in other ways to the bombings. Once Spain issued an arrest warrant for Daoud, U.S. authorities dismissed Mayfield's case and released the attorney from jail without charges.

Fallout from the Case

- In March 2006, a 273-page report was released by the office of Inspector General (IG) Glenn Fine regarding the misidentification of Mayfield. The IG's office acknowledged that there were similarities between the latent print found near the crime scene in Spain and Mayfield's fingerprint but concluded that the FBI, in identifying a match, may have been over-reliant on its own expertise and overconfident in IAFIS.

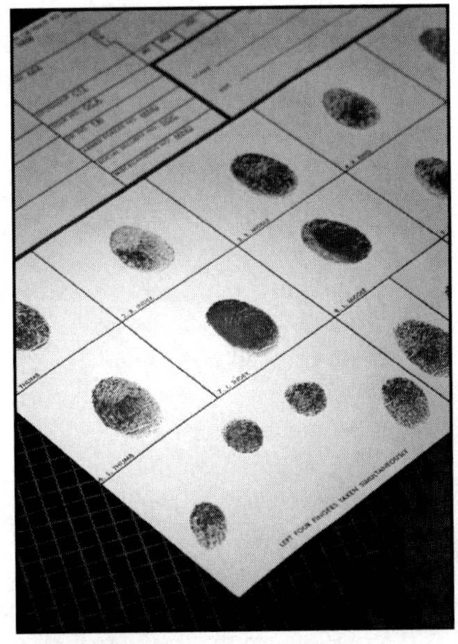

A fingerprint match must be 100 percent certain; any discrepancy that can't be explained away by the examiner is grounds for exclusion.

- The IG also speculated whether Mayfield's religion had played a part in the misidentification. The report concluded that at the time of their examination, FBI fingerprint experts were not aware that Mayfield was Muslim, was married to a Middle Eastern woman, or had legally represented a member of the Portland Seven terrorist cell. But the issue wasn't as clear with regard to the rest of the FBI's investigation.

- The IG's report noted that the latent print examiners ignored a full quadrant in the unknown print that did not match Mayfield's pattern. According to the report, the multiple alternative theories the FBI had to construct to conclude that LFP 17 belonged to Mayfield rose to such a level of unlikelihood that the potential match should have been dismissed.

- The report further argued that once the examiners thought the print belonged to Mayfield, they began to work backward, using Mayfield's clear print to look for similarities in the questioned print.

- The IG's report recommended several changes to FBI laboratory procedures with regard to fingerprint identification but found no misconduct on the part of FBI employees. To its credit, the FBI had already implemented changes in its practices.

- In 2004, Mayfield filed a lawsuit against the U.S. Department of Justice, the FBI, and several specific FBI employees for violations of his civil rights. Mayfield won his suit and was subsequently awarded $2 million.

Suggested Reading

Borchart, *Convicting the Innocent.*

Murray, *Overturning Wrongful Convictions.*

Petro and Petro, *False Justice.*

Scheck, Neufeld, and Dwyer, *Actual Innocence.*

U.S. Department of Justice, National Institute of Justice, *Latent Print Examination and Human Factors.*

Questions to Consider

1. Were you aware of the Gilchrist malpractice case? What safeguards do forensic scientists have in place to prevent mistakes in analyzing evidence? What safeguards should be in place?

2. What are some of the major causes of false convictions in the U.S. legal system?

3. What are some other possible lines of forensic evidence to seek truth in false convictions in cases where DNA testing is not possible, such as a drive-by shooting or a case of mistaken identity?

Guilty until Proven Innocent
Lecture 16—Transcript

Beginning around 1980, and spanning over 21 years, forensic scientist Joyce Gilchrist worked for the Oklahoma City Police Department. During that time, she earned the nickname Black Magic for her uncanny ability to make connections between evidence and perpetrators. She was also very convincing on the witness stand. Her contributions to cases and trials sent literally thousands of people to prison, with 23 of those individuals ultimately sentenced to death. Tragically, 11 of those people were already executed by the time an FBI review of Gilchrist's work demonstrated not only serious incompetence, but exposed her out-and-out lies, both in the lab and in the courtroom.

Gilchrist worked in several capacities within the Oklahoma City forensic lab, including trace evidence and forensic chemistry. She performed hair analysis, fiber examinations, and chemical and serological testing of body fluids, among other forensic procedures. Until her fraud was exposed, Gilchrist was considered an exemplary scientist. She was even named Employee of the Year in 1985 and was promoted to supervisor by 1994. But throughout her career, other forensic scientists were questioning her reliability; as early as 1987, a colleague reported her wrongdoings to a regional forensic association. But at that time, nothing was done to stop or even seriously investigate Gilchrist.

But sometime just before the spring of 2001, a secret FBI review was being conducted on her work. They looked at just eight Gilchrist cases to start with and found that in at least five of them she had made either serious mistakes or, in their words, went beyond "acceptable limits of forensic science." One of the problems with evidence like hair, fiber, and even blood type, is that they can only assign evidence to a relatively large category, not to a particular individual.

For example, although there are different hair forms and colors, a tremendous number of people would share the same patterns. That's also true of blood types; there are only four major blood groups, A, B, AB, and O, and billions of people share each of those types. The same goes for fiber evidence; manufacturers mass-produce clothing and carpeting, so the fibers from those sources are what we call in forensic science class evidence, not individuating evidence.

Still, that didn't stop Joyce Gilchrist from using hairs, fiber, and blood to place suspects at crime scenes with a confidence that was, scientifically, completely unwarranted. Other accusations of her malpractice included not doing the tests that might demonstrate a suspect's innocence and withholding evidence from defense counsel that would help counter the prosecution's case. In that sense, Gilchrist is what I would call a hired gun. She was willing to say or do whatever it took to get a conviction for the police department she worked for. Gilchrist let her bias and ego get in the way of the objectivity that is the foundation of good science.

Let me give you just a couple disturbing examples of Gilchrist's long reign of terror. A 20-year-old named Mark Fowler was arrested in 1985 for an Oklahoma City grocery store robbery in which three people were killed. He was put to death by lethal injection in January of 2001, partly based on the testimony of Joyce Gilchrist. Now Fowler was no angel; he admitted to robbing the store along with a man named Billy Ray Fox, but Fowler blamed Fox, a disgruntled former employee of the grocery store, for herding the victims into the back room and killing them.

A year after Mark Fowler's trial, his own grandmother, Anne, was raped and murdered in Oklahoma City. A few months later, another elderly woman was raped and killed in the same neighborhood. While Fowler sat on death row, a man named Robert Lee Miller Jr. was convicted of the murders of both Anne Fowler and her neighbor, again, largely owing to the testimony of Joyce Gilchrist. Miller was also given a death sentence. But 10 years later, in 1998, DNA evidence proved, beyond a shadow of a doubt, that Miller was not guilty, and showed that another man, one who Gilchrist earlier said could not have been the rapist and murderer, well he was actually the perpetrator. Mark Fowler's father was not only faced with the murder of his mother and the execution of his son, but also had to live knowing that Joyce Gilchrist could have been wrong in his son's case, too.

Let's look now at the two cases Miller didn't commit, but was found guilty of. Keep in mind this predated the common use of DNA technology in forensics and was before it was accepted in Oklahoma courts, although serological testing was already providing glimpses of a person's genetics. Evidence collected in the two crime scenes included semen, blood, saliva, and hair. The perpetrator was determined to be blood type A, when testing both blood and semen from both crime scenes. These individuals are called *secretors* because they also express their blood type markers in other body fluids, like semen and saliva. Plus, the blood contained a particular genetic marker that at the time was said to be more common in African Americans than people from other ancestries.

Given this information, police literally walked the neighborhood and requested blood samples from black males. That's how they came up with Robert Miller as their prime suspect. He not only lived where both victims had been attacked, but was also shown to be blood type A and a secretor. So police brought Miller in for questioning, and then he endured a 12-hour interview, despite the suspicion he was high on drugs at the time. Using this interrogation, police led Miller to a false confession. Much of that interview was recorded and was later used to demonstrate that Miller was delusional during the questioning. Lots of the detail Miller presented was not at all consistent with the crimes committed. Despite the inaccuracies, Miller's trial hinged largely on that coerced confession, combined with the extensive testimony about blood, saliva, and hair that was delivered by Joyce Gilchrist.

Gilchrist correctly testified that she could not exclude Miller as being the murderer, since he did have type-A blood. But even though the crime scene body fluid testing revealed four genetic markers that were not present in Miller's blood, Gilchrist claimed that discrepancy did not exclude him. She said the crime scene samples could have been tainted by the victim's own blood, even though the victim wasn't type-A.

The hair analysis conducted by Gilchrist included both human and non-human hairs. She stated that human hairs recovered from the crime scene were Negroid hairs, and Robert Miller is African-American. But just as with the inheritance of blood type, African ancestry is shared by a significant chunk of the population. And in case you're thinking Joyce Gilchrist may be a racist, she is also African-American. With regard to the non-human hair that investigators found at one of the crime scenes, Gilchrist said it greatly resembled hair taken from a roving neighborhood dog that Miller was fond of. But how many other people in that general vicinity cared for and may have petted that dog, maybe even including the victim?

As I mentioned earlier, analyses by Gilchrist were said to exclude another suspect who the police had also been originally looking at hard in this case. That was an African-American guy named Ronald Lott. Turns out he's the man the 1998 DNA testing ultimately proved raped and murdered those two women, for which he was convicted in 2002. Lott turned out to be a serial rapist with additional victims, and was executed in December of 2013. Robert Lee Miller Jr. now finally walks free after sitting in prison from 1988 to 1998, a decade of his life lost, and irreparable damage done to him and his family.

Reanalysis of the hairs showed that Gilchrist's work was, in the words of the second analyst, "meaningless and completely unjustified." Due to the FBI investigation, which initially reopened several cases, the governor demanded that thousands of cases in which she was involved be re-evaluated, and of course, she was fired.

Believe it or not, Gilchrist had the nerve to file a lawsuit against her former employer for over $20 million, claiming that she was really fired as retribution for a sexual misconduct charge that she made against a supervisor. Ultimately, it's been Gilchrist that has had to hand over money to former inmates who are now suing her for her part in their wrongful convictions. Sadly, though, some of them, maybe even Mark Fowler and the 10 others who were executed, partly based on the work and testimony of Joyce Gilchrist, well, they never lived to see justice served on her.

Now, while some types of evidence, like hair and blood type, cannot be conclusively linked to one specific person, fingerprints, like DNA, can be, right? During the morning rush hour on March 11, 2004, 10 bombs almost simultaneously detonated on four different trains in, or heading to, Atocha Station in Madrid, Spain; 191 people were killed and nearly 2,000 injured, making this one of worst acts of terror in Europe since World War II.

Investigators fairly quickly discovered that the bombs had been placed in backpacks and detonated by mobile phones.

Because the incident occurred three days before Spain's general election, the attack was originally thought to be the work of a Basque separatist organization, known as the ETA. But suspicions quickly turned toward the al-Qaeda terrorist group, since at that point, Spain's President, José Maria Aznar, was allied with the U.S. President, George W. Bush, and Spain was involved in the Iraq war. As a side note, Aznar was defeated in that general election by José Luis Rodríguez Zapatero, who removed Spanish troops from Iraq by May of 2004, just two months after the bombing incident.

Another potential link to al-Qaeda was that the bombing was two years and six months, to the day, after the 9/11 attacks on the World Trade Center in New York City, and would have been precisely 911, in other words, 9-1-1, days after it, if not for the leap year that had just occurred in February 2004, technically making the bombing 912 after 9/11.

The attack didn't end with the train bombings, though. A few weeks later, 12 kilograms, that's almost 26.5 pounds, of a nitro-based high explosive was found on another train track in Spain. Although there were over 100 meters of wire connected to it, the other end of that wire was not connected to anything else, so, as a bomb, the device was incomplete and non-functional.

The day after discovering this inactive device, the Spanish National Police figured out who they thought was responsible and closed in on an apartment in west Madrid. At that point, the suspects in the building detonated another suicide bomb, killing 7 of their members and 1 policeman, and wounding another 11 officers. Some of the suspects got away. Spanish authorities determined the explosives used at the apartment were of the same type as the earlier devices used in both the actual and potential train bombings.

While authorities had intelligence information about the group responsible, they initially didn't have much physical evidence to definitively link the crimes to any specific individuals. But while investigating around the scene of the train bombings, Spanish police recovered a blue plastic bag.

It contained seven copper detonating devices, and as luck would have it, a single usable fingerprint was on the bag's surface.

Over 5,000 miles away, just outside of Portland, Oregon, a 37-year-old attorney named Brandon Mayfield and his wife, Mona, started noticing some really strange happenings around the home they shared with their four children. They'd go out, and when they came back, would notice things weren't as they had left them. For instance, they knew they had closed a certain window blind, only to find it slightly open. The couple's door had two locks on it, and they typically used only one, and not the other, but when they came home, they'd find the door locked differently than the way it was when they left. And once on her freshly vacuumed carpet, Mona saw a shoeprint that she absolutely knew was not from a member of the family. It was their custom not to wear shoes in the house, as they were Muslims. They suspected a burglar had somehow gotten in, but nothing they owned was missing. Even more eerie was that this happened to the Mayfields on more than one occasion. Where is this mystery going, you might ask?

Well, Brandon Mayfield was born in Oregon in 1966. At age 20, he met his future wife, Mona, on a blind date. She was an Egyptian national and a Muslim, and as their relationship proceeded, he converted to the Islamic faith. When he met Mona, Mayfield was already in the United States Army Reserve, which he was a member of from 1985 to '89. In 1992, he joined the U.S. Army, and was an officer until 1994. In 1999, Mayfield earned his law degree, and in addition to practicing family law, he worked for an Oregon State Bar Association organization that was dedicated to helping clients who couldn't afford legal representation.

In 2003, Mayfield represented a man named Jeffrey Battle in a child custody matter. Battle was one of a group of American Muslims known as the Portland Seven. They had been convicted of conspiring against the United States with the Afghani Taliban Islamic fundamentalist movement. So despite Mayfield's 10 years of military service to his country, other associations, and possibly his religion, apparently made him suspicious to authorities. They ultimately connected Brandon Mayfield to the Madrid train bombings. How on earth did that happen?

Well, remember the fingerprint on the plastic bag containing the detonators? The Spanish National Police, working through the Paris-based international agency known as Interpol, made a request to the United States FBI. They asked that the Federal Bureau of Investigation compare that fingerprint to other prints, using IAFIS technology. IAFIS stands for the Integrated Automated Fingerprint Identification System. That's the FBI's software database of digitized fingerprint records that can be searched against each other. Because Mayfield had been in the United States Armed Services, his fingerprints were on file at the national level. On March 19 of 2004, eight days after the Madrid train bombing, the FBI's Latent Print Unit identified Brandon Mayfield as the source of the fingerprint they recovered from the plastic bag; they cataloged that fingerprint as LFP 17, which stands for latent fingerprint, number 17.

In actuality, the IAFIS search had generated a list of about 15 to 20 most likely fingerprint candidates; Mayfield was only one of those possibilities. Now, IAFIS is automated, but its results are not definitive proof. Any system-suggested matches are always followed up with a comparison by the eyes of a qualified fingerprint examiner. In the course of his side-by-side examinations, FBI agent Terry Green reported over 15 points of similarity between Mayfield's known print and the latent print in question, concluding that Brandon Mayfield was the source of the unknown fingerprint. Not only that, but a second examiner concurred with the match. The Chief of the FBI Latent Print Unit also reviewed the comparison. All three confirmed that LFP 17 came from Mayfield.

So that's what prompted the FBI to start 24-hour surveillance on Mayfield, and conduct the clandestine physical searches of both his residence and his law office. The Mayfield family wasn't imagining things at home. The Foreign Intelligence Surveillance Act of 1978, commonly known as FISA, had been used as the legal authority to investigate Brandon Mayfield as potentially being one of the perpetrators of the Madrid train bombings. Further justification came from the USA Patriot Act of 2001, which most people don't realize is an acronym that stands for Uniting and Strengthening America by Providing Appropriate Tools Required to Intercept and Obstruct Terrorism. Whew, I'll bet somebody stayed up all night coming up with that one!

The focus on Mayfield not only included his home and office, but also the family farm on which he grew up in Kansas. The search resulted from what's called a sneak and peek warrant. It allowed FBI agents to photograph papers at the Mayfield house, download their computer hard drives, and install sound-monitoring devices to eavesdrop on the family's conversations. The FBI also gathered Mayfield's bank and telephone records. Despite the Fourth Amendment, which requires probable cause to get a warrant, FISA allows these sneak and peek activities when such covert operations will let authorities collect information on who they consider foreign agents; that includes U.S. citizens, so long as the main purpose of the surveillance is to gather foreign intelligence. The Patriot Act of 2001 allowed even greater freedom for law enforcement to monitor U.S. citizens than did FISA did.

I know a retired FBI intelligence agent, and he told me about some of these procedures during those covert sneak and peek operations. First, the agents monitor the suspect. Once the person's habits and whereabouts are reasonably nailed down, they then bring in a team of specialists who are trained to get into a residence or office. Then they look around, set up monitoring devices, and leave without a trace. The team consists of lock experts, guys that install listening devices, people that can plant hidden cameras, computer forensics experts, and even what they call a closer, who brings plaster, wood putty, and a painting kit, to disguise any structural damage done by installing surveillance equipment. It really is pretty much like what we see in TV or on the movies. They post a lookout, which is probably why the Mayfield's window blinds had been disturbed; then the rest of the FBI team quickly goes about setting up their gear and then gets out. Mayfield ultimately told the media that the Portland FBI agents seemed pretty inept to him, since the couple immediately recognized that they were being watched.

OK, back to the fingerprint. Within about three weeks of identifying the print as Mayfield's, the FBI was contacted by the Spanish National Police. The Spanish authorities had independently determined that the print was, "conclusively negative" as a possible match to Mayfield. Nonetheless, the FBI prepared an affidavit that was presented to a U.S. District Court, and an arrest warrant was issued for Brandon Mayfield as what they termed a material witness in the Madrid bombing case. The FBI even went so far as to send an agent to Madrid to suggest that the Spanish National Police reexamine the fingerprint, in light of all the things the FBI knew about Mayfield.

See, the fingerprint wasn't the only evidence the FBI was acting on to arrest Mayfield. It's been suggested they also considered Mayfield's regular attendance at his local mosque, as well as the fact that his wife had placed a phone call to an Islamic charity that was being watched by the government for possible terrorist activities. Mayfield also advertised his law practice in a newsletter run by a guy that the FBI was already looking at for suspected links to terrorism.

Brandon Mayfield was arrested at his law office on May 6, 2004. Simultaneously, FBI agents came to Mona Mayfield's door and said her husband's fingerprints had been found on a bag in Madrid near the site of the bombings. She said she had two thoughts; either this is some kind terrible mistake, or the FBI is trying to frame my husband. They detained and questioned Mona and conducted a six-hour search of the couple's property. This time they didn't just download the family computer hard drives, they took their computers and modem, as well as the key to Mayfield's safety deposit box at his local bank. Among other things, they also confiscated several copies of the Koran from the home and papers containing Spanish writing, which later turned out to be one of the children's homework from a Spanish class.

In the affidavit they used as justification for the arrest and search, the FBI claimed "100 percent positive identification" of Mayfield from the fingerprint. The affidavit also had Agent Green reporting the Spanish National Police agency as being satisfied with the FBI's conclusion, which was not true. At his hearing Mayfield stated, "That's not my fingerprint, your Honor." He said he hadn't been out of the country in several years, over 10. And he produced his passport that had expired in 2003. An investigation of airline record systems showed no record of a Brandon Mayfield traveling overseas. The FBI countered, he could have used false documents and a pseudonym, even though Mayfield's whereabouts could be accounted for by family, friends, and clients during the timeframe it would have taken him to get to Spain and back, around the time of the bombing.

Spain denied the FBI's story and said the Americans simply refused the Spanish National Police's conclusions. In fact, Commissioner Carlos Corrales said he was surprised by the FBI's focus on Mayfield and stated, "It seemed as though they had something against him and wanted to involve us." After further investigation of the fingerprint, on May 19, Spanish authorities arrested an Algerian named Daoud, who not only was a definitive match to the print, but was also linked in other ways to the bombings. Once Spain issued an international arrest warrant for Daoud, U.S. authorities dismissed Mayfield's case. He was released from his two weeks of jail time, without being charged.

So what went wrong here, and how did this mistake happen within one of the most sophisticated forensic laboratories in the world? In March of 2006, a 273-page report was released by the office of Inspector General Glenn Fine, regarding the misidentification of Brandon Mayfield as being involved in the Madrid train bombing. While the report concluded the FBI investigators did not abuse their powers, it did specify that by providing flawed information to multiple law enforcement agencies that the U.S. Patriot Act might have complicated an already bad situation. The IG's Office acknowledged that there certainly were similarities between the latent print found near the crime scene and Brandon Mayfield's fingerprint. But, the report stated the FBI may have been over reliant on its own expertise, especially by ignoring the opposing conclusion of the Spanish authorities, and that perhaps the FBI had been over confident in IAFIS. Ultimately, though, FBI investigators, not the database, positively linked Mayfield to the fingerprint. IAFIS technology merely suggested him among a group of possibilities.

The IG also speculated whether Mayfield's religion played a part in the misidentification of the fingerprint. The report concluded that at the time of their examination the FBI fingerprint experts were not aware that Mayfield was Muslim or that he was married to a Middle Eastern woman and had represented a member of the Portland Seven terrorist cell in a legal matter. But the issue wasn't as clear with regard to the rest of the FBI's investigation or its failure to revisit the fingerprint comparison once the Spanish National Police concluded the print didn't come from Mayfield, since by that time the FBI was aware of all Mayfield's particulars.

The IG report stated there were conflicting accounts among FBI personnel they interviewed, and others, as to whether religion played a role in the Mayfield investigation. Perhaps it was more his association with other suspected terrorists that was to blame for the FBI's focus on Mayfield. In the end, though, the IG's report concluded it was the belief that the fingerprint belonged to Mayfield—not his religion—that caused the nation's highest policing agency to go after him.

The FBI defended itself by suggesting the misidentification was because they had been working from a relatively poor digital copy of the fingerprint, not the original. They also claimed the print wasn't really suitable for identification, that there were gaps within the fingerprint caused either by creases in the surface of the bag or multiple touches to the bag, or maybe both. But to counter that claim, the Inspector General's report argued that the Spanish National Police, and later the FBI itself, concluded the print belonged to Daoud. So how could it not be of sufficient quality for an ID?

It was also noted in the IG's report that the latent print examiners gave too much weight to what we call minutiae in the fingerprint comparison. That refers to the smaller aspects of the print, like sweat gland pores and how and where print ridges end or split. This is like losing the forest in the trees, since once investigators were vested in the minutiae, they didn't back up again for the big picture. Mayfield and Daoud's prints were strikingly similar in the smallest of details, but they were clearly different in other, more meaningful ways.

A print match has to be 100 percent certain; any discrepancy that can't be explained away by the examiner is grounds for an exclusion. It was later shown that the FBI realized that a full one-quarter, the upper left quadrant, of LFP 17, it didn't match Mayfield's pattern. But they ignored this in favor of assuming that someone else touched the bag in that exact place, creating a different pattern in that quarter of the fingerprint. All told, the IG's report said the multiple alternative theories that the FBI had to construct and string together in order to conclude the LFP 17 belonged to Mayfield, simply rose to such a level of unlikelihood that the potential match should have been dismissed. The Inspector General's report also argued that once the examiners thought the latent print belonged to Mayfield, they started to work backward, using Mayfield's clear print to look for its similarities to the poorer quality unknown print, even though the latent print lacked detail and clarity. But that's just not a sound scientific way to go about that kind of high-stakes comparison.

The IG's report recommended several changes to the FBI laboratory procedures with regard to fingerprint identification. As part of its own self-analysis, the FBI likewise found no misconduct on the part of its employees, and, to its credit, had already implemented changes in its practice, even before the Inspector General's report was declassified and released to the public.

In 2004, Brandon Mayfield filed a lawsuit against the United States Department of Justice, the Federal Bureau of Investigation, and several specific employees of the FBI. The suit claimed that Mayfield's civil rights were violated by these agencies, and persons, in their pursuit of him as a suspect in the Madrid train bombings, and, that his arrest was directly related to the fact that he is a Muslim. Mayfield won his suit, and was awarded $2 million, along with a rare official apology from the FBI, in exchange for dropping his claims against the government and its agents.

Without a doubt, the IAFIS fingerprint database and comparison system is extremely useful as a forensic tool. But even the most sophisticated technology in the world cannot completely save us from human error or bias. And while the U.S. Inspector General did not implicate either the technology or the personnel in the Brandon Mayfield case, as we saw with Joyce Gilchrist, there have been forensic scientists who were clearly guilty of not mere mistakes, but rather serious, inexcusable malpractice.

Political Assassinations

Lecture 17

In this lecture, we'll look at some cases from around the world that involve the intersection of political intrigue and forensic science. We'll travel to Bulgaria, Great Britain, Sweden, the Gaza Strip, and Switzerland, among other stops, to look at three political assassinations that remain worldwide mysteries. Specifically, we'll investigate the poisoning of the anticommunist journalist Georgi Markov, the shooting of controversial Swedish prime minister Olof Palme, and the poisoning of PLO leader Yasser Arafat.

Georgi Markov and the Poisoned Umbrella

- Georgi Markov was born near Sofia, Bulgaria, in 1929 and, after studying chemical engineering, became a teacher at a technical college; he also dabbled in writing. After 1954, when communist Todor Zhivkov became the leader of Bulgaria, Markov's avocation as a playwright and author began to expand into more subversive and dissident writings, particularly against communism.

- Increasingly, Markov's political and satirical writing garnered disapproval from the strict Bulgarian government. In the early 1970s, the author left Bulgaria and settled in London with a job as a commentator on BBC radio. Markov was ordered to return to his home country, but he refused and was granted political asylum by Great Britain. He continued his work at the BBC and Radio Free Europe, despite being convicted in absentia in Bulgaria.

- To stop Markov's anticommunist broadcasts, Zhivkov enlisted the help of his minister of the interior, General Dimitar Stoyanov. In June 1977, Stoyanov submitted a lengthy report to the Soviet Union's Politburo about the "enemy emigration" occurring out of Bulgaria and its damaging ramifications to state security. A secret agreement was made in which Stoyanov was given permission to contact Yury Andropov, head of the KGB, to provide "technical support" in the killing of Markov.

- On September 7, 1978, while Markov waited for his bus to work, he suddenly felt a stinging pain in the back of his right thigh. He turned around and saw a heavyset man behind him bend over and pick up an umbrella. The stranger muttered "sorry" with a thick accent before getting into a taxi.

- That evening, Markov noticed redness and swelling on the back of his thigh. He developed a high fever and was admitted to the hospital the next day. Doctors were stumped by his symptoms, initially diagnosing him with blood poisoning. Markov died three days later of massive heart and organ failure, leaving behind a wife and a 2-year-old daughter. He was 49.

- Because Markov had reported earlier death threats he'd received to his coworkers, Scotland Yard ordered a full autopsy. Forensic pathologists noted fluid in his lungs and small hemorrhages dotting many of his organs. His liver showed signs of poisoning, but toxicology uncovered no obvious lethal substance in his body. An examination of Markov's blood showed extremely high white cell counts, which are often seen in serious bacterial or viral infections, some cases of drug toxicity, and other causes of severe physical stress.

- During tissue sampling, a block of flesh was removed from the back of Markov's right thigh, around the site of the initial injury. In it, forensic scientists discovered a metal pellet about 1.5 millimeters in diameter. The pellet had no clear traces of a poison, but doctors suspected that it had contained some type of natural toxin that dissipated in Markov's body. Ricin, a slow-acting toxin whose symptoms mimic other illnesses or infections, was thought to be a likely candidate.

- After the fall of the Soviet Union in 1991, it was discovered that Soviet technicians had developed a modified umbrella that could be used as a type of syringe to inject something like the pellet used on Markov. It's also possible that the umbrella may have served as a distraction while the poison pellet was delivered into Markov's thigh by a syringe.

- The London police have tried to solve the Markov murder for decades. In January 1993, a likely suspect was discovered: Francesco Gullino, a Danish nationalist of Italian descent who was living and working as an antiques dealer in Copenhagen. Gullino admitted to working for the Bulgarian secret police but denied involvement in Markov's killing. The case remains open in England.

The Assassination of Olof Palme

- Olof Palme was born in Stockholm in 1927 to a wealthy, conservative family. He earned his law degree from the University of Stockholm and, afterwards, became an active member of the Social Democratic Workers' Party of Sweden. He steadily moved up in the ranks of the party and was ultimately elected leader. In 1969, he was appointed prime minister.

- Throughout his career, Palme was dedicated to "socialism, peace and solidarity" and was a frequent critic of both Russian and U.S. policies. He became a supporter of the Palestinian Liberation Organization (PLO), a group the United States considered a radical terrorist organization until the early 1990s. Needless to say, Palme had strong political opinions and made many enemies, as well as close allies.

- As prime minister, Palme enacted numerous governmental changes with the welfare of the people in mind, but in order to adopt these reforms, he significantly raised taxes. This move probably led to his loss in the 1976 election, ending 40 years of continual rule by the Social Democratic Party. But Palme remained involved with the party and was reelected prime minister in 1982.

- Given that he was a fairly controversial and outspoken leader of a large European country, Palme was not as concerned with security as he should have been.

- On February 28, 1986, he and his wife, Lisbet, went out to a movie; on the way home, a man wearing dark clothes and a ski mask shot the prime minister in the back at point-blank range. He fired a second shot at Mrs. Palme, but she sustained only a superficial wound.

- Some witnesses reported that a car carrying two other men picked up the assailant, but no one got the vehicle's license number. An ambulance rushed Palme to the hospital, where he was pronounced dead.

- The only pieces of forensic evidence recovered in the case were bullets removed from the victim's bodies. No bullet casings were recovered at the scene, which led police to believe that the gun used was a revolver.
    - As part of the investigation, police searched Palme's apartment and offices, trying to locate wire-tapping equipment that might have tipped off the assassins to the couple's evening plans. No surveillance equipment was found.

    - There have been claims that police work at the scene was sloppy and inadequate, and it has even been alleged that right-wing members of the Swedish police force could have been involved in the assassination.

- In 1988, a man named Christer Pettersson, a known drug addict and criminal, was named as a suspect. Lisbet Palme picked Pettersson out of a police lineup, and he was convicted primarily on her testimony. Less than a year later, Pettersson was released for lack of evidence. At one point, Pettersson confessed to shooting the prime minister but later recanted. He died in 2004 and remains the only person ever arrested for Palme's murder.

- More than 100 suspects have been brought up during the investigation of the Palme assassination, including supporters of apartheid in South Africa, Israeli Mossad assassins, right-wing

Chilean fascists, the Yugoslavian secret service, and even the CIA. The case has never been solved.

The Poisoning of Yasser Arafat

- Yasser Arafat was born in Cairo in 1929. His father was a Palestinian from the Gaza Strip, and his mother was from Jerusalem. She died when her son was only 4 years old, and he was sent to the Old City of Jerusalem to be raised by relatives.

- As a young man, Arafat studied Judaism and Zionism but ultimately became an Arab citizen and began to help smuggle arms into Palestine, which at that time was under British control. During the Arab-Israeli War of 1948, Arafat aligned himself with the Muslim Brotherhood. As a newly graduated civil engineer, he was called to fight in the 1956 Suez crisis, in which Egypt stood up to British, French, and Israeli control of the Suez Canal.

- Arafat was part of the guerrilla force in Gaza that was ejected when President Nasser of Egypt allowed the United Nations to take over the disputed area. Arafat went to Kuwait, where he and other Palestinians founded the Fatah movement. Fatah's ideology was to have Palestinians take back their land on their own, not through support from any established Arab government. The movement financed its efforts by making alliances with major businessmen, particularly those in the oil trade.

- In the early 1960s, Arafat moved to Syria and began to persuade fighters from the PLO's military arm to join him; from there, he crossed into the Jordanian-occupied West Bank to recruit more supporters. By 1969, Arafat had become chairman of the PLO, and for many decades, he served as the international symbol of the Palestinian struggle for independence.

- Arafat alternated between negotiation and violence in his politics. After the signing of the Israeli-Palestinian Declaration of Principles, he shared the 1994 Nobel Peace Prize with two Israeli officials. But he later participated in further violence against Israeli occupation of the West Bank, Gaza Strip, and east Jerusalem. Ultimately, the Israelis put his compound under siege, and Arafat was held there in confinement beginning in 2002.

- In October 2004, at age 75, Arafat suddenly fell ill with what was thought to be stomach flu and was airlifted to a military hospital in France. He went into a coma and died about a month later. The official cause of death was a massive brain hemorrhage that resulted in a stroke. Despite questions surrounding his death, his wife would not allow an autopsy.

- Eight years after his death, in the summer of 2012, an Al Jazeera television station apparently convinced Mrs. Arafat to provide some of her husband's belongings for forensic analysis. The Swiss Institute of Radiation Physics found levels of the radioactive poison polonium-210 in Arafat's clothing. Polonium is one of the rarest elements in nature; if ingested, an amount the size of just one grain of sand can kill.

- After the discovery of the polonium, Palestinian officials reopened the investigation into Arafat's death. In November 2012, forensic experts took samples of Arafat's remains to be analyzed independently by teams from Switzerland, France, and Russia. A year after the exhumation, Swiss scientists reported that evidence supports the likelihood that Arafat was poisoned by polonium. Just as with the assassinations of Georgi Markov and Olof Palme, Arafat's cause and manner of death are finally known, but the identity of the perpetrators remains a mystery.

Suggested Reading

Blum, *The Poisoner's Handbook.*

Bondeson, *Blood on the Snow.*

Emsley, *Molecules of Murder.*

Sifakis, *Encyclopedia of Assassinations.*

Questions to Consider

1. Can you name other political assassinations than the three covered in this lecture?

2. Do you know of other cases where remains of potential or known assassination victims have been exhumed for further testing and analysis?

Political Assassinations

Lecture 17—Transcript

Georgi Markov was born near Sofia, Bulgaria in 1929, and after studying chemical engineering, became a teacher at a technical college; he also dabbled in writing. This was just after World War II, in a time when Bulgaria, as well as most of Eastern Europe, came under the Communist rule of the Soviet Union. There was great fear of Fascism and the Nazis, as well as the widespread civil and social destruction that marked post-war Europe that apparently led many people to think that Communism, rather than capitalism, would help the European economy. After 1954, when Communist Todor Zhivkov became the leader of Bulgaria, Georgi Markov's avocation as a playwright and author began to expand into more subversive and dissident writings, particularly against Communism.

Markov won literary awards in his native Bulgaria, but increasingly, especially during the Cold War era of the 1960s, the author's political and satirical writing won disapproval from the strict government. Markov felt pressure to leave Bulgaria, and ultimately settled in London in the early 1970s with a job as a commentator on BBC radio. His weekly program, *Distant Reports about Bulgaria*, caused him to be censored in his home country and led Zhivkov to order Markov to return back to Bulgaria. Markov refused and was granted political asylum by Great Britain, where he continued his work at the BBC and Radio Free Europe, despite being convicted, in absentia, in Bulgaria.

Zhivkov made a decision that he had to find a way to stop Markov's anti-Communist broadcasts, so he enlisted the help of his Minister of the Interior, General Dimiter Stoyanov. The pair gave Markov the code name Wanderer, sometimes seen translated as Vagrant. And in June of 1977, General Stoyanov submitted a lengthy report to the Soviet Union's Politburo about, "the enemy emigration" that he saw occurring out of Bulgaria and its damaging ramifications on state security. A secret agreement was made giving General Stoyanov permission on behalf of Bulgarian leader Zhivkov to contact Yuri Andropov, who was at that time head of the KGB, which was the Soviet Union's security agency. The KGB was to give, "technical support" in the killing of Georgi Markov.

In early 1978, Markov began receiving phone threats, and although he didn't seem too concerned, he did mention them to his coworkers at the BBC. By the summer of 1978, Markov felt the threats were becoming more serious and increasingly sinister. The alleged final threat stated, "You will not become a martyr. You will simply die of natural causes. You will be killed by a poison that the West cannot detect or treat."

On Bulgarian leader Zhivkov's 67th birthday, September 7, 1978, while Markov waited at London's Waterloo Bridge Bus Station, headed to work, he suddenly felt a stinging pain in the back of his right thigh. He turned around and saw a heavyset man behind him bend over and pick up an umbrella. The stranger muttered "sorry" with a thick accent, before getting into a taxi. When Markov got to work, he mentioned what had happened to his coworkers.

That evening, Markov noticed redness and swelling on the back of his thigh where he had felt that stinging sensation earlier. He developed a high fever throughout the night, and Markov was admitted to the hospital the next day. The doctors were stumped by his symptoms; their original diagnosis was blood poisoning. The thinking was he had developed septicemia from some kind of bacterial infection. But Markov died three days later, on September 11, 1978, age 49, of massive heart and organ failure, leaving behind a wife and two-year-old daughter.

Because of what Markov had reported to coworkers and police, Scotland Yard ordered that he receive a full autopsy, which was conducted at a public mortuary in London the next day. Forensic pathologists noted fluid in his lungs, commonly seen when the heart is failing, and small hemorrhages dotted many of Markov's organs. His liver showed the signs of poisoning, but toxicology uncovered no obvious lethal substance in his body. An examination of Markov's blood showed extremely high white cell counts, the kind often seen in serious bacterial or viral infections, or in cases of drug toxicity and other disorders that cause severe physical stress.

During tissue sampling, a block of flesh was removed from the back of Markov's right thigh, around the site of the initial injury. In it, forensic scientists discovered a tiny metal pellet, about 1.5 millimeters in diameter, that's slightly smaller than a pinhead. The pellet itself was pretty sophisticated; it was made of 90 percent platinum and 10 percent iridium, and was later discovered to be a watchmaker's bearing. It had two tiny, round tunnels drilled through it by a laser in an X-shaped pattern that intersected in the middle. The pellet had no clear traces of a poison, but the doctors suspected it had contained some type of natural toxin that dissipated in Markov's body.

By process of elimination, ricin seemed the most likely poison. Investigators figured that the ricin was put into the holes in the pellet and then the platinum was coated in something. The coating had to be a substance that melted when it reached body temperature, allowing the toxin to enter Markov's bloodstream. Ricin is a naturally-occurring substance produced in castor beans and can be isolated during the extraction of castor oil from the beans. Although the oil itself is easily extracted from the plant, the ricin is much more difficult to purify into a form that can be used as a poison. An amount of ricin as tiny as a few crystals of table salt can kill an adult. Ricin is a slow-acting toxin, and it mimics other illnesses or infections. At the time of Markov's poisoning, there was no known antidote, although since then, some antitoxins have been developed by the U.K. and U.S. militaries, but those have had very limited human testing so far.

In terms of how it acts on living cells, ricin first penetrates the cell membrane and then prevents the cell from manufacturing proteins. These proteins include necessary enzymes, some hormones, and other important biological molecules. Due to its effects on cells, which make up all body tissues, the eventual result of ricin poisoning is complete organ failure and death. German scientist Peter Stillmark first found ricin to be toxic in 1888. Later, during World War I, ricin was actually considered as a possible weapon of mass destruction, but was discovered not to be a good choice for two main reasons; it can't tolerate extremely high temperatures of over about 176 degrees Fahrenheit, and, at that time, there was absolutely no antidote.

Ricin poisoning has been used, or attempted, in small-scale terrorist and assassin operations, like the incident that killed Georgi Markov. In fact, in 2013, ricin-laced letters were sent to U.S. President Barack Obama, New York City Mayor Michael Bloomberg, and a gun-control group. Small-time actress Shannon Guess Richardson allegedly sent the letters, hoping to frame her estranged husband as having sent them. Now, back to the Markov case.

Another reason investigators suspected ricin as the poison in the Markov case, and why they believed a coating of some type had to be used, relates to another Bulgarian exile, named Vladimir Kostov. He was living in Paris and also working as a Radio Free Europe correspondent. Like Markov, the Bulgarian secret service was after Kostov and had given him the code name "Judas." That's because Kostov had once worked as a Bulgarian intelligence officer. About two weeks before Markov's assassination, Kostov was getting off of an escalator in a Paris Metro station, when he felt something sting him on the back. He had no idea what happened, but went to the doctor within hours. Kostov's physician told him it might have been a wasp, since he couldn't feel anything under the man's skin.

Later that day, Kostov developed a high fever, and the area on his back got painful and swollen. He went to a different doctor who told him the wound didn't look like a wasp sting, but he had no idea what it was. Fortunately for Kostov, his fever went away and he got better. But when fellow Bulgarian exile Georgi Markov was killed and Kostov heard the story, he went to the French authorities and requested they investigate whether what had happened to him two weeks before Markov's death could somehow be connected.

At that point, Kostov was taken into protective custody and given a more thorough medical exam. X-rays revealed a small metal pellet lodged between his rib and the skin on his back. Under local anesthetic, a surgeon removed a thumb-sized chunk of skin surrounding the foreign object. A detective from Scotland Yard took the tissue sample containing the pellet back to England, where forensic scientists analyzed it. The pellet was identical to the one that had earlier been found in Markov's body and had traces of a sugary wax on its surface, which, since it hadn't melted, may have saved Kostov's life. Blood tests on Kostov showed traces of antibodies to ricin, meaning his white blood cells had created defenses to fight the poison; he just hadn't been exposed to enough of it to kill him, as it had Markov.

After the USSR fell apart in 1991, it was discovered that Soviet technicians, perhaps working in the KGB poison laboratory known as The Chamber, had developed a modified umbrella that could be used as a type of syringe to inject something like the pellet used on Markov. But some sources think the umbrella may have just served as a distraction, while the poison pellet was delivered into his thigh by some kind of syringe. There was no real damage to the jeans Markov wore and no evidence of any type of residue indicating the pellet had been injected, say, with an explosive of some type. Maybe the man behind Markov just bent down so his face wouldn't be seen and injected Markov with some kind of syringe while picking up the umbrella.

In Kostov's case, he didn't see an umbrella, but did say that when he felt the sting in his back, he heard the clattering sound of something small hitting the escalator; the sound was also reported by his wife. In either case, in a crowded setting like a public transportation hub, a certain amount of bodily contact is expected. Just a simple jostle could easily conceal whatever injection method was used to introduce pellets into both Markov and Kostov. Plus, such close proximity would ensure no mistaking or missing the target, and it wouldn't harm bystanders.

The London police have been trying to solve the Markov murder for decades. Of course, they got no help from Bulgaria until the end of Communist rule there, in 1989. But Bulgaria conducted widespread destruction of records in early 1990, before the new government took over, which would have included any paperwork related to Markov's assassination. In January of 1993, though, a likely suspect was discovered, a guy by the name of Francesco Gullino, a Danish national of Italian descent who was living and working as an antiques dealer in Copenhagen.

Gullino admitted to working for the Bulgarian secret police under the code name Agent Piccadilly but denied being involved in Markov's killing. Authorities recruited Gullino after he was caught twice smuggling drugs into Bulgaria. They apparently set him up with an alias and a job in Denmark in exchange for conducting espionage for them in Western Europe. Gullino was the only Bulgarian spy known to be in London at the time of the assassination, but investigators haven't been able to gather enough evidence to indict him. The Markov case is still open in England, but in 2008, the statute of limitations expired in Bulgaria, so officials there haven't pursued the case any further.

Now, I don't know about you, but I really didn't recognize the name Olof Palme until recently. If you're a political history buff, you may be quite familiar with the man, but if you're a reader of modern fiction, you may have heard his name first, or at least became more curious about Palme, after reading *The Girl with the Dragon Tattoo* by Stieg Larsson. That novel and its sequels were set in Sweden and include historical references, including events involving Swedish Prime Minister Olof Palme.

Palme was born in Stockholm in January of 1927 to a wealthy conservative family. After attending the University of Stockholm for a few years, he received a scholarship to finish his Bachelor's degree in 1948 at Kenyon College in my home state of Ohio. The topic of his undergraduate thesis was the United Auto Workers Labor Union, known as the UAW. After graduation, Palme spent time in Detroit working with the head of the UAW. He also traveled throughout the United States, Mexico, and later visited several third-world countries, where he witnessed many forms of social injustice that greatly influenced his political life.

After returning to the University Of Stockholm to earn his law degree, Palme became a very active member of the Social Democratic Party of Sweden. In 1953, the Prime Minister of Sweden recruited Palme as his secretary. He steadily moved up the ranks and was ultimately elected leader of the Social Democratic Party. When Prime Minister Erlander resigned in 1969, King Gustaf the VI appointed Palme the new Prime Minister of Sweden. Incidentally, several years later, under Palme's rule, a parliamentary form of democratic government was established, and the constitutional power of the king was removed. How's that for a thank you?

Throughout his career, Olof Palme was dedicated to "Socialism, Peace, and Solidarity" and was a frequent critic of both Russian and U.S. policies. He was against war in general, including the U.S.-Vietnam War, and he accepted U.S. Army deserters into Sweden. Despite that, Palme became a supporter of the Palestine Liberation Organization, known as the PLO, led by Yasser Arafat. That was a group the U.S. considered a radical terrorist organization until the early 1990s. Needless to say, Palme had strong political opinions and made many enemies, as well as close allies. In fact, by way of some titillating trivia, one of those close associates was U.S. actress and political liberal Shirley MacLaine. She met Prime Minister Palme at an anti-Vietnam war conference in New York, and the two began a long-term love affair, despite the fact that each was married.

While he was Prime Minister, Palme enacted lots of governmental changes with the welfare of the people in mind, but in order to adopt those reforms, he significantly raised taxes. This probably led to his loss in the 1976 election, ending 40 years of continual rule by the Social Democratic Party. Still, Palme remained very involved with the party and also served as a United Nations mediator in the early 1980s during the Iran-Iraq war. In 1982, Palme was reelected Prime Minister of Sweden. He is quoted as having once said, "I know that the Thatchers and the Reagans will be out in a few years. We have to survive till then."

Given he was a fairly controversial and outspoken leader of a large European country, Palme was not as concerned with security as he probably should have been. He often went out without bodyguards, and not only to sneak around with Shirley MacLaine. On the afternoon of February 28, in 1986, Palme's wife, Lisbet, called their grown son Marten from her office and talked about seeing a movie together that night. That evening, at 8:30, Olof and Lisbet left their apartment in Stockholm without their bodyguards for a short walk to the metro station, where then, they would travel to the theater and meet Marten and his wife.

After the show, the couples separated, and Olof and Lisbet started walking back to the station. At 11:21, a man wearing dark clothes and a ski mask came up behind them and shot the Prime Minister in the back, at point-blank range. He fired a second shot at Mrs. Palme, but she sustained only a superficial wound, and the gunman ran off. Some witnesses reported seeing a car carrying two other men that picked up the assailant, but no one got the vehicle's license number. An ambulance rushed Palme to the hospital, but he was pronounced dead at 12:06 on the morning of March 1, 1986, aged 59 years.

The only pieces of forensic evidence recovered in the case were bullets of the Winchester-Western .357 variety that were removed from the victims' bodies. The composition of the bullets was consistent with lead fragments in the clothing of both Prime Minister Palme and his wife. No bullet casings were recovered from the scene, which led police to believe the gun was a revolver, since they don't eject casings. The most common guns that use those types of bullets are several varieties of U.S.-manufactured Smith and Wesson .357 Magnums.

As part of the investigation, police searched Palme's apartment and both Lisbet's and Marten's offices to try to locate any wire-tapping equipment. If someone had been listening in on their conversations on the day of the murder, they would have known where and when to find the Prime Minister without his bodyguards. But no surveillance equipment was found at any of the locations they searched. There have been claims, though, that the police work at the scene was sloppy and inadequate. It's even been alleged that right-wing members of the Swedish Police Force could have actually been involved in the assassination.

In 1988, a man named Christer Pettersson became a suspect. Pettersson was a known drug addict and criminal who had been previously convicted of manslaughter in a 1970 stabbing death. Lisbet Palme picked Pettersson out of a police lineup, and he ended up being convicted, mainly on her testimony. But less than a year later, Pettersson was released for lack of evidence, especially since no weapon had been found. At one point, Pettersson actually confessed to shooting the Prime Minister, but then recanted, saying he may or may have not done it; he just really couldn't really remember. In 2004, Pettersson died of a brain hemorrhage after an epileptic seizure. He remains the only person ever arrested for the murder of Olof Palme.

It's been claimed that over 500 Smith and Wesson revolvers have been test-fired in order to make bullet comparisons to the slugs recovered from the Palme's bodies. In 2006, a tip led police to a lake in Central Sweden where they found a .357 Smith and Wesson. Based on its serial number, they figured out the gun had been stolen in a home burglary and then used in a 1983 post office robbery. Authorities had high hopes when they sent it to Sweden's National Lab of Forensic Science, but in 2007 it was announced the gun was too corroded to provide any type of confident comparison.

Well over a hundred theories and suspects have been brought up during the investigation of the Palme assassination. He vehemently opposed racism in South Africa, and several apartheid supporters have been named as suspects, but no evidence has been found to link any of them to the crime. There's also been speculation that Israeli Zionist Mossad assassins killed Palme due to his longstanding support of the Palestine Liberation Organization. Palme granted asylum to Chilean refugees after a coup overthrew President Salvador Allende in 1973, so right-wing Chilean fascists have also been implicated in Palme's death. The Yugoslavian Secret Service was known to have been following Palme and have also been suspected, same with the Kurdish PKK separatist movement from Turkey.

There's even been discussion of possible U.S. Central Intelligence Agency involvement, since Palme spoke out against the events that developed into the U.S. Iran-Contra scandal. Palme clearly made a collection of enemies over the years. Some groups have even claimed responsibility for his killing, but nothing's been proven. Sweden used to have a 25-year statute of limitations on murder, but that's been lifted because of the Palme case, which remains open to this day.

Now, a few minutes ago, I mentioned Yasser Arafat, who was the leader of the Palestine Liberation Organization from 1969 to 2004. Arafat was born in Cairo, Egypt in 1929, the same year as Georgi Markov, and just two years after the birth of Olof Palme. So all three of those men were contemporaries. They were all about the same age during events like as World War II, the rise of Communism, the Cold War between the U.S. and the Soviet Union and its allies, as well as the political conflicts in the Middle East that still boil today.

Although Arafat was born in Cairo, his father was a Palestinian from the Gaza Strip. His mother was from Jerusalem, and after she died when Arafat was only four years old, he was sent to the Old City of Jerusalem to be raised by relatives. He was in Jerusalem for several years before returning to Cairo. There, Arafat dug into studies of Judaism and Zionism to try to understand the conflicting viewpoints, but ultimately he became an Arab nationalist and began to help smuggle arms into Palestine, which at that point was under British control. During the Arab-Israeli War of 1948, Arafat aligned himself with the Muslim Brotherhood. And as a newly-graduated civil engineer,

Arafat was called to fight in the 1956 Suez Crisis, in which Egypt stood up to Britain, France, and Israel over control of the Suez Canal.

Arafat was part of the Fedayeen guerrilla force in Gaza that was thrown out when President Nasser of Egypt allowed the United Nations to take over the disputed area. Arafat went to Kuwait, where he and other Palestinians, some of whom were members of the Egyptian Muslim Brotherhood, founded the Fatah movement. Fatah's ideology was to have Palestinians take back their land on their own, not through support from any established Arab government. They financed their efforts by making alliances with major businessmen, particularly those in the oil trade.

In the early 1960s, Arafat moved to Syria and began to persuade fighters from the Palestinian Liberation Organization's military arm to join him; from there, he crossed into the Jordanian-occupied West Bank to recruit even more supporters. Regardless of where they lived, Palestinians were increasingly being drawn to Arafat and the Fatah's Palestinian solution, that's the one that would take back the land they saw as theirs.

By 1969, Arafat had become the Chairman and leader of the PLO, and for many decades, he moved throughout the Middle East serving as the international symbol of the Palestinian struggle for independence. Arafat alternated between negotiation and violence in his politics and by the early 1990s had made his way back in the Palestinian territories of the West Bank. After the secret Oslo Accord led to the official signing of the Israeli-Palestinian Declaration of Principles in Washington, DC, Arafat shared the 1994 Nobel Peace Prize with two Israeli officials. But within years, he participated in further violence against Israeli occupation of the West Bank, the Gaza strip, and east Jerusalem. Ultimately, the Israelis took his Ramallah compound under siege, and Arafat was held there in confinement, beginning in 2002.

In October of 2004, at 75 years old, Arafat suddenly fell ill with what was thought to be stomach flu, so he was airlifted to a military hospital in France. He went into a coma and died about a month later. The official cause of death was ruled to be a massive brain hemorrhage that resulted in stroke. Despite questions surrounding his death, including rumors of a possible poisoning, his wife, Suha, didn't allow an autopsy.

But eight years after his death, in the summer of 2012, an Al Jazeera television station apparently convinced Mrs. Arafat to provide them some of her husband's belongings. The items had been used by Arafat during his one month illness, and included a hat containing some of his hair, Arafat's toothbrush, and clothing with some of his urine and bloodstains on it. Reference samples of some of Arafat's belongings used prior to his illness were also gathered. All this evidence was sent to a forensic laboratory in Switzerland.

The Swiss Institute of Radiation Physics found levels of the radioactive poison polonium-210 in Arafat's clothing, that's the same toxin that was used in the highly publicized 2006 poisoning of former KGB spy Alexander Litvinenko. Polonium is one of the rarest elements in nature; it's usually made with a particle accelerator or in a nuclear reactor, and although it can't pass through skin, an amount the size of just one grain of sand can kill a person if they ingested.

Toxicologists agreed that the timeframe typical of polonium poisoning correlated well with Arafat's one-month illness before his death. Some were skeptical, though, saying polonium's short half-life would have made it undetectable within several years after exposure. Some said the polonium was planted in Arafat's belongings long after his death to try to implicate the Israelis. The lab itself admitted there was no good chain of custody for the items brought to them through Arafat's widow.

But Palestinian officials had always suspected the Israelis poisoned Arafat. So after the polonium discovery, the investigation into his death was reopened with the full blessing of Suha Arafat and current PLO leader Mahmoud Abbas. Mrs. Arafat went on the record saying only a country with access to nuclear elements could have carried out this poisoning. Although Israel is not one of the eight nuclear-state nations, it's widely believed that Israel does have nuclear capabilities. Still, given that Arafat was holed up in his West Bank compound for over two years before he died, only interacting with fellow Palestinians, it's hard to imagine how the poison could have been slipped into his food or drink.

Despite cultural taboos and great statements of protest, in November of 2012, several teams of forensic experts went to Yasser Arafat's mausoleum in the West Bank city of Ramallah. It was decided that the remains would be sampled, and teams from Switzerland, Russia, and France would independently analyze Arafat's tissues in their respective forensic labs. The grave was closed the same day after the sampling of his body took place.

In November of 2013, a year after the exhumation, the Swiss scientists reported that Arafat was likely poisoned by polonium. Traces of the poison were found both in the PLO leader's body and in the soil around his remains. The Swiss reported amounts of polonium found in his rib were in the range of 18 to 36 times greater than incidental values.

Then, a month later, the results of the Russian testing were revealed. Their lab found absolutely no traces of polonium in Arafat's tissues. As for the French lab, well, it did find traces of polonium but ruled they were at natural levels, likely due to environmental influences, not poisoning. So despite exhuming Arafat's body and testing by three different laboratories, his death has so far not been proven an assassination, and his cause of death still remains a forensic mystery. But maybe that depends on who you ask.

Mysteries of the Romanov Family

Lecture 18

Few political assassinations have involved targeting an entire family, but when power is handed down within bloodlines from generation to generation, one way to effect political change is to eliminate the family to which the power belongs. That was the case in 1918 for the House of Holstein-Gottorp-Romanov in Russia. In this lecture, we'll look at the execution of the Romanov family, the mysteries surrounding the location of the bodies, and new evidence discovered about the case in the 1990s. We'll also see the scientific process by which anthropologists identified the remains of the Romanovs after almost 75 years.

Execution of the Romanovs

- In 1917, Russia was not only fighting in World War I, but it was also heading into its own civil war. Czar Nicholas II had been in power since late 1894, but in 1917, he was on the front and seemed to be losing touch with the suffering of the Russian people back home. Nicholas had left his wife, Alexandra, in charge, but many Russians despised her because she was part German and part English; some even believed that she was a German spy.
    - Alexandra was also a carrier of hemophilia, a disease she had passed on to the couple's only son, Alexei.
    - Alexandra became close to the Russian mystic and healer Grigory Rasputin, believing that he could cure Alexei. Rasputin himself was murdered in 1916.
- Internal unrest of the Russian people ultimately resulted in the February Revolution in 1917, which led to the house arrest of the Romanov family and the forced abdication of Nicholas. The Bolshevik, or "Red," Party, headed by Vladimir Lenin, was rising in Russia. The anti-Bolshevik supporters of the czar were known as the "Whites."

- In October 1917, the Bolsheviks moved the czar and his family to a facility in Yekaterinburg, which had become the Red capital. Initially, the Bolsheviks planned to bring Nicholas to trial, but the White Army gained power and, in the summer of 1918, began to close in on Yekaterinburg; thus, the Reds changed their plan. There was too much to lose if the White Army somehow managed to free the Romanovs and reinstate Nicholas as ruler.

According to later accounts, Czar Nicholas and Alexei were shot first, before a firing squad began killing the other family members and servants.

- On the night of July 16, 1918, Yakov Yurovsky, the chief Bolshevik guard at the house where the Romanovs were being held, ordered the royals and their servants to the basement.
    - In total, there were seven members of the Romanov family: Nicholas and Alexandra; their four daughters, Olga, Tatiana, Maria, and Anastasia; and the 13-year-old Alexei. The four servants included the family physician, maid, valet, and cook.
    - Historical references don't agree on all the details, but the consensus is that once in the basement room, Yurovsky told the assembled Romanovs and their servants that they had been gathered together to be executed. Yurovsky shot Nicholas first, then turned his gun on Alexei.
    - A waiting firing squad entered the room and began shooting the others. Although the executioners were told to aim for the heart, bullets ricocheted around the room madly. Because of

the danger to the executioners themselves, a decision was made to cease fire and bayonet the survivors. The dead were carried out of the building about 20 minutes after the first shots.

Mysteries in the Aftermath

- Although the execution of the czar was almost immediately publicized around the world, the gory details were held back. Early accounts made no mention of the murders of the rest of the family. Alexandra and Alexei were supposedly being held in "a place of security." One of the reasons for the confusion was that Yurovsky ordered the disposal of the bodies in a secret location.

- Over the years, many versions of what happened to the bodies circulated. One 1935 account came from the American adventurer and author Richard Halliburton, who traveled to Russia to interview one of the alleged firing-squad members, Peter Ermakov. Ermakov told Halliburton that the bodies had been taken to an abandoned mineshaft near Yekaterinburg. There, the clothing and bodies had been burned.

- Despite the Bolsheviks' best efforts, the White Army overtook Yekaterinburg about a week after the murders. When troops found the basement room, the crime scene had been scrubbed clean, but there were what looked to be bullet holes in the walls. The Whites also sifted through the ashes at the mine and found clothing and other items belonging to the victims. The only trace of human remains found there was a finger thought to belong to a middle-aged woman; perhaps it had been cut off to remove one of Alexandra's rings.

- Among the spinoff mysteries related to the inability to locate the Romanov remains were rumors that some members of the imperial family had survived, especially the children. Over the decades following the murders, many imposters arose, including those claiming to be Alexei or Anastasia, the youngest of the four Romanov daughters.
  - Following a failed suicide attempt in Berlin in 1920, a woman known as Anna Anderson claimed that she was Anastasia. A

private investigator who did a background check on the woman identified her as a Polish factory worker named Franziska Schanzkowska, who had suffered from mental illness after a head injury.

- Anderson spent the rest of her life insisting that she was Anastasia. She traveled around the United States and Germany, gaining notoriety and media attention.

- In the late 1960s, a Charlottesville, Virginia, history professor who took an interest in Anderson's story married her to allow her to permanently relocate to the United States. The pair lived an eccentric life, and Anna died in Virginia in 1984; her body was cremated the day after her death.

Discovery and Identification

- In 1991, after the collapse of the Soviet Union, some previously unreleased government documents offered a different story of the events surrounding the disposal of the Romanov bodies. Apparently, the plan to burn the remains at the mine was changed at the last minute. But while the bodies were being driven to the new location, the truck got stuck in the mud, and the Bolshevik soldiers dug a pit and hastily burned the remains along the roadside.

- Following this lead about the disposal site, searchers discovered a 6-by-8-foot common grave that they immediately suspected contained the Romanov family. About 1,000 bones and bone fragments were recovered from the shallow pit. However, the burial contained only 9 skeletons—not the 11 that would have accounted for the Romanov family and their servants.

- Initial anthropology assessments suggested that the group consisted of five females and four males. Anthropologists use pelvic bones, skulls, and bone measurements to judge whether a skeleton is male or female, but that assessment can be made only for the remains of individuals who have undergone the typical skeletal changes associated with puberty.

- To determine which five of the six executed women were present and which four of the five men, anthropologists looked for signs of age, using both skeletal and dental development. Based on this analysis, all four males were adults and were assumed to be Nicholas (age 50), the family doctor (age 53), the valet (age 61), and the cook (age 48). None of the skeletons appeared to be from a child near age 13.

- To further identify the men, scientists turned to dental work—about which at least some facts were known—and photographic superimposition. In this process, known images of a person are compared to a skull to identify similarities between specific dimensions of the bones and the photo. Such comparisons focus on features that don't tend to change over the course of life, such as the distance between eye orbits, the width of the face, and so on.

- The five female skeletons were all from adults, but two were obviously older than the others, probably Alexandra (age 46) and the maid (age 40). One had platinum dental work, and Alexandra's diaries mentioned visits to the dentist. The other had substantial wear facets on the bones of her ankles, a sign of work done while on her hands and knees; she was assumed to be the maid.

- The four Romanov daughters were just five years apart in age, ranging from 22 to 17. Two of the young skeletons looked older than the third and, combined with photographic superimposition, were thought to be Olga (age 22) and Tatiana (age 21). The third skeleton showed features suggesting that it belonged to either Anastasia (age 17) or Maria (age 19), but scientists disagreed on that point. One of the sisters, along with Alexei, was definitely missing from the grave.

- The next step was DNA testing, the nuclear form of which was first used in forensics in 1986; it was still a relatively new technique for identification at the time the skeletons were discovered. Particularly new was mitochondrial DNA testing, which is especially useful on hair, bones, and other badly degraded samples.

- In 1992, samples were taken from fragments of teeth and bones of the nine skeletons from the suspected Romanov grave to be compared with known royal family reference samples. The results, released in 1993, showed that one set of remains belonged to Nicholas and another to Alexandra, and three were from the couple's daughters. The four other skeletons were not related to the Romanovs and were presumed to be the servants.

- The reference samples for Alexandra were taken from Prince Phillip, husband of Britain's Queen Elizabeth and the maternal grandson of Alexandra's sister. Because mitochondrial DNA is passed from a mother to all her children—whether female or male—and through the women down their entire bloodline, Prince Phillip's mitochondrial DNA exactly matched Alexandra's and those of her daughters.

- Confirming Nicholas's identity required the exhumation of his brother. Grand Duke George was the original heir to the Russian throne, but he died in 1899 at age 28. DNA samples from living relatives also confirmed that one skeleton belonged to Nicholas II.

- Richard Halliburton's 1935 publication provided graphic details of the assassination, relayed by Peter Ermakov, including the sequence of the killings and the weapons used. All nine skeletons showed facial fracturing, confirming that they had probably been hit with rifle butts. Some of the gunshot and bayonet wounds were also consistent with Ermakov's account.

- After the forensic analyses, the Romanov family was buried in St. Petersburg without the two missing children. But researchers continued to dissect Yurovsky's writings, and learned that two of the bodies had been buried separately from the others. In 2007, near the site discovered in 1991, metal detectors led investigators to a grave containing about 44 bone fragments and numerous bullets. In 2008, results of DNA testing showed that the two remaining Romanov siblings had at last been found.

Suggested Reading

Farquhar, *A Treasury of Deception.*

Maples and Browning, *Dead Men Do Tell Tales.*

Massie, *The Romanovs.*

Murray, *Forensic Identification.*

Primorac and Schanfield, eds., *Forensic DNA Applications.*

Rathbun and Buikstra, *Human Identification.*

Sifakis, *Encyclopedia of Assassinations.*

Slater, *The Many Deaths of Tsar Nicholas II.*

Steadman, *Hard Evidence.*

Questions to Consider

1. Do you know of other cases in which entire families have been killed for political reasons?

2. Had you heard of the Romanovs prior to this lecture? What did you know about the case, particularly with regard to the Anastasia imposters?

3. The Russian mystic and faith healer Grigory Rasputin was an advisor to the Romanov family. Can you think of other examples of a ruler or ruling family that had such a "spiritual" guide?

Mysteries of the Romanov Family

Lecture 18—Transcript

Few political assassinations have involved targeting a single, specific, and entire family. But when power is handed down within bloodlines, from generation to generation, one way to effect political change is to eliminate the family to which the power belongs. That was the case in 1918 for the House of Holstein-Gottorp-Romanov in Russia.

In 1917, Russia was not only fighting in World War I, it was also heading into its own civil war. Tsar Nicholas II had been in power since late 1894, titled Emperor and Autocrat of All the Russias. However, in 1917, he was on the military front and apparently losing touch with the tremendous suffering of his Russian people back home. Nicholas had left his wife, Alexandra, in charge of ruling from the capital, Petrograd, better known as St. Petersburg. But many Russians despised Alexandra because she was from Germany. She had been born Victoria Alix Helena Louise Beatrice, or Alix of Hesse for short, and was part German and part English; in fact, she was one of the granddaughters of England's Queen Victoria.

During World War I, Russia was part of the Allied Forces, along with the British Empire and France, later to be joined by Japan, Italy, and ultimately the United States by 1917. Germany was part of the opposing Central Powers. For this reason, some Russians even thought Alexandra was a German spy in their midst. She was also one of the female carriers of the so-called royal hemophilia that afflicted several European ruling families that were descendant from Queen Victoria.

After providing the imperial family four daughters, Alexandra finally gave birth to the couple's only son, Alexei, who was hemophiliac. Although they tried to keep Alexei's disease a secret, once it was known, it proved to be yet another reason Alexandra was not liked by the Russians, having brought the, "English disease" to their throne. She also became very close to the unusual Russian mystic and faith healer, Grigori Rasputin, whom she apparently believed could cure Alexei. Rumor was Alexandra and Rasputin were lovers, or at least that he was too closely controlling her political actions, another reason for Russians to distrust their Tsarina. Rasputin himself was murdered in 1916, a mystery in itself.

Internal unrest of the Russian people eventually resulted in the February Revolution of 1917. This led to the establishment of a weak provisional government, the house arrest of the Romanov family, and the forced abdication of Tsar Nicholas the Second. The Bolshevik, or Red Party, headed by Vladimir Lenin, was rising in Russia. The anti-Bolshevik supporters of the monarchy and the Tsar were known as the Whites.

The Bolsheviks took over and imprisoned the Tsar and his family in the old capital of Siberia in the Ural Mountains. But in October of 1917, they moved the prisoners to a facility that the Reds termed The House of Special Purpose in Yekaterinburg. This had become the Red capital. Initially the Bolsheviks planned to bring Nicholas to trial, but the White Army gained power, and in the summer of 1918 began to close in on Yekaterinburg, so the Reds changed their plan. There was just too much to lose if the White Army somehow managed to free the Romanovs and reinstate Nicholas as the ruler of Russia.

So on the night of July 16, 1918, the chief Bolshevik guard at the house where the Romanovs were being held, a man named Yakov Yurovsky, ordered the family doctor to gather the royals and their servants and lead them all to the basement. The family was told that the fighting White and Red armies were getting dangerously close to the house, so they were to be moved by truck in order to protect them.

It took a while for the royals to all dress, but shortly, they were all assembled into a small basement room with their servants. In total there were seven members of the Romanov family, Tsar Nicholas, who was 50 years of age; his 46-year-old wife, Alexandra; their four daughters, the Grand Duchesses, Olga who was 22, Tatiana age 21, 19-year-old Maria, and Anastasia who was 17, as well as the couple's 13-year-old son, and would-be heir, Tsarevich Alexei. Their four servants included the family physician, maid, valet, and cook.

Historical references don't agree on all the details, but the general consensus is that once in the basement room, a couple hours after midnight, so now

July 17, over the rumble of an idling truck out back, Chief Yurovsky told the assembled Romanovs and their servants, they were actually there to be executed. The truck engine was a ruse to help muffle the sounds. Yurovsky shot Tsar Nicholas first and then turned his gun on young Alexei. Quickly, a waiting firing squad entered the small room and began shooting the others.

Although the executioners were told to aim for the heart, bullets were ricocheting around the room, and in some cases seemingly off the bodies of the Romanov daughters. Due to the dangers to the killers themselves, a decision was made to cease fire and bayonet the survivors. When the bodies were checked for signs of life, it was discovered that about 18 pounds of diamonds and other jewels had been sewn into the bodices of the girls' dresses, which was likely what caused the deflected ammunition. About 20 minutes after the first shots, the dead were carried out of the building.

Although the execution of the Tsar, in particular, was quickly publicized in newspapers in Russia and then around the globe, the gory details were held back. Early accounts made no mention of the murders of the rest of the family. The front page of the July 20 New York Times stated, "Nicholas Romanoff, ex-Czar of Russia, was shot July 16, according to a Russian announcement by wireless today. The former Empress and Alexis Romanoff, the young heir, have been sent to a place of security."

Yurovsky ordered the bodies disposed of where no one could find them, especially not the White Army or other supporters of the Tsar. That way the Bolsheviks could tell whatever story they wanted. Plus, there would be not a single bone found to be used as a martyr's relic by the opposition. The location of the gravesite was to become a well-guarded secret. In fact, almost 20 years later, a Bolshevik official still bragged, "The world will never know what we did with them."

As you can imagine, though, many different versions of what became of the victims' bodies circulated for years. One 1935 account came from American adventurer and author Richard Halliburton, who traveled to Russia to interview one of the alleged firing squad members, named Peter Ermakov. Ermakov told Halliburton he was the man Yurovsky put in charge of finding a good place to get rid of the bodies. He chose an abandoned mineshaft a dozen miles away from Yekaterinburg. Ermakov said that the day before the killings were to occur, soldiers brought sulfuric acid, gasoline, and firewood and took them deep into the mine in preparation to completely destroy the bodies.

This is how Ermakov described this crime scene after the bodies were moved to the truck. "There was blood everywhere and slippers and pillows and handbags and odds and ends swimming around in a red lake." Ermakov said it took a couple hours to drive to the disposal site. It was there the clothing and jewels were removed from the victims. The valuables were to be sent to Moscow, but the clothing was burned right there at the mine. By that time, though, it was already getting light, so they decided to stash the 11 bodies deep in the mine and wait for sunset that evening to burn them.

According to Ermakov, that next night the bodies were moved to the mouth of the mine, where layers of logs had been used to create a funeral pyre. They put the bodies on the wood, poured sulfuric acid on the victims, then added the gasoline and set it all ablaze. Ermakov claims to have personally watched to be sure not a piece of physical evidence of the bodies was left. He said if something came off the pile, he pushed it back on, and kept adding more gasoline to make sure the fire continued until there was nothing left but ash. He then claimed the ashes were gathered into buckets, which were taken up to a high place and flung into the wind. Then the next day it rained. Here's a quote that Halliburton attributed to Ermakov, a little over 15 years after the killings. "So, if anybody says he has seen a Romanoff or a piece of a dead one, tell him about the ashes, and the wind, and the rain."

Despite the Bolsheviks' best efforts, the White Army was able to overtake Yekaterinburg, but sadly for the Romanovs, that was about a week after the murders. When troops found the basement room, the crime scene had been scrubbed clean. Remember, the world didn't yet know the entire family had been killed, and that amount of blood described by Ermakov could not have come from only executing the Tsar. But at that basement crime scene, there were reportedly what looked like bullet holes in the walls, as well as scratches on the wallpaper, possibly from bayonet blades.

Members of the White Army had heard about the possible use of the old mine outside of town as the dumpsite for the bodies. So they rode out there and found the remnants of two bonfires. Sifting through the ashes, they found 65 pieces of physical evidence. Among them were two belt buckles, fragments from two military hats, and other regalia known to be worn either by the Tsar or his son. From the women they found some jewelry, shoe buckles, hooks and buttons from corsets, and an eyeglass case that was known to be Tsarina Alexandra's. They also recovered an eyeglass lens and an upper denture, thought to belong to the doctor. So the clothing was clearly burned there. The remains of a small dog were also found; the family had at least one dog, a little black dog called Ortino, while in Bolshevik captivity.

But the only trace of human remains they found at the mine was a finger they thought belonged to a middle-aged woman; perhaps it had been cut off to remove a ring from Alexandra's finger. Or maybe it was blown off of her body, since at least one source indicates grenades were used in a failed attempt to seal off the mine. So although the White Army found evidence at the mine, the bodies themselves were not discovered.

Among the spinoff mysteries related to the inability to locate the Romanov remains were rumors that some members of the imperial family had survived, especially the children. In the decades following the murders, many imposters arose claiming to be the surviving children. The most frequent contenders said they were Alexei, heir to the throne. But the most famous imposters were several alleged to be Anastasia, the youngest of the Romanov's four daughters, who was just 17 at the time of the assassinations.

One widely-held Anastasia possibility was a woman who, for at least for part of her life, was known as Anna Anderson. Following a failed 1920 suicide attempt in Berlin, and after recovering in a mental hospital, Anna claimed that despite some amnesia, she believed she was Anastasia Romanov, sole survivor of the Romanov family. Now this woman could not speak any Russian, even though the killings had only happened just a few years prior, but some said she did look a lot like the youngest Romanov daughter. So much so, in fact, that some people began to be convinced, blaming any inconsistencies in her story on amnesia related to the trauma of witnessing the killings.

Many who knew the family well kept insisting that the woman was an imposter, but just to be sure, Tsarina Alexandra's brother organized a background check. The conclusion of the private investigator was that the impersonator was a Polish factory worker named Franziska Schanzkowska, who had suffered from mental illness after a head injury at a munitions factory where she worked.

Regardless, the woman spent the rest of her days trying to prove, and insisting she was, the Grand Duchess Anastasia. She traveled around the U.S. and Germany, gaining notoriety and media attention, while living off the kindness of supporters and going in and out of asylums and hospitals. Books, plays, a ballet, and movies were written about the possibility, or impossibility, that Anna Anderson was actually the Grand Duchess Anastasia.

In the late 1960s, a Charlottesville Virginia history professor who took an interest in Anderson's story, married her to allow her to permanently relocate to the United States. The pair lived an eccentric life, including a late 1983 episode in which the woman's husband sprung her from a mental facility, resulting in a police alert that spanned 13 states. Anna Anderson, now Manahan, died in Virginia in 1984, aged 82. Her body was cremated the day after her death.

Now, let's fast forward to 1991, when the Union of Soviet Socialist Republics, or U.S.S.R., collapsed and some previously-unreleased government documents were made public. These offered a different story of the events surrounding the disposal of the Romanov bodies, 73 years prior. Apparently, the day after the killings, perhaps because of Ermakov's bragging, Yurovsky, remember, he's the Bolshevik guard chief, he was afraid too many people had heard about the plan to burn the remains of the Tsar and his family out at the old mineshaft that night.

So although the clothing had already been burned at the mine, Yurovsky made a last-minute change of plans to get rid of the bodies themselves at an alternate location. But while the remains were being driven from the mine to the new site, the truck got stuck in the mud. So the Bolshevik soldiers were forced to dig a pit and hastily bury the bodies right there, along the roadside.

They poured sulfuric acid on the victims' faces and covered the bodies with rubble, wooden planks, and the backfill from the hole they had dug.

Following the new lead about the disposal site, searchers were looking in one potential area and came across a six-by-eight-foot common grave they immediately suspected contained the Romanov family. Under the direction of Russian authorities, about 1,000 bones and bone fragments were carefully recovered from the shallow pit. However, the burial contained a total of only 9 skeletons, not the 11 that would have accounted for the seven members of the Romanov family and their four servants.

Initial anthropology assessments suggested the group consisted of five females and four males. We anthropologists use pelvic bones, skulls, and other bony landmarks, as well as lots of bone measurements, to judge whether a skeleton is male or female, but we can only do that for adult remains, well, at least for those individuals who have undergone the typical skeletal changes associated with puberty. The hormones of puberty cause lifelong changes in the skeleton that result in diagnostic features related to a person's sex.

So which four of the five executed males were present, and which five of the six women were present in that group of nine? For that assessment, the anthropologists looked for signs of age. We can use both skeletal and dental development, in particular, to gauge the age of younger individuals. We can come up with fairly tight age range for subadults. For instance, in fetal stages, we can come within months of the age. By the late teens and early '20s, we can often give an age estimation spanning just a few years. Once mature, though, sometimes remains can only be assigned to an overall category, like young adult, middle-aged adult, or older adult, although often those generalizations can be narrowed down to a couple decades. When skeletal maturation is complete, we have to look to degeneration and other typical changes that reflect advancing age.

Based on the anthropological analysis, all four males were adults, so assumed to be Tsar Nicholas age 50, the 53-year-old family doctor, the valet who was 61, and the 48-year old cook. None of the skeletons appeared to be from a child, whether male or female, near age 13, which was the age of

Tsarevich Alexei. Because only about 15 years separated the actual ages of the males in the group, to try to figure out which man was which, scientists turned to dental work, which apparently some facts were known about from these men, as well as photographic superimposition.

A photo superimposition uses known images of a person and compares them to the skull in question to look for similarities between specific dimensions of the photo and the bones. The skull can be manipulated until it's viewed from the same angle that the subject in the picture was positioned. Then the skull is photographed and an overlay of the two images can be analyzed. The comparison focuses on features that don't tend to change over life, like the distance between the eye orbits, the width of the face, and the slope of the forehead, even how well the ear placement in the photograph of the known subject matches the auditory openings in the unknown skull.

As for the five female skeletons representing the six possible women, all were from adults, but two were obviously older than the others. Two of the missing women were known to be approximately middle-aged, 46-year-old Tsarina Alexandra and the family's maid, who was 40. One had platinum dental work, and Alexandra's diaries made mention of visits to the dentist. The other skeleton showed substantial wear facets on the bones of her ankles, a sign of working while on her hands and knees, so assumed to be the maid.

But the four Romanov daughters, the Grand Duchesses, were a mere five years apart in age, ranging from 22 down to 17. Fortunately, though, during those years both skeletal and dental maturation are still being completed. So given a small group of remains, anthropologists can often seriate the individuals in order of age. Two of those skeletons looked older than the third, and, combined with photographic superimposition, were thought to be 22-year-old Olga and 21-year-old Tatiana. The third skeleton showed features suggesting it belonged to either 17-year-old Anastasia or 19-year-old Maria, but scientists brought in from around the world disagreed on that point.

A colleague of mine, the late Dr. Bill Maples, was one of the experts who analyzed the remains. Maples said his height estimate put the third young woman at about the same stature as the other two. A photograph of the girls taken about a year before the killings showed Anastasia to be noticeably shorter than either Tatiana or Marie, and slightly shorter than her oldest sister, Olga. Maples found this entry in their mother's diary, written less than a year before the murders. "Anastasia is very fat, like Marie used to be, big, thick waisted, then tiny feet. I hope she grows more." This supports that the third skeleton was not Anastasia, but the older Marie. In any event, though, the Romanov's 13-year-old son, Alexei, was definitely missing from the grave, as was one of his four sisters.

The next step, as you can probably imagine, was DNA testing. Nuclear DNA was first used in forensics in 1986, so it was still a relatively recent technique for identification at the time the skeletons were discovered. Particularly new was mitochondrial DNA testing, which is especially useful on hair, bones, and other badly degraded samples, where few, if any copies of nuclear DNA might be found. Although mitochondrial DNA testing wasn't used in court cases until 1996, it was well into studies of its usefulness in identification by the time the mass grave was found.

So in 1992, samples were taken from fragments of teeth and bones from the nine skeletons in the suspected Romanov grave, and that was to be compared with known royal family reference samples. The results, released in 1993, showed one set of remains did belong to Nicholas, another to Alexandra, and that three were from the couple's daughters. The four other skeletons were not related to the Romanovs, so presumed to be their servants.

Where did they get the family reference samples? Well, the current Prince Philip, husband of Britain's reigning Queen Elizabeth, is the maternal grandson of Tsarina Alexandra's sister. In other words, he is the grandnephew of the Tsarina, and his mother would be Anastasia's first cousin. Since mitochondrial DNA is passed from a mother to all of her children, whether male or female, and then through the women down their entire bloodline, Prince Philip's mitochondrial DNA exactly matched Alexandra and those of her daughters. But Nicholas would not have shared the some mitochondrial DNA pattern as his wife and daughters, so confirming his identity required the exhumation of his brother, Grand Duke George, buried in St. Petersburg. George had died suddenly from complications of tuberculosis when he was 28, while out riding his motorcycle in 1899. In addition to George's

DNA, samples from living relatives also confirmed one skeleton belonged to Nicholas. That skeletal DNA also matched blood on one of Nicholas's shirts that had been tucked away in a museum since an 1891 assassination attempt that happened in Japan.

But one of the imperial couple's daughters and their only son were still missing. Could the two youngest children have survived? Could Anastasia-wannabe Anna Anderson have really been the Grand Duchess after all? Anderson's body had been cremated, and she had no known children, so how could they get DNA for her? As luck would have it, scientists found a hospital tissue sample that had been kept from an intestinal surgery Anderson had in 1979. They were also able to locate a lock of her hair found in a book belonging to her husband. Not only did those two samples match each other, Anderson's DNA showed she was not in any way related to the Romanov family. And remember a private investigator had been hired by Tsarina Alexandra's family, and he said Anderson was really munitions factory worker, Franziska Schanzkowska. Researchers located the maternal grandnephew of that missing Polish woman, and his mitochondrial DNA was an exact match to the so-called Anna Anderson.

Here's another interesting thing about the Romanov grave found in 1991; it had already been discovered by a small group of Russians in May of 1979, a movie producer, an ethnographer, and a couple of geologists. They had creatively petitioned to take soil studies in the area to cover their real intention of locating the grave. They had access to some of guard Yurovsky's writings, and using those, had located the gravesite over 10 years earlier. They even removed three skulls and photographed them at the site, but then, thinking better of it, reburied them at the same place. At that time, even speaking of the Romanov execution was taboo, so amazingly, the group kept their findings a secret until the U.S.S.R. collapsed.

OK, now let's consider what's known or suspected about the victims' injuries. Halliburton's 1935 publication describes Ermakov's graphic account of the assassinations, including the sequence of the killings and the weapons used. Ermakov said there were initially only three executioners in the room and said that Yurovsky shot Nicholas in the face, and the bullet passed right through the Tsar's brain. Then Yurovsky turned and shot Alexei in the head once. Ermakov claimed that he, himself, shot Tsarina Alexandra in the mouth at a distance of only six feet. He then turned his gun on the doctor, who turned away and was shot in the neck. Next, Ermakov said he shot the cook in the body and then in the head, while the man was crouching in a corner.

Someone else in the trio besides Ermakov had already shot the valet by that time, and the third killer, a guy named Vaganov, was repeatedly firing at the girls. Ermakov said no one had apparently yet shot the maid. She was in another corner, covering herself with a couple of pillows that she had brought for the family's alleged truck evacuation. Well, the pillows were also filled with jewelry, deflecting the bullets, so Ermakov said a guard came in from the hallway and used a bayonet on her throat. At that point, the rest of the men rushed in and were repeatedly beating and bayoneting the victims, dead or alive. Rifle butts were used to smash their faces.

When they found young Alexei still moaning on the floor, Yurovsky shot him twice more in the head. Ermakov went on to claim that a total of 38 shots had been fired; I don't have any idea why he would be able to be so exact. He even knew the type of ammunition and firearms used. One of the men turned 17-year-old Anastasia over onto her back, and she cried out, so he finished her off with the butt of his rifle. It makes me almost sick to have to speak those details, but it's important to examine how Ermakov's account matches the trauma analysis for at least the nine skeletons recovered in 1991, especially given how he lied about the disposal of the bodies at the mine.

The facial skeletons of all nine showed fracturing, confirming they had likely been bashed with rifle butts. The skulls of only three of the individuals showed through-and-through gunshot wounds. The doctor was shot from left to right through his skull; perhaps Ermakov only thought he hit the man's neck. Olga was shot in the left side of her jaw and the bullet exited her forehead, so the gun would have had to be below her chin, or she was lying on the ground at the time. Tatiana was shot in the back of the head.

One definite bayonet wound went through the breastbone of who the Russians claimed was the valet. But Dr. Bill Maples believed that sternum was inventoried to the wrong body, and actually belonged to Tsar Nicholas.

A total of 14 bullets were found in the grave, two of which were still lodged in the doctor's remains. The Russian forensic experts agreed that the ammunition matched the types of weapons mentioned in the historic accounts. They also found a grenade detonator in the grave, probably from the activity at the mine. The victims' tooth enamel and skull surfaces also showed evidence of exposure to acid, just as Ermakov's story claimed.

After the forensic analyses, the Romanov family was buried in St. Petersburg in the imperial crypt of Saints Peter and Paul Cathedral, on July 17, 1998, 80 years to the day from the execution, without the two missing children. Researchers continued to dissect Yurovsky's writings and learned two of the bodies had been burned and buried separately from the others, in an attempt to throw off the body count if the White Army discovered the main grave. In 2007, about 70 meters from the site discovered in 1991, metal detectors led investigators to a grave that contained about 44 burned bone fragments and numerous bullets. In 2008, results of DNA testing showed the two remaining Romanov siblings had at last been found.

Forensics of Genocide

Lecture 19

The term *genocide* is used for the deliberate extermination of a targeted group of people—whether for political, racial, economic, or other social reasons. A key word in this definition is *deliberate*. To be considered genocide, a massive loss of life must have been intentional, unlike other types of mass fatalities, such as those stemming from natural disasters, disease, or other factors There have been many tragic examples of genocide throughout world history; in this lecture, we'll learn about some of the technology being used to locate and study mass graves, and we'll look at a case study of genocide and the exhumation of mass graves in Guatemala.

Genocide in the 20th Century

- The 20th century witnessed a number of genocides. Among these was the Turkish decimation of as many as 2 million Armenians during and after World War I, in what some say were the first organized concentration camps. Shortly after that, in the Soviet Union, about 20 million died at the hands of Joseph Stalin—from both his failed socialist agricultural projects and as targets of his regime.

- The agricultural, industrial, and social changes of Mao Zedong's Great Leap Forward also led to mass deaths from starvation throughout China in the late 1950s and early 1960s. In the Cultural Revolution that followed, anyone believed to be anti-government was imprisoned or murdered. Estimates of the number of people killed in China between 1949 and 1976 range from 50 to 70 million. We find intent in both Stalin's and Mao's agricultural genocides because the practices continued despite the knowledge that so many were dying as a result.

- Probably the most well-known genocide came at the hands of the German Nazis, who killed about two-thirds of Europe's Jews in the 12-year Holocaust between 1933 and 1945. Total estimates are that

In addition to Europe's Jews, others targeted in the Holocaust included Jehovah's Witnesses, Roma and Polish people, Russian prisoners of war, homosexuals, and people with physical or mental disabilities.

11 million people died—most by poison gas administered in Third Reich death camps.

- No one really knows the number of people who have died in North Korean prison camps since 1945 or from the longstanding practice of refusing or diverting aid from starving Korean peasants. And in just the three years from 1975 to 1978, Pol Pot's Khmer Rouge exterminated about 2.2 million in the Killing Fields of Cambodia.

- The Kurdish genocide in northern Iraq that occurred in the late 1980s under Saddam Hussein killed almost 200,000 people, including other minority groups in the region. And the Hutu killing spree in Rwanda in 1994 left somewhere between 500,000 and 1 million of the Tutsi people dead, while the United Nations pulled out its troops and abandoned the country.

Mass Graves

- According to the United Nations, a mass grave contains three or more executed people but could include many more. In addition, genocides may involve a tremendous number of separate burial locations. For example, in Cambodia, mapping teams have already identified more than 20,000 gravesites from the Khmer Rouge's three-year killing spree.

- Locating clandestine graves can be difficult. Memories of witnesses fade and topography changes over time. Forensic scientists have developed new technologies or adapted existing ones to help find the graves of genocide victims before they are forever lost.
  - For example, scientists working with the forensic arm of Physicians for Human Rights and the International Commission on Missing Persons have used ground-penetrating radar to find subsurface disturbances. This technology can even potentially show where bodies were after they've been removed, as was common after the Bosnian War.

  - Since about 2005, forensic investigators have also used before-and-after satellite images to try to find evidence for mass graves from the air.

- In early 2013, forensic anthropologists at the University of Tennessee's Anthropological Research Facility—known as the Body Farm—began a project with the goal of learning more about identifying mass burials. Researchers there have buried 10 donated human bodies and plan to monitor the burials over the course of three years to study changes on the surface and underground at the sites.
  - Light detection and ranging (LIDAR) technology will be used to map the surface topography. Multispectral imagery will be used to identify soil and vegetation changes that occur after a grave is constructed.

  - In addition, entomologists have been brought in to look at insect succession; botanists will study plant changes; soil experts will analyze chemical isotopes; and molecular genetics researchers will examine whether bodies in multiple graves can contaminate one another's DNA profiles.

- The goal of the project is to identify some combination of techniques that will ultimately provide a faster, safer, and more definitive approach to locating mass graves from genocides, perhaps using drones.

Case Study: Genocide in Guatemala

- Between 1960 and 1996, more than 200,000 people were murdered in Guatemala. The majority were of Mayan descent. They were victims caught in a political crossfire between ruthless government military forces and Marxist guerrilla rebels who sought to overthrow the corrupt administration. This genocide was mostly about race and land use, and the bulk of the killings occurred in the highlands where the majority of Mayans live.

- Some non-Mayan citizens, particularly Latinos of high socioeconomic status who held anti-government views, also quietly disappeared during this time. Many of these people were from the city centers and universities, mainly college faculty, students, and others who sought political reform. The fate of nearly all these victims is still unknown.

- In the summer of 2004, forensic anthropologists, social anthropologists, and archaeologists were working in and around Guatemala City and in the mountains near the town of Chichicastenango. Those working to investigate the genocide had received many death threats, and the entrance to the laboratory of the Guatemalan Forensic Anthropology Foundation was heavily guarded.

- The hallways of the foundation's offices and lab were lined with wooden coffins of the size that would be used for infant burials. These are used to house the decomposed remains of victims exhumed from mass graves by the archaeologists because fragmentary skeletons don't require adult-sized caskets. After graves were exhumed, the remains were brought to the lab, where anthropologists attempted to identify the dead and document injuries observed on the bones in the hopes of using the information to prosecute war crimes.

- In the foundation's lab, scientists document wounds to the skeletons that can often be corroborated by eyewitness accounts of the murders gathered by social anthropologists. Gunshot wounds attest to the typical method of execution, but some remains show evidence of machete wounds or smashing of the skull.

- Most of the Mayan victims in the highland graves are known individuals. They were executed, either alone or in small groups, and their bodies were left to rot as examples to their family members and fellow villagers. Often, the murderers did not permit the families to bury their dead.
    - After the killers moved on, family members frequently snuck back to the scene to quickly and secretly bury their loved ones, usually in unmarked graves. Thus, in many cases, the villagers could actually lead anthropologists to the graves.

    - The reason for disturbing the dead was to make legal records of their deaths and to allow family members to properly rebury their loved ones.

- To find the graves, trained social anthropologists visit villages to interview witnesses to these old crimes.
    - In a village near Chichicastenango, a middle-aged man named Don Diego described the murder of four of his friends and his brother. The men were assumed to be conspiring against the government simply because they were gathered together outside a small store one afternoon.

    - Diego and the other men were chased into the woods by some soldiers; Diego ran in a different direction than the others. As he hid, he heard gunshots. The next day, he and some of the villagers went into the woods and found the bodies of the five men riddled with bullets. Diego and the others buried their friends deep in the woods near the top of a mountain.

  - Twenty-two years later, the graves were uncovered by forensic anthropologists as the families and friends of the dead men gathered around the exhumation site.

- In another village near Chichicastenango, two graves were hidden in the cornfields, near the edge of the forest. In one of the graves, three bodies had been covered with beautiful Mayan blankets. Under the cloth was discovered the remains of what appeared to be an adult female and two small children, one about 5 years old and the other, an infant. When they were buried, the youngest had been wrapped in the female's arms. The importance of establishing the ages and identities of these victims is to show that they were not enemy combatants.

- In a typical scenario, guerrillas would come through a village and ask for food, shelter, and clothing. If the Mayans refused, the guerrillas would kill them, but if the village cooperated and the government forces found out, the military would execute the whole village or just some as an example to others. Loved ones were often made to watch—and sometimes even forced to participate in—the torture and killing of their family members.

- Eventually, as a result of the bravery of those who came forth and told their stories, graves have been exhumed, injuries have been documented, and identifications have been made. Following that, the bones of the dead can be bundled into small coffins; the government can issue death certificates; and preparations can be made to return the remains to their families.

- In early 2012, Efraín Ríos Montt—a former general in the Guatemalan army who became the country's de facto president during the worst of the atrocities—was finally tried for genocide and crimes against humanity. He was found guilty on May 10, 2013, and sentenced to 80 years, but just 10 days later, the decision was annulled on a technicality. A retrial has been set for January of 2015.

Suggested Reading

Kiernan, *Blood and Soil.*

Mallet, Blythe, and Berry, eds., *Advances in Forensic Human Identification.*

Montejo and Perera, *Testimony.*

Murray, *Forensic Identification.*

Rathbun and Buikstra, *Human Identification.*

Steadman, *Hard Evidence.*

Totten and Parsons, eds., *Centuries of Genocide.*

Questions to Consider

1. What types of forensic methods are used to identify and/or analyze the remains of victims of genocide?

2. What are some of the worldwide efforts and/or organizations that work to help prevent these types of atrocities?

Forensics of Genocide

Lecture 19—Transcript

The term *genocide* is used for the deliberate extermination of a targeted group of people, whether for political, racial, economic, or other social reasons. A key word here is *deliberate*. To be precise in calling a massive loss of life "genocide," the deaths must have been intentional, which differentiates genocide from other types of mass fatalities, whether natural, or even man-made.

So, for example, the tremendous number of Native Americans who died during the colonization of North America would technically not be considered genocide, since the vast majority lost their lives to European diseases, not European warfare. That doesn't deny how terrible and world-altering the decimation of the Native-American population was, especially to an anthropologist like me. It's simply to say that the precise definition of genocide doesn't apply to the bulk of the lives lost, since contagion wasn't well understood at the time and the deaths from disease weren't intentional. A specific event, however, like the approximately 4,000 Cherokee who died during the Trail of Tears migration between Florida and Oklahoma in the 1830s, well that is genocide, because the U.S. government forced the relocation, and there was every reason to believe that many would not survive that journey.

There have been so many tragic examples of genocide throughout world history. Here are just a few major events, from only the 20th century, alone. The Turks decimated as many as 2 million Armenians throughout and after World War I, during which some say the first organized concentration camps were used. Shortly after that, in the U.S.S.R., about 20 million died at the hands of Joseph Stalin; his failed socialist agricultural projects caused great famines in the new Soviet nation. Stalin's regime targeted former Russian military and the upper echelon of Russian society, including a variety of, "enemies of the people."

Likewise, the agricultural, industrial, and social changes of the so-called Great Leap Forward of China's Mao Zedong, led to mass death from starvation throughout the Chinese countryside in the late 1950s and early

’60s. Mao’s Cultural Revolution that followed caused the imprisonment or murder of anyone believed to be against the government. Some estimate 50 to 70 million people were killed in China between 1949 and 1976. Now, the intent comes into play in both Stalin’s and Mao’s agricultural genocides because the practices continued despite the knowledge that so many were dying from them.

Probably the most well-known genocide came at the hands of the German Nazis, who killed about two-thirds of Europe’s Jews in the 12-year Holocaust between 1933 and 1945. Others targeted include Jehovah Witnesses, Roma and Polish people, Russian prisoners of war, homosexuals, and those who were mentally or physically handicapped. Total estimates are that 11-million people died, mostly by poison gas administered in Third Reich death camps.

No one really knows the number of people who have died in North Korean prison camps since 1945, or from the long-standing practice of refusing or diverting aid from starving Korean peasants, which many believe is still happening today. And in just the three years of 1975 to 1978, the Killing Fields of Cambodia resulted in about 2.2 million, that’s over 20 percent of the country’s population being exterminated by Pol Pot’s Khmer Rouge; victims were minority ethnic and religious groups, as well as intellectuals and the professional class.

The Kurdish genocide in Northern Iraq that occurred in the late ’80s under Saddam Hussein and his cousin, known as Chemical Ali, killed almost 200,000 people, including other minority groups in the region. And the Hutu killing spree in Rwanda, which happened over only 100 days in 1994, left somewhere between 500,000 and a million of the Tutsi people dead, while the United Nations pulled out their troops and abandoned the country. Unbelievably, this long list is just a small sample. At the time I was writing this lecture, mass graves were being discovered in other parts of Africa and in Syria.

Now, let’s talk about how human rights groups locate genocide graves and use the evidence found in them to understand critical details about the atrocities committed and to help prosecute those responsible. According to the United Nations, a mass grave contains three or more executed people, but obviously could contain many more. In addition, these genocides can involve a tremendous number of separate burial locations. For example, in Cambodia, mapping teams have already identified over 20,000 gravesites from the Khmer Rouge's three-year killing spree.

The systematic archaeological excavation of mass graves and the forensic analyses of the remains and artifacts in them are evidence of crimes against humanity. But I can tell you, locating a clandestine grave can be like trying to find a needle in a haystack. The memories of witnesses fade, and the topography changes over time. So forensic scientists have developed new techniques or adopted existing ones to help find the graves of genocide before they're forever lost to time.

Scientists, like those working with the forensic arm of Physicians for Human Rights and the International Commission on Missing Persons, have used techniques like ground-penetrating radar to look for sub-surface disturbances. Using this method, radio waves reflected in the soil are picked up by an antenna to help detect underground anomalies. This can potentially even show where bodies were after they've been removed, as was a common practice after the Bosnian War. The bad guys would move remains once they thought investigators were closing in on the graves.

But ground-penetrating radar has to be used on the surface. So since about 2005, forensic investigators have also used before-and-after satellite images to try to find evidence for mass graves from the air. That's only effective, though, for graves with hundreds of bodies in them, not smaller graves of, say, 10 people, which are far more common, especially in recent genocides.

In early 2013, some of my forensic anthropology friends from the University of Tennessee's Anthropology Research Facility, commonly called the Body Farm, began a project with the goal of learning more about identifying mass burials. The researchers excavated four holes along a lake at Fort Loudoun State Park and then buried 10 donated human bodies. One site was filled back in with only soil to serve as an experimental control; into a second hole, a single body was placed; three bodies were buried in the third grave; and the other six corpses were literally piled into the fourth pit. Bullets and shell casings were also placed into some of the graves, as well as personal effects, like wallets.

The plan is to monitor the burials over the course of three years to look for surface changes, but also to use a variety of non-invasive methods, so-called remote-sensing techniques, to look at what's going on inside the graves. One is LIDAR technology; that stands for Light Detection and Ranging, where lasers are used to map an area's surface topography. That's because burials typically start as a mound, but become a depression as the soft tissues of the bodies decay. The other technique is multi-spectral imagery, which monitors different wavelengths of light as they bounce off the ground's surface, including those outside the range visible to humans, like infrared. This can show soil and vegetation changes that occur after a grave is constructed and nature begins to fill back in.

But the project is even more multidisciplinary, especially at ground level. Entomologists have been brought on board to look at insect succession; botanists will study plant changes; soil experts will analyze chemical isotopes; and molecular genetics researchers will examine whether bodies in the multiple graves can contaminate one another's DNA profiles. The scientists completed baseline studies of the plants, soil, and topography before they dug the graves for comparison to anticipated changes. At the conclusion of the project, they're going to bring in international human rights anthropologists; they'll be invited to a week-long workshop to learn how to properly excavate mass graves.

From the data collected during the study, the Oak Ridge National Laboratory will try to generate computer models and algorithms. They're hoping that some combination of the techniques used, and the information gained, will ultimately provide a faster, more definite, and safer way of locating mass graves from genocide, maybe using drones. But the methods could also be helpful in any other type of forensic situation where a clandestine grave is suspected, like even one of my own cases.

Having the research project last three years will give a sense of how time alters the graves and their contents. That way, even if forensic archaeologists and anthropologists can't get to genocide graves immediately, say, due to dangerous conflict that's still going on in the area, sites can be marked on satellite-generated maps, and images can be continuously monitored until it's safe to go in. All in all, this high-tech University of Tennessee study might prove invaluable when forensic teams are called upon to locate and exhume mass graves.

Now I want to tell you about some fairly low-tech, but very personal experiences I've had learning about mass graves and genocide. In the 36 years between 1960 and 1996, over 200,000 people were murdered in Guatemala. Most of them died at the hands of their own government. The majority were of Mayan descent. They were victims caught in a political crossfire between ruthless government military forces and Marxist guerrilla rebels who sought to overthrow the corrupt administration. This genocide was mostly about race and land use, and the bulk of the killings occurred in the highlands where the majority of Mayans live, having been driven there centuries ago by the Spanish conquest.

In addition to the suffering and loss of the Mayan people, some non-Mayan citizens, particularly Latinos of high socioeconomic status who held anti-government views, also quietly disappeared during that time. The majority of these *desaparecidos*, as those who simply disappeared are called, were from the city centers and universities, mainly college faculty, students, and others who sought political reform. The fate of nearly all these victims is still unknown.

Guatemala sits at the north end of Central America beneath the Yucatan Peninsula of Mexico. It's modern capital, Guatemala City, is just a few hours by plane from Cincinnati. In the summer of 2004, as part of a television project for the National Geographic Channel, I was invited to observe and participate in the work of human rights anthropologists in Guatemala, including forensic anthropologists, social anthropologists, and archaeologists. Much of our work took place in and around Guatemala City, extending up into the mountains into the northwest, near the town of Chichicastenango, at elevations up to nearly 9,000 feet above sea level.

In Guatemala City, I was picked up at the airport to be taken to the laboratory of the Guatemalan Forensic Anthropology Foundation, known as the FAFG.

During this drive through the city, I was in a vehicle with tinted windows and told by my armed driver that I needed to stay close by him for my protection, not only because I was a female traveling alone in a strange country, but because the anthropologists working to resolve these murders had received many death threats. On the outskirts of Guatemala City, we reached the walled, gated, and heavily guarded entrance of the laboratory.

The Foundation's offices and lab were in a renovated house; the former Japanese ambassador's home, I believe. There were small wooden coffins, the size that would be used for baby burials, lining its hallways, as testimony to the Foundation's work. These are used to house the decomposed remains of victims that are exhumed from mass graves by the archaeologists, since fragmentary skeletons don't require adult-sized caskets. After graves are exhumed, the remains are brought to the lab, where anthropologists attempt to identify the dead and document injuries observed on the bones in the hopes of using the information to prosecute war crimes.

The Director of the Foundation, Fredy Peccerelli, told me, "The main mission of the Foundation is to provide dignity to the family members through the identification of the bodies, and also create a scientific historical record." But just as importantly, the Foundation operates to document to the world, in a scientific and public way, that these events did happen in Guatemala's recent history, and that the people who died mattered to each other and to the past, present, and future of Guatemala—in essence, that they are not forgotten.

The roots of the Foundation predate 1996, when the United Nations brokered a peace accord and created a commission to investigate the human rights violations in Guatemala. Even before that time, a core group of anthropologists was already secretly investigating the atrocities at great danger to themselves. In the early '90s, my dear, departed friend, the famous Dr. Clyde Snow from Oklahoma, retired from academia and began making trips to Central and Latin America to train local anthropologists in this kind of work. Clyde, who died in May of 2014, was one of the granddads of forensic anthropology and the virtual founder of human rights anthropology.

His Guatemalan team began with a staff of five, and when I visited in 2004, the FAFG had about 60 staff members. Some worked in the offices, others studied remains in the laboratory, and still more worked in the field interviewing witnesses or locating and excavating graves.

In the lab, scientists document wounds to the skeleton, which can often be corroborated by eyewitness accounts of the murders that are gathered by the Foundation's social anthropologists. Gunshot wounds, like the one I saw on the back of a 62-year-old Mayan woman's skull, were the typical method of execution, but there were many others.

One 15-year-old boy had 88 non-lethal machete wounds to his arms and legs, which probably caused him to bleed to death. To save ammunition, young children and babies were often killed by holding their legs and swinging them against trees or rocks to smash their skulls. There probably isn't a means of torture or murder that the anthropology team has not seen in one case or another. The horrors are almost unspeakable.

Regarding the fieldwork, I should first point out that most of the Mayan victims in the highland graves are known individuals. They were either executed, alone, or typically in small groups, and had their bodies left to rot as an example to their family members, and fellow villagers. Often, the murderers did not permit the families to bury their dead. But typically, within a day or two of the killings, after those responsible moved on, the family members snuck back to the scene to quickly and secretly bury their loved ones, usually deep in the woods and in unmarked graves, and at the risk of being caught and killed themselves. So in many cases, the villagers often knew who died and when, where they're buried, and could actually lead the anthropologists to the graves.

So you might wonder, then, why the dead would even be disturbed. Well from a practical standpoint, if a woman buried her husband 20 years ago, there's no record of his death, or the events surrounding it, except in her memory. From a legal standpoint, no death certificate means she can't remarry, and unrecorded deaths mean children can't inherit land from their deceased parents. But from a humanitarian and psychological perspective, it's hugely important that the victims be reburied with a proper funeral ceremony so family members can bring final dignity to their dead. No one should be killed and then thrown out to rot like garbage before being hastily interred in a secret grave by fearful relatives. It's an essential part of the healing process to give names and proper burials to these dead.

So then the first step in bringing the anthropology team in to assist has to be that the family members come forward and request help from the government. But when I was there, many were still very untrusting of the authorities. So trained social anthropologists—what some of us might call cultural anthropologists or even social workers—well, they go visit the villages to mediate the process and interview witnesses to these old crimes.

In the village of Chujulimul near the town of Chichicastenango, I participated in one of these interviews with a middle-aged man named Don Diego. He told us that he and five other men, one of whom was his brother, were gathered outside the small store in their tiny village one afternoon. He said some guys from the army came by, and when they saw the six men gathered, assumed they were somehow conspiring against the government. So they chased the men into the woods, but Don Diego ran in a different direction. Then he heard gunshots from the forest, but he hid. The next day he got some of the villagers, and they went into the woods and found the bodies of the five men scattered among the trees and riddled with bullets. He had tears in his eyes as he told us the story, even though the incident happened 22 years before. He said he felt guilty for running away and sorry that he got to live his life while his friends and his brother didn't.

The clandestine graves where Don Diego and other villagers buried the men were deep in the woods, near the top of a mountain. It took us the better part of an afternoon to get up there. Near the end of the trip, we had to leave the vehicles and climb the steep slopes by foot. Local men had previously helped clear the sites in the woods where they knew the graves were. By the time I got there, Foundation anthropologists had already uncovered two of the three burial sites that held Don Diego's brother and his friends. One of the graves contained two men, one of whom was probably only about 16 years old. The other grave, in which I worked, contained the remains of a man who was estimated to be in his mid-40s. The third grave was a short distance away, and not yet opened. Over the years, I've worked at lots of crime scenes as a forensic anthropologist, but I had never experienced anything like that. The family and friends of the dead men were all around the exhumation site. They talked softly, sang, some of them cried. Their little children, who weren't even alive when these awful things happened were laughing and playing in the woods nearby.

In the grave where I worked, we found the remains of a man. His wallet was buried with him. In it were some photographs, a small mirror, and comb, and a calendar and coins, all dated before 1982 when the burials had taken place. The wallet was passed around the crowd, and one woman began to sob and cry out, "Esta se llama Pedro Morales Morales, mi hermano." It was her brother who had gone to the store that day and been killed with his friends. Twenty-two years ago, as a young girl, she had watched the villagers bury him at that site. The social anthropologists, who are also trained as grief counselors, tried to comfort her, while we continued to excavate.

When the clothing was cut away from the man's skeleton, we were able to see he had a gunshot wound that shattered the bone of his thigh and another gunshot wound to his hand, which he had probably held up in defense. But the cause of death was obvious; he had a single gunshot wound to the side of his head.

On another day, along with Dr. Clyde Snow, the Foundation Director, Fredy Peccerelli, and Chrystyna Elych from the United Nations, I visited another exhumation site. This site was in the village of Panimache, also near Chichicastenango, about three hours northwest of Guatemala City; it was high in the mountains. Some of the village women met and led us, along with their sheep, up the mountain to the site. There were two graves at this location, both hidden in the cornfields near the edge of the forest that surrounded the tiny community.

As I walked past the first gravesite, I saw villagers watching the progress of some of the Foundation anthropologists. I was assigned to a different team that went deeper into the cornfield to assist at the second burial site. The entire community was around the area; the village school had given the kids a holiday, and everybody else had suspended their work that day so they could see what the anthropologists were doing.

The bodies in the grave where I worked had been covered with beautifully-colored Mayan blankets when they were buried. Under the cloth covering, which had held up well, you could easily see a row of three round mounds that looked like they were probably skulls. We pulled back the blanket and found the skull of the first person, who appeared to be an adult female. As we continued to peel the blanket away, we found the remains of two small children, one about five and the other one an infant. When they were buried, the youngest had been wrapped in the female's arms; she was probably the child's mother or grandmother.

The importance of establishing the ages of these victims, as well as their identities, is to show that these were not enemy combatants. How can a five-year-old child or an infant be considered harmful to the military? How could that 62-year-old woman I mentioned earlier have been a threat to a man with a high-powered rifle? These graves help establish, to any court, that there could be no legitimate purpose for the executions of the women and children in them. Unfortunately, though, even though the Foundation had recovered some 3,000 sets of remains from over 400 sites, by the time I visited in 2004, there had only been three successful trials.

Also at Panimache, I was surprised to see a group of uniformed school kids who had come a long way on a bus from a larger town. They had been brought by their history teacher to see the work of the anthropologists. As they quietly watched us uncovering the bones of the children in the grave, I wondered if those school kids considered, had they lived at another time, that could be them in that grave.

These aren't just lessons for Guatemalan school children, but the rest of the world, as well. In my memory, the conflicts in Guatemala had been only vague news stories during the time they happened, but I was shocked when one of our drivers, a man in his early 40s who grew up in Guatemala City, told me that even many people who lived in Guatemala at the time weren't aware of the genocide. He said, if you were an upper or middle-class teenager during those years, your parents protected you from the stories, not only to shield you from the horrors, but also because young boys would run off and join the guerilla fighters out of what he called romantic notions, since the guerrillas lived like bandits up in the mountains. But in many ways, the guerrilla fighters were no better than the military.

A typical scenario was that the guerrillas would come through a village and ask for food, shelter, and clothing. If the Mayans refused, the guerrillas would kill them; but if the village cooperated, and the government forces found out, the military would execute the whole village, or maybe just some as an example to the others. Loved ones were often made to watch, and sometimes even forced to participate in the torture and killing of their family members.

Eventually, though, due to the bravery of those who came forth and told their awful stories, graves have been exhumed, injuries have been documented, and identifications have been made. Following that, the bones of the dead can be bundled into those baby coffins I mentioned, the government can issue death certificates, and preparations can be made to return the remains to their families. I was privileged to attend one of the Foundation's repatriation ceremonies in the remote mountain town of Cotzal, where just over two decades before, a bloody confrontation between the army and guerrillas led to the mass murder of 20 villagers, mostly women, children, and the elderly.

The journey to Cotzal was long and difficult. We traveled nearly four hours on dusty dirt roads, following a truck that held 10 coffins and several cardboard boxes of fragments that were too badly damaged for identification. As we neared the top of the mountain, we came upon a group of village women returning from their agricultural fields. The forensic anthropologists pulled the truck over so the women could ride in the back, with the coffins, on our way up to their town.

The village center consisted of just two buildings, a one-room town hall, and a one-room schoolhouse. The truck was unloaded and the caskets were carried into the town hall. Once we were inside, the anthropologist opened the containers so that families could pay their last respects. The Foundation had been able to identify 7 of the 10 dead, a mother, father and their two sons; and a grandmother and her two grandsons. Without dental or medical records, and with DNA technology being too expensive at the time, the families have to content themselves with whatever results the Foundation can provide. The remains were to be left all day for visitation, so flowers and candles were put on each casket, and they put fresh pine needles on the floor so the mourners would stir up the fragrances as they moved among the coffins. Local men were playing horns and drums, and we could hear the music as we got into our vehicles to leave that grieving village.

Despite years of death threats to the Foundation's anthropologists and their families, the work has continued and prospered. I remember telling Fredy, as strange as this sounds, that each and every one of those late-night threatening phone calls, or each time his vehicle had to be searched for a bomb, well, that was testament to the Foundation's success. Those threats mean that the anthropologists have done such powerful work in identifying remains and documenting war crimes, that those responsible were now running scared, afraid of being caught and made accountable for the awful things they had done.

In early 2012, I emailed a note of congratulations to both Fredy Peccerelli and Clyde Snow when I heard that Efraín Ríos Montt, a former general in the Guatemalan army who became the country's de-facto president during the worst of the atrocities, well, he was finally going to be tried for genocide and crimes against humanity. He was found guilty on May 10, 2013 and sentenced to 80 years, but just 10 days later, the decision was annulled on a technicality. What a disappointment for the people seeking justice, including the forensic anthropologists. A retrial has been set for January of 2015, but at that point, Montt will be 88 years old.

Still, the work of human rights anthropology will continue in fields and in laboratories around the globe. We can only hope that groups like the Forensic Anthropology Foundation of Guatemala, and many others, can find the graves of genocide, document the dead, and bring the perpetrators to justice. That might help some of the world's many war-torn countries to heal the deep wounds that scar their history and their people.

The Nazis and the Witch of Buchenwald
Lecture 20

One of the reasons that the Nazis stand out in the West as well-known perpetrators of atrocities is that before Hitler's rise to power, Germany didn't have a history of brutality against its own people. Paradoxically, at the time Nazism arose, Germany was among the most highly cultured and educated societies of its time, but the Third Reich stood in stark contrast to that. In part, it is because of the sophistication of German society—particularly in medicine, science, and technology—that we have a record of evidence regarding what happened under Nazi rule. In this lecture, we will discuss some of the forensic evidence involving the Nazis and how it was used at trial.

Background on Buchenwald

- Buchenwald was opened in 1937 and, with its satellite labor sites, was the largest of the concentration camps on German soil. Over the period of its operations, the site was used to house 240,000 people that the Germans considered undesirable because of mental or physical disabilities, sexual orientation, or religious differences. The camp also housed political detainees, including as many as 350 Allied prisoners of war.

- Although Buchenwald was not specifically a Nazi death camp, accounts indicate that 33,462 prisoners died there. An additional 8,483 Russian POWs were executed upon arrival at Buchenwald, and about 13,500 people died during the death march to evacuate prisoners as the Allies closed in the camp. In total, about 56,500 people died at or while leaving Buchenwald.

- The majority of the deaths at Buchenwald were not random or unanticipated events. Some prisoners were murdered—either by shooting in the stables or hanging in the camp crematorium—but most died from starvation or were simply worked to death;

some of those deemed too weak to work were transferred to extermination camps.

- Buchenwald was a training site for the Schutzstaffel (SS), whose members learned the incarceration, torture, and prisoner work-detail methods of the Third Reich at the camp. Buchenwald was also the site of Nazi medical trials and other experiments on prisoners, after which most died or were killed. Life expectancy at Buchenwald was estimated at just three months.

Human Experimentation

- The Nazis kept meticulous records of their human experimentation and preserved human specimens. In some studies, they removed portions of prisoners' nerves, muscles, and bones to examine tissue regeneration. Healthy people were infected with hepatitis and malaria or dosed with mustard gas. People were subjected to extreme cold or low oxygen levels to simulate high altitudes. All manner of wounds were created, then fragments of glass, wood shavings, and bacteria were rubbed into them to test antibiotics and other treatments.

- The primary intent of these experiments was to look for ways that the German military could counter frostbite, altitude sickness, and common wartime injuries and infections. But there were other studies, too, such as the use of X-rays, surgery, and drugs to sterilize people.

- At Buchenwald, the laboratories carried out experimentation in three particular areas: poisoning, chemical burns, and vaccines for typhus. These were of interest to the Third Reich military because of their usefulness in either murder or warfare-related injuries.
    - In the poison studies, Buchenwald prisoners were given various toxic substances in their food, after which they would either die and be autopsied or be killed for autopsy. In addition, some inmates were shot with poison bullets to study the effectiveness of administering toxins at long range.

- Non-anesthetized Buchenwald prisoners were also burned with phosphorous-containing compounds—the type used in incendiary bombs to set fire to buildings and other targets during warfare. The purpose was to research different types of medical treatments, such as salves and ointments developed by the Buchenwald scientists.

- The third major area of experimentation at Buchenwald involved typhus. Healthy prisoners were infected with typhus to create a living reservoir of bacteria for study. More than 90 percent of those inmates died. Researchers used another population of prisoners for typhus experiments, administering several types of potential vaccinations before infecting the subjects with typhus several weeks later. Another group served as a scientific control, receiving no vaccine, only the infection.

- In 1946, at the Nuremberg trials, a specific set of hearings was held about Nazi medical experimentation, known as the Doctors Trial.
  - Witness testimony, scientific journals, medical records, pathology specimens, and photographs were all used as forensic evidence to document the cruel experiments conducted on the prisoners at Buchenwald and elsewhere. Some of those involved were acquitted; others were sentenced to prison terms; and some were put to death.

  - Specific decisions made at Nuremberg helped refine and codify ethics for the use of human subjects in research.

The Witch of Buchenwald

- One of the commandants at Buchenwald was Karl Otto Koch; together with his wife, Ilse, Koch was arrested in 1943 by his own Nazi superiors for criminal behavior. It was alleged that the pair used prisoners for private gain, murdered those prisoners to keep them from talking, and embezzled from the Third Reich.

- Koch's superiors discovered that he had several of the medical personnel at Buchenwald shot, then listed them on death records as political prisoners. It was later discovered that the executed men had treated Koch for syphilis, and he didn't want his illness known.

- Koch was tried in an SS court, found guilty, and executed by a Nazi firing squad in 1945. The prior year, his wife had been acquitted on all charges of corruption owing to lack of evidence, after which the 38-year-old Ilse went to live with family away from Buchenwald.

- Karl Koch had become commandant of Buchenwald in 1937, but a couple of years before his arrest in 1943, he was transferred to occupied Poland to help establish a Nazi camp there. In his absence, Ilse—known as the Witch of Buchenwald—gained prominence and may even have filled a supervisory role at the camp. She is alleged to have become quite predatory on prisoners and promiscuous with other camp leaders.

- Ilse enjoyed watching all manner of punishment doled out to inmates. She's alleged to have engaged in random acts of violence, including riding around on horseback and lashing out at prisoners at whim. Ilse teased the prisoners and guards with her sexuality. She is said to have watched male prisoners strip naked upon their arrival at camp and made lewd comments about the prisoners' bodies to other SS wives.

- Ilse was particularly friendly with one of the Buchenwald physicians, who was preparing a dissertation on tattoos as indicators of criminal behavior. It is alleged that Ilse came up with the idea of having the tattoo study done at the camp.
    - About 40 inmates with particularly interesting tattoos wound up at the infirmary and never returned. After that, pieces of tattooed skin were found in the Buchenwald pathology lab. It was alleged that the skin was used to make lampshades, book covers, knife sheaths, and handbags and that fellow prisoners were forced to manufacture the artifacts.

    - It was also said that the Nazi pathologist at Buchenwald tried his hand at making shrunken heads from prisoners. Some of these objects were given to SS members as gifts, while Ilse kept others.

Liberation

- On April 4, 1945, the first Nazi camp to be liberated by U.S. forces was one of the Buchenwald work camps. In response, the Nazis at the main Buchenwald facility began the prisoner evacuation march. A week later, some of the remaining prisoners—who knew by clandestine radio communications that Allied liberators were on their way—rushed the watchtowers and killed the guards, using guns they had been stealing and stashing for years.

- Later that afternoon, the U.S. army arrived, and soon, the 21,000 camp survivors received food, water, and medical attention. The liberators were shocked by the appearance of the inmates. The next day, journalist Edward R. Murrow visited Buchenwald to report its horror to the world. He saw stacks of bodies by the crematorium and tried to count at least some of the dead but gave up after he reached more than 500.

- General Patton ordered military personnel to force local German citizens to witness the Nazi atrocities that had been going on nearby. They made both men and women exhume victims from mass graves where the signs of torture and violent death were clearly observable. Civilians were forced to march to Buchenwald and parade past piles of dead bodies and a table that held alleged human artifacts.

Ilse's Trials

- Among many stories the Allies heard from the survivors at Buchenwald were accounts of Ilse Koch's legendary brutality and sadism. After the war ended, she was among the first of the Nazis to be tried in a U.S. military court martial. The proceedings began on April 1, 1947, at Dachau. Ten days later, despite her claims that none of the accusations was true, the Witch of Buchenwald was found guilty of "violation of the laws and customs of war" and sentenced to life in prison.

- The trials at Dachau were later reviewed, and Ilse's sentence was commuted to four years—essentially, the time served since her arrest.
    - During the review process, Ilse's American defense attorney criticized the prosecution based on the lack of any forensic evidence solidly connected to her, especially related to the accusations that ornamental goods had been made of human skin at her request.

    - No lampshade had been produced at trial, despite its presence in many photographs taken during the liberation. American authorities admitted that it had somehow gone missing.

- After Koch's sentence was commuted, an irate public pushed for a retrial in the German courts. She was rearrested in 1949 and charged by the prosecutor's office in Bavaria. The second trial began in 1951 and lasted seven weeks.
    - This time, four witnesses testified that Koch was observed to be personally involved in choosing prisoners with tattoos or engaging in other actions related to the manufacture of lampshades from human skin. Again, however, no lampshades or other ornamental objects made from skin could be produced. Charges related to those allegations were dropped.

    - But in January 1951, the Witch of Buchenwald was found guilty of incitement to murder, incitement of attempted murder, and incitement of bodily harm. She was again sentenced to life in prison.

- We will never really know what crimes Ilse Koch was guilty of, but from the time she was picked up by the American military in the summer of 1945—after being recognized by a former prisoner at Buchenwald—she spent the rest of her life incarcerated, either by sentencing or in awaiting trial. Whether we believe she was as hideous as accounts say or was a scapegoat or surrogate for Nazi atrocities, she never walked free again. In 1967, she hanged herself in a German prison.

Suggested Reading

Kiernan, *Blood and Soil.*

Przyrembel, "Transfixed by an Image."

Totten and Parsons, eds., *Centuries of Genocide.*

Whitlock, *The Beasts of Buchenwald.*

Questions to Consider

1. What are some of the technological advances of the mid-20th century that were misused by the Nazis to commit mass killings?

2. What technological advances of the time allowed the collection of evidence against the Nazis?

3. Do you think Ilse Koch was a scapegoat for the Nazis? If so, what is the basis for your conclusion?

The Nazis and the Witch of Buchenwald

Lecture 20—Transcript

There are many reasons why the Nazi cruelties are probably the best-known atrocities of a country against its own people, at least in the Western world. But to me, one stark difference between what happened in Germany and other inhumane regimes of the first half of the 20th century, such as the Ottoman Empire and the Soviets, is that prior to Hitler's rise to power, Germany didn't really have a recent history of brutality against its own people.

Paradoxically, at the time Nazism arose, Germany was among the most highly cultured, educated, and some would even argue, enlightened societies of the time. But the Third Reich sure stood in stark contrast to that. In part, it's due to the sophistication of German society at the time in areas like medicine, science, and technology that we have a strong record of forensic evidence for the evils that happened under Nazi rule.

Any crime is essentially a piece of history that must be understood by uncovering and interpreting forensic evidence. That evidence might come in the form of testimony, perhaps from perpetrators, witnesses, or survivors of an experience, like the Nazi death camps. If investigators are fortunate, there may also be physical evidence to analyze, like the photographs or specimens from medical experiments conducted at concentration camps. Physical evidence is particularly important, since we know that perpetrators often lie about criminal activity and eyewitness memories are not always reliable. The Third Reich kept near meticulous records of most of their actions, and so left behind a legacy of their reign of terror. In this lecture, we'll discuss some of the forensic evidence against the Nazis and how it was used at trial.

I'm going to focus on one of the concentration camps, in particular, called Buchenwald, and discuss some of the medical experimentation and other Nazi practices that took place there, some of which were, allegedly and surprisingly, under the direction of a woman. Buchenwald was opened in 1937, and when you count its satellite labor sites, was the largest of all the concentration camps on German soil. Over the period of its operations, the site was used to house 240,000 people the Germans considered undesirable.

Their criteria included mental or physical disability, sexual orientation, or religious differences, Jews, of course, but other groups as well.

The camp also housed political detainees, including as many as 350 Allied prisoners of war. Although not a specific Nazi death camp, the Buchenwald accounting indicates that 33,462 prisoners died there. But there were an additional 8,483 Russian POWs who were executed upon arrival at Buchenwald, and so not included in that count, as they weren't considered housed at the camp. Another about 13,500 emaciated people died during the death march to evacuate prisoners, as the allies closed in on Buchenwald. When you add up all who died at or upon leaving Buchenwald, the toll rises to over 56,500, and it wasn't even considered one of the Nazi extermination camps.

I tell you those numbers particularly so you can see the careful accounting that took place, but also, the extent of the killing at what was alleged to be a work camp. The majority of the deaths at Buchenwald were not random or unanticipated events. Some prisoners were out-and-out murdered, either by shooting in the stables or hanging in the camp crematorium, but most died from starvation or were simply worked to death. Some of those deemed too weak for labor were transferred to extermination camps. Buchenwald was a training site for the Schutzstaffel, or so-called SS members, so they could learn the incarceration, torture, and prisoner work-detail methods of the Third Reich. It was also one of the places that conducted Nazi medical trials and other terrible experiments on prisoners, after which most died or were killed. Life expectancy at Buchenwald was estimated at just three months.

The human experimentation that occurred at the hands of the Nazis was known not only from their meticulous records and countless photographs, but also preserved human specimens, some of which are still curated today at the U.S. Holocaust Memorial Museum. In some studies conducted at various camps, they removed portions of prisoners' nerves, muscles, and bones to examine tissue regeneration. The tissues taken were sometimes surgically implanted into other inmates to study transplantation.

Healthy people were infected with hepatitis and malaria, or dosed with mustard gas. People were subjected to extreme cold, such as three hours in a tank of ice water. Prisoners were put into compression chambers and subjected to low oxygen levels to simulate altitudes up to nearly 70,000 feet. Men were forced to eat nothing and drink only seawater, some of which was chemically processed to try to desalinate it. All manner of wounds were created, including head injuries, and then fragments of glass, wood shavings, and bacteria rubbed into them to test a variety of antibiotics and other treatments. It makes me sick to speak of it.

The primary intent of those experiments was to look for ways the German Army, Navy, and Air Force could counter frostbite, altitude sickness, and common wartime injuries and infections. But there were other studies, too, such as the use of X-rays, surgery, and drugs to sterilize people, so the Nazis could limit the reproduction of those they thought inferior. The infamous Dr. Josef Mengele at the Auschwitz death camp specifically experimented on children, especially twins, and the handicapped. There was no consent on the part of any of the victims of these experiments. Many historic photographs show what appear to be doctors and nurses either experimenting on or "caring for" all types of experimental subjects. How any of those people could rationalize calling themselves scientists or medical professionals is beyond me.

With specific regard to Buchenwald, three particular areas of experimentation were assigned to their laboratories—poisons, chemical burns, and vaccines for typhus. These were of interest to the Third Reich due to their usefulness in either murder or warfare-related injuries. The fourth area of research involved both surgical and hormonal experiments to try and "cure" what they considered deviant sexual orientation, that's the reason Buchenwald housed homosexuals.

In the poison studies, Buchenwald camp prisoners were given various toxic substances in their food. Following that, they would either die and be autopsied or be killed for an autopsy so the mechanisms of the toxin could be examined. In addition, some inmates were shot with poison bullets in a non-lethal place on their body to see if that could be effective means of administering toxins at long range, that way, if a Nazi gunshot didn't kill the enemy in battle, the poison bullet would.

Non-anesthetized Buchenwald prisoners were also deliberately burned with phosphorous-containing compounds, the type used in a variety of incendiary bombs used to set fire to buildings and other targets during warfare. The purpose was to research different kinds of medical treatments, such as salves and ointments the Buchenwald scientists were developing.

The third major area of experimentation at Buchenwald involved typhus, which is a relative of spotted fever. Rather than use Petri dishes to culture the microbes, healthy prisoners were infected with typhus to create a living reservoir of bacteria for study. More than 90 percent of those inmates died. Researchers used another population of prisoners for the actual typhus experiments, administering several types of potential vaccines before infecting the subjects with typhus several weeks later. Another group served as a scientific control, receiving no vaccine, only the infection.

In 1946, the famous post-war Nuremberg trials held a specific set of hearings about Nazi medical experimentation, known as the Doctors' Trials. Witness testimony, scientific journals, medical records, pathology specimens, and photographs were all used to forensically document the cruel experiments on the prisoners at Buchenwald and elsewhere. Some of those involved were acquitted; others were sentenced to prison terms; and some were put to death.

Although medical experimentation on humans certainly happened before that time, and has continued since, the studies conducted by the Nazis were so atrocious they ultimately changed research directives for the entire civilized world. Specific decisions made at Nuremburg helped refine and codify ethics for the use of human subjects in research, like regarding consent, degree of risk to participants, and the right of a person to withdraw from a study at any time.

Now, let's return to Buchenwald and discuss other crimes that took place there, particularly under the direction of one of the camp commandants, Karl Otto Koch, and his notoriously brutal wife, Ilse, who became known as the Witch of Buchenwald. This pair was apparently so terrible that their own Nazi superiors had them arrested in 1943. It was alleged they used the camp's prisoners for their own private gain, murdered those prisoners to keep them from talking, and embezzled from the Third Reich. In line with their precision, the Nazis definitely had rules about how a camp should be run and who was to be killed, and they believed the commandant and his wife were corrupt and insubordinate.

Here's one reason Herr Koch was arrested: His superiors discovered he had several of the medical personnel at Buchenwald shot and then listed them on the death record as political prisoners. It was later discovered the executed men had treated Koch for syphilis, and he didn't want word of his illness getting out. He may have contracted the infection from the camp brothel, where female prisoners were forced to work as prostitutes, or perhaps from his wife, who was alleged to be pretty promiscuous, but some also contend he caught the disease from another man. That would have been most scandalous for the Commandant, since homosexuals were one group singled out for incarceration at Buchenwald. A report commissioned by the U.S. Army after the war names Koch a homosexual and his wife a nymphomaniac.

Anyway, Karl Koch was tried in an SS court, found guilty, and executed by a Nazi firing squad in 1945, at age 47. The prior year, his wife had been acquitted on all charges of corruption due to lack of evidence. Following that, the 38-year-old Ilse Koch went to live with family away from Buchenwald. But that's definitely not the end of her story or her legal troubles.

Why was the Witch of Buchenwald so hated, feared and reviled? We need to go a bit further back in time to answer that. Karl Koch became Commandant of Buchenwald in 1937, with his wife Ilse at his side. But a couple of years before their 1943 arrests by the Gestapo, Karl had been transferred to occupied Poland to help develop and staff a new Nazi camp. In his absence, his wife Ilse gained prominence at Buchenwald, and some sources say she even filled a supervisory role. It was during this time she's alleged to have become both predatory on prisoners and promiscuous with other camp leaders, including high-ranking SS officers and even the medical staff. There's even been suspicion that Frau Koch engaged in sexual relationships with inmates.

Ilse Koch was particularly sadistic, and especially enjoyed watching all manner of punishment doled out to inmates, while other SS wives would avert their eyes. She's alleged to have engaged in her own random acts of violence, including riding around on horseback and lashing out at prisoners at whim. Ilse teased the prisoners and guards with her sexuality, including wearing clothing that would be considered provocative in that day. And if a man looked at her, it's said she would have him punished. It's alleged she would watch the men stripped naked upon their arrival at camp and make lewd comments about the prisoners' bodies to other SS wives.

All the while, Ilse Koch had a lovely home at Buchenwald and bore three children there, though one died in infancy. She enjoyed fine clothes and many luxuries while thousands starved just a short walk away. Although the couple was paid well as Nazi elite, other SS members began to resent the Kochs' excesses and wondered how they were affording them. It was later discovered—and this was part of the reason they were arrested by the Gestapo in 1943—that the Kochs were having personal property, even gold teeth, taken from the prisoners. They were putting the funds into Swiss bank accounts.

Ilse Koch was particularly friendly with one of the Buchenwald physicians. He was preparing a dissertation on tattoos as indicators of criminal behavior. Even today, in the forensic ID of unknown persons, we recognize tattoos can be marks of cultural affiliation, like gang membership, sexual identity, or religion. Anyway, it was said that Koch herself came up with the idea of having the tattoo study done at the camp, based on watching men work with their shirts off as she rode around Buchenwald on horseback.

About 40 inmates with tattoos, particularly interesting ones, wound up at the infirmary and never returned. After that, pieces of tattooed skin were found in the Buchenwald pathology lab. It was alleged the skin was used to make lampshades, book covers, knife sheaths, and even handbags, and that fellow prisoners were forced to manufacture those artifacts. It was also said the Nazi pathologist at Buchenwald tried his hand at making shrunken heads from prisoners, like he'd seen done in other cultures. Some of these objects were given to SS members as gifts, while others, the Witch of Buchenwald kept for herself. So notorious was Ilse Koch, that long after she was arrested in 1943 and imprisoned pending her SS trial, she was still legendary at Buchenwald.

On April 4, 1945, the first Nazi site liberated by U.S. forces was one of the Buchenwald work camps. This caused the Nazis at the main Buchenwald facility to begin the prisoner evacuation march I mentioned earlier, which killed over 10,000 people. A week later, some of the remaining prisoners, who found out by clandestine radio communications that Allied liberators were on their way, well, they rushed the towers and killed the guards using guns they had been stealing and stashing for years.

Later that afternoon, the U.S. Army arrived, and soon the 21,000 camp survivors received food, water, and medical attention. The liberators were shocked by the appearance of the inmates. One 85-pound French POW said he once weighed 170. A stable that was intended for 80 horses housed 1,200 men. Stacks of corpses lay by the crematorium. It was all captured on film.

The next day, journalist Edward R. Murrow visited Buchenwald to report its horror to the world. He was shown a group of children, some as young as six, who an adult inmate introduced with a sarcastic, “Here are our children, the enemies of the state.” Murrow saw the stacks of bodies, he saw them and tried to count at least some of the dead; there were more than 500 in two piles that he tried to inventory, after which he gave up counting. One line in his radio broadcast, which I’ve listened to several times, is, “If I’ve offended you by this rather mild account of Buchenwald, I am not in the least sorry.”

General Patton ordered the U.S. Army to force local German citizens to witness the Nazi atrocities that had been going on nearby. They made both men and women exhume victims from mass graves, where the signs of torture and violent death were clearly evident. It’s said the locals had already once dug up some of the original Nazi gravesites, and then reburied some of the dead farther from town, because they couldn’t stand the stench of death.

Twelve-hundred local civilians were made to march to Buchenwald and parade past piles of dead bodies. They were also forced to pass a table that held alleged human artifacts, including two shrunken heads, a lampshade said to be made of human skin, framed and unframed pieces of tattooed skin, and jars of human organs recovered from the camp’s pathology lab. All this was captured on film that I’ve seen. The smiling faces of the German people on their way to the camp are a stark contrast to the looks on their faces once they got there.

Among many stories the Allies heard from the survivors at Buchenwald were accounts of Ilse Koch's legendary brutality and sadism. So after the war ended, she was one of the first one of the Nazis rounded up, and among the first to be tried by the U.S. military. These were not the famous Nuremberg trials, but rather, a military court martial involving 31 Nazis from the Buchenwald camp, including Ilse Koch. The proceedings began on the first day of April in 1947 at the Dachau concentration camp. Ten days later, on the anniversary of the Buchenwald liberation, despite her claims that none of the accusations were true, a panel of eight American military judges found the Witch of Buchenwald guilty of "violation of the laws and customs of war," and sentenced her to life in prison.

One exceptionally scandalous thing about the trial was that 41-year-old Koch was eight-months pregnant as she took the stand in 1947, despite having been in American custody at Dachau for several years awaiting trial. The father of her child had to be either a fellow inmate, one of her American jailers, or one of the members of the military tribunal helping prepare the case. To this day, no one knows.

After the trial, a review process was demanded, and within a year, all 31 of the Buchenwald convictions were either overturned, or at least had their sentences reduced. The final conclusion of the review commission regarding Koch was that she had been a victim of, "propaganda and mass suggestion." So in 1948, the American military governor and administrator of occupied Germany, General Lucius Clay, commuted Koch's sentence to four years, essentially time served since her arrest. This caused international outrage. Some even alleged Clay had to be the father of her child, based on his clemency. Later, in his biography, Clay claimed he took more abuse over his leniency on Ilse Koch than anything else he did in Germany.

During the review process leading to Clay's decision, Koch's American defense attorney criticized the prosecution based on the lack of any forensic evidence solidly connected to her, especially around the accusations that ornamental goods had been made of human skin at Frau Koch's request.

No lampshade had been produced at trial despite its presence in many photographs taken during the Buchenwald liberation. American authorities admitted it had somehow gone missing.

I've seen a forensic pathology report that was produced for the Judge Advocate General in May of 1945 by a military medical lab in New York City. It describes three pieces of tattooed human skin recovered at Buchenwald. One of the pieces is described as containing two nipples and an umbilicus, that's the scientific name for the bellybutton. A microscopic examination showed it to be consistent with human skin. However, none of the specimens were part of any lampshade or other decorative object, just three separate pieces of what the forensic report calls, "tattooed skin hides." Ultimately, there just wasn't enough concrete evidence in the Buchenwald trial. There was not a single eyewitness account by any of the over 450 inmates that testified to show that Ilse Koch played any part in either choosing inmates for tattoo harvesting, skinning any victims herself, or even requesting that items be made out of human skin.

After Clay commuted Koch's sentence, an irate public, including many veterans' organizations, the Jewish World Congress, and ultimately a U.S. Senate subcommittee, pushed that Koch be retried in the German courts. This was particularly to avoid double jeopardy; that's the rule that a person can't be tried twice for the same offense. She would have to be prosecuted for what she had done to German nationals, rather than citizens of other countries.

In addition, no one could produce any official documentation outlining the role of SS wives in concentration camps. Plus, since Koch wasn't technically on staff at Buchenwald, some questioned whether a military tribunal really had any authority over her in the first place. The general perception became, even among German citizens, that in some ways, Koch had been more depraved than the SS guards, since they were acting on military orders, while she was acting by choice. So Ilse Koch was rearrested in 1949 and charged by the Prosecutor's Office in Bavaria, West Germany.

The Witch of Buchenwald's second trial—well, really her third, considering the first charges brought against her and her husband by the SS—this trial began in 1951 in Augsburg and lasted seven weeks. This time, four witnesses for the prosecution testified that Ilse Koch was observed personally involved in either choosing prisoners with tattoos or other actions related to the manufacture of human skin lampshades. But again, no lampshades or any other ornamental objects made from human skin could be produced, so that part of the charge against her was dropped. But in January of 1951, the Witch of Buchenwald was found guilty of incitement to murder, incitement of attempted murder, and incitement of bodily harm. She was again sentenced to life in prison.

Now, I'm not trying to say Koch was innocent in any way, but she was certainly vilified in both the German and international media despite clear lack of evidence in any of her trials of specific, diabolical, personal behavior on her part. For this reason, some scholars have said that, in a way, Ilse Koch had become larger than life and really came to symbolize the depravity of the entire Nazi regime, perhaps second only to Hitler himself, and primarily because she was a woman—especially when you consider many of the cultural norms surrounding femininity during the mid-20th century.

Ilse Koch had been a pretty young woman, who rose from the sophisticated and enlightened German society, but joined the Nazis and ultimately wound up at Buchenwald. That she, who, at the time had been the mother of two small children, saw the horrors at Buchenwald first-hand and not only did nothing to stop them, but became part of the madness, well that seemed to epitomize how truly horrible and depraved things had gotten in Nazi Germany.

What Ilse Koch was guilty of or not, we will never really know, but the fact remains that from the time she was picked up by the American military at her sister-in-law's home in the summer of 1945 after being recognized by a former prisoner of Buchenwald, she spent the rest of her life incarcerated, either by sentencing or awaiting one trial or another. Whether you believe she was as hideous as accounts say, or was a scapegoat or surrogate for the Nazi atrocities, she never walked free again. In 1967, Ilse Koch hanged herself in a German prison.

Shortly before her suicide, Koch was visited by her son, Uwe, born when she was in prison. He had been taken from her at birth, but learned at age 19 that Koch was his mother. If she ever told her son his father's identity, it's never been made public. During her incarceration, Koch wrote, including to her children, of the utter cruelty that she had endured in 10 long years of difficult prison, due to Jewish slander and perjury. In my opinion, even if she wasn't guilty of all the things she's said to have done, even if the Buchenwald accounts of her were of mythic proportions, surely, what she had seen happen to the Jews and others could in no way be compared to her incarceration.

Now, in the few minutes we have left, I'd like to talk a little more about alleged Nazi objects made of human tissue. Some of the artifacts, seen in the countless images and films from the Buchenwald liberation have been housed for many years at the National Museum of Health and Medicine in Silver Spring, Maryland. That's an interesting facility with a history that dates back to the Civil War, when it first served as the Army Medical Museum. It was also affiliated with the former Armed Forces Institute of Pathology, a group I taught with several times at annual forensic anthropology courses. I've visited the museum, too and have a couple of forensic anthropology colleagues who work there.

Among the museum's collections from World War II are five pieces of tattooed leathery skin from Buchenwald. There's also a sixth one housed at the National Archives, in DC, likewise confiscated from Buchenwald. Reports indicate that as of 2004, three had been positively identified as human. A couple of the pieces have small, round cutouts on their left side that appear to be from a hole punch, so perhaps they were kept in a binder at one time. The one from the National Archives is larger and has pinholes around its edges. In photos from Ilse Koch's 1947 military trial at Dachau, what look like these same three pieces are tacked up as courtroom exhibits. Perhaps that's where the pinholes originated. At one time, the piece at the National Archives had a label calling it part of a, "human skin lampshade," but the shade on the table in the Buchenwald liberation images doesn't show any tattooed design of any kind; it's just a solid-colored beige lampshade.

One of my forensic anthropology colleagues at Michigan State University, Dr. Norm Sauer, was called on not too long ago by the Holocaust Memorial Center in Michigan to try to authenticate some alleged Nazi souvenirs.

Among these were two chess sets made of bone, a bar of soap said to be made with human fat, some bone fragments and ashes, and a lampshade, fairly similar to the one in the Buchenwald photos. Forensic analysis showed none of the items were of human origin, with the possible exception of some of the tiny bone fragments. In 2010, a swastika-embossed bar of soap for sale in a Montreal curiosity shop was also tested, but no human DNA was found.

Six months after Hurricane Katrina hit the U.S., a man found a lampshade at a garage sale; the seller said it was made from "The skin of Jews." After paying $35, the man took the lampshade home, examined it and was shocked that he could see pores and wrinkles in it. A year later he sent the shade to a New York magazine writer who paid $5,000 for forensic testing, which showed it was made of human skin. Labs in Germany and Israel confirmed the DNA testing, but nobody knows for sure where the shade originated or whether it's connected to the Nazis. The owner has tried to donate the lampshade to museums in both Israel and the United States, but all of them have turned him down.

In November of 2013, the auction website eBay apologized for putting over 30 items, alleged to have belonged to Holocaust victims, on their site. They quickly removed the items from listing, saying eBay doesn't allow such postings, but somehow those slipped through their review process. Although the horrors of the Nazi atrocities should never be forgotten by the public, they also should not be publicly marketed.

The Spies Have It

Lecture 21

Espionage is as old as human history itself. Clandestine observation and reporting have affected political affairs around the globe and caused battles to be won and lost. When you think about it, we probably don't even know about history's best spies or most interesting cases of espionage because they were never exposed. In this lecture, we'll look at a few famous examples of undercover activities of the last 100 years.

The Duquesne Spy Ring

- William Sebold had served in the Imperial German Army during World War I but moved to the United States after the war's end. By 1936, he was a naturalized U.S. citizen and worked as an engineer in various industrial plants. In 1939, during a trip to Germany, Sebold was approached by a high-ranking member of the Gestapo to persuade him to spy on America.

- Fearing danger to family members still living in Germany, Sebold reluctantly agreed. Under German guidance, he received training in radio communications and microphotography. But while he was still in Germany, Sebold's passport was stolen, forcing him to make a trip to the American consulate in Cologne, where he revealed his recent encounter with the Gestapo.

- Sebold told U.S. officials of the Nazis' plans for him but said that he wanted to work with the FBI as a double agent against Germany. He had been told by the Germans that he was to play the role of Harry Sawyer, a diesel engineering consultant. He would receive and respond to encoded messages from the Nazis.

- In February of 1940, Sebold returned to New York, and within a 16-month period, an FBI agent posing as German spy Harry Sawyer sent and received more than 500 phony messages. The FBI outfitted Sebold himself with an office in Manhattan where he could meet with other German spies; the office was bugged and had two-way mirrors through which Sebold's encounters with German agents could be filmed.

- Among Sebold's German contacts was Frederick Duquesne, who was born in South Africa but became a U.S. citizen in 1913. Duquesne particularly hated the British because several of his family members had been killed in the Boer Wars; as a result, he had spied for Germany during both world wars.
    - Duquesne's World War II spy ring consisted of 33 men and women. Nearly all the men were originally from Germany, and most had become naturalized U.S. citizens.
    - The purpose of Duquesne's spy network was to collect information about military and industrial weaknesses of the United States that could be exploited should America enter the war, which happened, of course, after the Japanese bombed Pearl Harbor.
- The Germans set up a meeting between Sebold—as Harry Sawyer—and Duquesne, during which Duquesne told Sebold about ways German agents could set fire to American industrial plants. Duquesne also shared plans he'd stolen from the DuPont Corporation that illustrated the new atomic bomb being developed by the United States. Sebold relayed the information to U.S. authorities.
- Based on the Sebold sting operation, in late June and early July of 1941, all 33 members of the Duquesne spy ring were captured. In January 1942, all 33 were sentenced to serve a total of more than 300 years in prison. Sebold disappeared after the trial, presumably into a witness protection program. He is thought to have died in California in 1970.

Velvalee Dickinson

- Another infamous spy of World War II was Velvalee Blucher, born in 1893 in Sacramento, California. After attending Stanford

University, Velvalee worked at an agricultural brokerage firm in San Francisco for Lee Dickinson, the man she would later marry. Her husband's company had many Japanese and Japanese-American clients, and as a result, Mrs. Dickinson became fascinated with Japanese culture. In 1937, the couple moved to New York City.

- In New York, the couple opened a shop selling rare and antique dolls. It was run by Velvalee and catered to wealthy collectors in the United States and around the world. At first, the shop barely stayed afloat, but the couple's fortunes mysteriously turned around after the United States entered World War II.

- Less than two months after the Japanese attacked Pearl Harbor, wartime censors began intercepting letters that had been sent to a Buenos Aires address but were returned marked "address unknown." The letters were from different senders in the United States.
    - Each letter mentioned dolls and contained bits of personal information about the supposed sender, who denied writing the letters.

    - More curiously, according to the alleged senders, the personal information about them in the letters was correct, and the signatures even resembled their handwriting, but they hadn't sent the letters.

    - Cryptographers easily determined that the references to dolls and other information in the letters correlated with the status and location of U.S. ships, especially those that had been damaged at Pearl Harbor.

- When trying to connect information from the letters and the unsuspecting U.S. citizens whose names and addresses had been hijacked, the FBI came up with a single commonality: All the victims of the scam had done business with Velvalee Dickinson's New York doll shop.

- The letters were all postmarked between January and June of 1942 from a variety of U.S. cities. The FBI found hotel records establishing that the Dickinsons had been in those cities around the dates the letters were mailed. Digging into Mrs. Dickinson's background, the FBI found that her contacts with a variety of Japanese organizations spanned major cities in the United States and dated back to the early 1930s. Apparently, none of these contacts had notified Velvalee that the woman in Buenos Aires to whom she'd been sending coded messages had moved.

- The FBI watched the Dickinsons for some time, hoping to gain more information about a possible spy ring. After Mr. Dickinson died of heart disease in March 1943, agents decided to close in on Velvalee. She was arrested on January 21, 1944; in August of 1944, she was convicted of evading censorship laws and sentenced to 10 years in federal prison. She got an early release in 1951 and disappeared from public view in the mid-1950s. No one apparently knows how, where, or when she died.

Mata Hari

- The woman known as Mata Hari was born in the Netherlands in August of 1876 as Margaretha Zelle, the only daughter of a wealthy businessman who eventually went bankrupt. At age 18, she married Rudolph MacLeod, a Dutch army captain stationed in Indonesia (known at the time as the Dutch East Indies). MacLeod was 20 years older than Margaretha and allegedly abusive to her. Their unhappy marriage led to two children, but the couple's son died at age two.

- While living on the island of Java, Margaretha is said to have become enamored with two things that would end up directing the rest of her life: Indonesian dance and men in uniform. In 1902, she and her husband returned to the Netherlands, separated, and began a custody battle for their young daughter. When her husband threatened to stop supporting the child, Margaretha gave over custody and moved to Paris to start a new life.

- In Paris, Margaretha capitalized on her exotic beauty; she marketed herself as Mata Hari, a Hindu princess from the Far East, and quickly became a successful exotic dancer. She began to have relationships with rich and powerful men, including military officers and politicians. In 1914, she agreed to a six-month contract in Berlin, but when World War I broke out in August of that year, the job fell through, and Mata Hari returned to the Netherlands.

- She was recruited as a wartime spy by the Germans but apparently did little to earn the money they paid her, except to resume her European travels. Around the same time, she also agreed to act as a double agent for French counterintelligence.

- As a result of her frequent European travels, her liaisons with men of all nationalities, and the fact that she could speak multiple languages, Mata Hari was constantly watched by numerous military intelligence agencies. In 1916, Scotland Yard detained her for questioning but couldn't pin anything on her.

- In January of 1917, the French intercepted a radio message from the German military that credited Mata Hari with transferring information to them. In February 1917, Mata Hari was arrested in Paris and put on trial, but neither the French nor the British could bring forth any definitive forensic evidence or testimony against her. Her defense attorney—a former lover—was not even allowed to cross-examine the prosecution's witnesses.
    - Despite her claims that her international liaisons were due to nothing more than her infamous exotic dancing, Mata Hari was found guilty after only 45 minutes of deliberation and executed by a French firing squad in October 1917.

    - Today, for all the infamy that's been associated with her name, sources disagree on whether Mata Hari actually ever spied for anybody.

Robert Hanssen

- Throughout the late 1980s, the FBI began to recognize that there was a dangerous double agent in its midst. The KGB was increasingly in possession of highly classified information that could only be coming from one or more of the FBI's own agents.

Before Hanssen made his final drop, FBI agents searched his car and found seven classified documents that he had downloaded to his palm device and apparently planned to pass to the Russians.

- When the USSR disbanded in 1991, the KGB's successor agency became known as the SVR. For a while after the transition, it looked as though the leaks had stopped, but soon, the double agent appeared to be working again. By the late 1990s, the FBI and the CIA were focused on one particular man, but after two years of investigation and surveillance, they realized their target was not the mole.

- The FBI next identified a former KGB operative who claimed he could find information on the double agent. The Russian agreed to surrender files he had stolen on the American mole in exchange for the FBI's promise of $7 million, along with safe passage to the United States and new identities for his entire family; to this day, that Russian's name remains classified.

- In November of 2000, the FBI identified the mole as Robert Hanssen, an expert in wiretapping, surveillance, and computers who had worked for the FBI for nearly 25 years. In December

2000, the FBI promoted Hanssen to a new position and gave him an assistant, who was assigned to watch Hanssen's every move. Agents also began 24-hour surveillance of Hanssen's home.

- Hanssen was finally caught making a drop on February 18, 2001. Since 1985, he had apparently passed thousands of pages of documents to the KGB and, later, the SVR. His motive was purely financial; over the two decades he was a double agent, he received nearly $1.5 million in cash and diamonds from the Russians. On July 6, 2001, Hanssen pled guilty to espionage and was ultimately sentenced to life in prison with no possibility of parole.

Suggested Reading

Butler and Keeney, *Secret Messages*.

Hynd, *Passport to Treason*.

Robenault, *The Harding Affair*.

Shipman, *Femme Fatale*.

Sylado, *My Name Is Mata Hari*.

Wise, *Spy*.

Questions to Consider

1. How has spying changed over the past 50 years, 100 years, 200 years, or more?

2. What are some of the lines of evidence investigators use to identify spies?

3. How common do you think spying is today, and where do you believe it is occurring? Do you think there's more spying today or less than there was in the 1900s?

The Spies Have It

Lecture 21—Transcript

Espionage is as old as human history itself. The clandestine watching of others and reporting on their actions has caused battles to be won and lost, as well as affected all types of relationships. When you think about it, we probably don't even know about history's best spies, since they were never caught or exposed. But, in this lecture, let's look at a few of the more notorious cases of espionage over the last 100 years.

World War II was a particularly active time for some fairly sophisticated spying, and the most successful agents, of course, were those with genuine connections to more than one country. William Sebold, born in 1899 in Mülheim, had served the Imperial German Army during World War I, but moved to the United States a couple years after the War's end. By 1936, he was a naturalized U.S. citizen. Sebold worked as an engineer in various industrial plants, including some involved in aircraft manufacturing. In 1939, during a trip to Germany to visit his sick mother, a high-ranking member of the German Secret Service approached Sebold to persuade him to spy on America.

Fearing danger to his family still living in Germany and recognizing the Germans knew about an early arrest he hadn't reported during his U.S. citizenship application, Sebold reluctantly agreed. Under German guidance, Sebold received training in both radio communications and microphotography and was given instructions for returning to the U.S. But while still in Germany, Sebold's passport was stolen, forcing him to make a trip to the American consulate in Cologne, where he revealed his recent encounter with the Gestapo.

Sebold told U.S. officials of the Nazis plans for him, but said he wanted to turn the tables and work with the Federal Bureau of Investigation as a double agent against Germany. He told the consulate, the Nazis had instructed him to return to the U.S. and serve as an intermediary between Germany and other German spies in America. He was to play the role of Harry Sawyer, diesel-engineering consultant. The Germans told Sebold he was to receive encoded messages from the Nazis and likewise respond to them in code.

In February of 1940, Sebold did return to New York, and within a 16-month period, using a Long Island shortwave-radio transmitter, not Sebold, but the FBI posing as German spy Harry Sawyer, sent over 300 phony messages to the Nazis and received over 200 messages from them in return. The FBI also outfitted Sebold with an office in Manhattan, where he could meet with other German spies while collecting the forensic evidence that would incriminate them. The FBI bugged the office with hidden microphones and used two-way mirrors to film Sebold's encounters with German agents.

Among Sebold's German spy contacts was ringleader Frederick Duquesne, who was born in South Africa, but became a U.S. citizen in 1913. Duquesne particularly hated the British, after the Boer Wars killed several in his family, and as a result, had spied for Germany during both World War I and II. The Duquesne World War II spy ring consisted of 30 men and 3 women, one of whom was Duquesne's Arkansas-born girlfriend. Nearly all the male spies were originally from Germany, and most had become naturalized U.S. citizens. At the point Sebold met Duquesne, the purpose of the Duquesne spy network was to collect information about the United States, including any military and industrial weaknesses that could be exploited should America enter World War II, which happened, of course, after the Japanese bombed Pearl Harbor in Hawaii.

Those in the Duquesne spy ring worked in a variety of jobs, including some low-key positions, like trans-Atlantic ship's barber, ship's butcher, ship painter, truck driver, and the owner of a restaurant where covert meetings were held. But other Duquesne spies held high-level technical jobs, including one employed by Carl L. Norden, Incorporated, manufacturer of the Norden bombsight, an aircraft device that ensures bombing accuracy, as well as other highly-confidential materials essential to U.S. national defense. The United States Army Air Forces considered the Norden equipment so top secret that American bombardiers were required to take an oath stating they would defend its secrecy with their own life, if needed.

Anyway, the Germans set up a meeting between Sebold, as Harry Sawyer, and Duquesne, during which Duquesne told Sebold about ways German agents could set fire to American industrial plants. Duquesne also shared plans he'd stolen from the DuPont Corporation in Delaware that illustrated the new atomic bomb being developed by the U.S. Another Duquesne spy, Paul Bante, provided Sebold with detonation caps and dynamite he assumed would be used to carry out an attack. Sebold was able to relay all this German information and forensic evidence into U.S. hands.

Based on the Sebold sting operation, on June 29, 1941, in a massive roundup, 93 U.S. Special Agents captured 30 members of the Duquesne spy ring; the other three were aboard ships, but were arrested at their next ports of call; 19 pled guilty to espionage, but the other 14 entered pleas of not guilty and were brought to trial in Federal District Court in September of 1941. On December 13, 1941, less than a week after Pearl Harbor was attacked, the jury found all of them guilty.

In January of 1942, all 33 members of the Duquesne espionage ring were sentenced to serve a total of over 300 years in prison. Sebold disappeared after the trial, presumably into a witness protection program. He's thought to have died in California in 1970. A key German-intelligence leader later said that Sebold's activity dealt what amounted to a death blow to German spy efforts in the United States. The FBI's first Director, J. Edgar Hoover, called the incident "the greatest spy roundup in the history of the United States." It remains that to this day.

Another infamous spy of World War II was a woman born Velvalee Blucher in 1893, in Sacramento, California. After attending Stanford University, Velvalee worked at an agricultural brokerage firm in San Francisco for Lee Dickinson, the man she would later marry. Her husband's company had many Japanese and Japanese-American clients, so, as a result, Mrs. Dickinson became fascinated with Japanese culture. The couple became quite the socialites; Mrs. Dickinson, dressed in a kimono, attended functions at the Japanese Consulate in San Francisco, and later in New York City, where the couple moved in 1937 after the Great Depression caused their California company to fail. By that time, Mr. Dickinson's health was failing as well.

In New York, the couple opened a doll shop, first in their home, and later at a storefront they leased on Madison Avenue. Velvalee ran the shop, which catered not only to New Yorkers, but other wealthy U.S. and foreign nationals who collected rare and antique dolls. But the Dickinsons' shop was barely staying afloat, often requiring loans to keep it in business. Mysteriously, though, the couple's fortune turned around after the United States entered World War II.

Less than two months after the Japanese attacked Pearl Harbor, wartime censors intercepted a letter that had been sent to Buenos Aires, but was being returned to the U.S. marked "address unknown." The letter had been sent to a woman named Inez Molinali, who no longer lived at that address and was actually named Inez Molinari. The letter's return address was Portland, Oregon, and the alleged sender's name was Sara Gellert. The censor's problem with the correspondence related to a cryptic message about three, "old English dolls" that had been left at a "doll hospital." FBI code breakers figured out the dolls were references to U.S. warships, and the doll hospital was code for a shipyard. When questioned, Gellert said she didn't send the letter, and she had no idea who did.

Over the next several months, mail censors sent the FBI four more similar letters they discovered. Each had been addressed to Buenos Aires in the misspelled name of Inez Molinali, and each had been returned to a different U.S. sender as undeliverable. Each letter mentioned dolls and contained bits of personal information about the supposed sender, who denied writing the letter, or knowing anything about Molinali. More curiously, the alleged senders told the FBI the personal information about them in the letters was correct, and the signatures even kind of resembled their own handwriting, but they hadn't sent the letters. Neither did the U.S. postmarks of all the letters match the cities provided in the return addresses.

One of the letters mentioned a doll dressed in a grass hula skirt being repaired in Seattle, while another talked about a lovely Siamese temple dancer damaged by a tear in its middle, but was being fixed. Cryptographers easily figured out that the doll with the hula skirt referred to a ship that had recently been moved from Hawaii to Seattle, and that the Siamese doll was really a damaged aircraft carrier under repair.

In addition to doll references, the letters contained other curious entries, like a letter allegedly sent from not far outside my hometown of Cincinnati, Ohio. It mentioned a Mr. Shaw, who was said to have destroyed a letter because he was ill and was currently having his damaged car repaired but would soon be back at work. The Cincinnati FBI office sent this to cryptographers, who determined the message was about the U.S.S. Shaw, a destroyer that had been hit by the Japanese but was being repaired. All these letters to Argentina could ultimately be decoded as veiled references to the status and location of U.S. Navy ships, especially those that had been damaged at Pearl Harbor.

When trying to connect the information about the dolls, the Japanese, and the unsuspecting U.S. citizens, all of those whose names and addresses had been hijacked, the FBI came up with one single commonality: All victims of this scam had done business with Velvalee Dickinson's New York doll shop. In fact, at one time, each had written to Velvalee from their home addresses, and as a result, had unknowingly provided her with samples of their signature, as well as the personal details about themselves that ended up in the letters.

An FBI document examiner determined that although the signatures on the letters were similar to each of the alleged senders' actual handwriting, all did show standard characteristics of forgery, like inconsistencies in letter formation and a lack of natural flow. The letters themselves were typed, and the keystroke hammers of old-style typewriters become increasingly unique over time due to wear. Investigators could tell that five different typewriters had been used to formulate the letters. One of the alleged senders was able to provide the FBI with a typed letter previously sent to her by Velvalee Dickinson. The forensic examiner could tell it had been created using the same typewriter as one of the letters addressed to Buenos Aires.

The five letters were all postmarked between January and June of 1942, and from a variety of U.S. cities, four on the West Coast, one from New York. So the FBI next focused on those places. They found hotel records establishing that the Dickinsons had been in each of those cities around the dates the letters were mailed. Investigators examined typewriters available to guests in those hotels and could match the other four forged letters to hotel typewriters. Digging into Mrs. Dickinson's background, the FBI found that her contacts included a variety of Japanese organizations and spanned San Francisco, New York, Washington, DC, dating back to the early 1930s. They further learned the Dickinsons had regularly entertained Japanese officials in their San Francisco home.

Apparently, none of Velvalee's Japanese contacts had notified her that the Buenos Aires woman, to whom she'd been sending coded messages, had moved. While Inez Molinari had lived at that address at one time and was married to a man who did business with the Japanese, when she was found and questioned, she claimed to know nothing about the letters the code breakers had intercepted. She had no connection with the Dickinsons and didn't even collect dolls.

And, why did Velvalee Dickinson choose those third parties whose return addresses she fraudulently used? The best the FBI could figure, some of those people were clients with whom the Dickinsons had a negative business interaction, like a longstanding unpaid balance with the doll shop. Velvalee probably would have done better to use the addresses of complete strangers who may not have been so easily traced back to her.

Now, the FBI watched the Dickinsons for some time, hoping to gain more information about a possible spy ring to which they belonged, but after her husband died of heart disease in March of 1943, they decided to close in on Velvalee. On January 21, 1944, two agents followed as the petite 50-year-old woman, conservatively dressed as usual, walked into her bank. When she opened her safe-deposit box, the men grabbed Velvalee and put her under arrest. Although she struggled with the agents, she was barely 5′ tall and weighed only about 100 pounds, so obviously was no match. The agents confiscated the bank box, finding cash and checks totaling nearly $16,000, about 10,000 of which was in $100 bills. That was quite a sum of money at that time and prompted a lot of explaining for a woman whose doll shop had been known to have a cash flow problem.

Velvalee told the FBI agents she had found the money under her husband's mattress after his death, but she had no idea where he'd gotten it. Agents were able to trace the serial numbers of some of the bills back to Japanese Naval Inspectors' Offices in New York. At that point, Velvalee admitted her husband had been spying for the Japanese. She stated a man, connected to the Japanese Navy, had paid her husband $25,000 in hundred dollar bills to send coded information to the Buenos Aires address. But as part of the investigation, agents questioned Mr. Dickinson's physician, who said the man had been so mentally incapacitated during the time in question, there was no way he could have carried out the plot.

Unable to confidently convict Velvalee with espionage, she was charged with violation of wartime censorship laws, which prohibited U.S. citizens from passing confidential information to foreign entities. While in jail pending trial, they pressed Velvalee not only to admit her guilt, but also provide counterintelligence about Japanese activities, since World War II was still in progress.

Knowing the FBI would continue to investigate and that espionage could carry the death penalty, Velvalee admitted that she herself had written the letters. She told investigators she and her husband would travel between the East and West coasts of the U.S. using their doll business as cover. On these visits, they would watch the shipyards to gain information; they'd hang around near the docks, talking to unsuspecting locals about the repair progress and get other details.

In August of 1944, Velvalee Dickinson was convicted of evading the censorship laws; she was sentenced to 10 years in federal prison. She earned an early release in 1951 and then disappeared from public view in the mid-1950s. No one apparently knows how, where, or when she died. One source I consulted points out an interesting fact, that for all the paranoia around Japanese-Americans that led to the internment camps, in which so many people were held, the only spy ever known to hand over U.S. military secrets after the attacks on Pearl Harbor was little old Velvalee Dickinson.

Here's a comparison I find interesting. Today, people are debating privacy issues related to the U.S. government's monitoring of American phone and email communications with foreigners. During World War II, though, people seemed to take for granted these types of surveillance efforts, which at least in the Velvalee Dickinson case, clearly did net important intelligence about spies in our midst.

When considering famous spies, especially female wartime agents, the name Mata Hari is likely to come up. She was born in the Netherlands in August of 1876, as Margaretha Zelle, the only daughter of a wealthy businessman, who eventually went bankrupt. After her mother's death, Margaretha's search for a better life caused her to answer a Dutch military officer's newspaper ad, soliciting a wife. A few months later, in 1895, at age 18, she became Mrs. Rudolph MacLeod, wife of an Army Captain stationed in Indonesia, at that time known as the Dutch East Indies. MacLeod was 20 years older than Margaretha, and allegedly abusive to her. Their unhappy marriage led to two children, but the couple's son died at age two, supposedly from treatment for congenital syphilis the children's father passed to Margaretha, before she gave birth.

It was while living on the island of Java in the Dutch East Indies that Margaretha is said to have become enamored with two things that would end up directing the rest of her life: Indonesian dance, and men in uniform. In 1902, the couple returned to the Netherlands, they separated, and began a custody battle for their young daughter. When her husband threatened to stop supporting the child, Margaretha gave over custody and moved to Paris to start a new life. Beautiful and 5′ 10″ tall, Margaretha had dark hair, dark eyes, and an olive complexion, perhaps related either to the days of Spanish domination over what's now the Netherlands, or the 17th- and 18th-century colonialism of the Dutch East India Company, centered in what's now Jakarta. Margaretha capitalized on her exotic looks and knowledge of Indonesian dance; she reinvented herself as Mata Hari, which means, "the eye of the day," in other words, the Sun. She marketed herself as a Hindu princess from the Far East and quickly became a very successful exotic dancer. Mata Hari's veil dances were notoriously unusual for the day, because they would include taking off all of her clothing, except for a jeweled bra she apparently refused to remove, being self-conscious of a small bust.

In great demand throughout Western Europe, Mata Hari began to have both short and long-term relationships with numerous rich and powerful men, including military officers and politicians. She was always able to find men to spend money on her, and many considered her little more than a high-paid prostitute, known in those days as a courtesan. In 1914, Mata Hari agreed to a six-month contract in Berlin, but when World War I broke out in August of that year, the job fell through, and at nearly 40 years old, she returned to the Netherlands.

A man from the German consulate in Holland contacted the well-connected Mata Hari to tell her the Germans were recruiting wartime spies. She was given the codename H21 and a cash payment, but apparently did little to nothing to earn the money, except resume her European travels. Returning to Paris, Mata Hari allegedly met the love of her life, a Russian military officer named Vladimir Maslov who was only 21, nearly two decades younger than the former dancer. After returning to the front, Vladimir was injured and taken to a hospital in northeastern France. Because the hospital was within the war zone, civilians had to get permission to travel there, and while Mata Hari was applying for a permit, she met Georges Ladoux, the captain of French counterintelligence.

Even though she had already accepted German espionage money, Mata Hari swore to Ladoux that, like her Dutch homeland, she was neutral, and due to her many years in Paris was really a French sympathizer. Ladoux asked if she would spy on the Germans for the French, and Mata Hari agreed to seduce a German officer to obtain information for Ladoux. All this, mind you, supposedly to get a pass to visit her young Russian lover. Some say the mission backfired on her, and that the French were actually setting her up so they could later claim she was passing French secrets to the Germans.

Due to her frequent European travels, her liaisons with men of all nationalities, and the fact that she could speak French, English, Italian, and some German, in addition to her native Dutch, Mata Hari was constantly watched by all manner of military intelligence agencies. In 1916, Scotland Yard detained her for questioning but couldn't pin anything on her. She allegedly told them she was working for the French Intelligence but was suspected of being a double agent for the Germans.

In January of 1917, the French intercepted a radio message from the German military that credited Mata Hari, their agent H21, with transferring information to them. Many now believe the German military may have actually set the former dancer up so the French would no longer trust her. In February of 1917, Mata Hari was arrested in Paris and put on trial, but neither the French nor the British could bring forth any definitive forensic evidence or testimony against her. Her defense attorney, a former lover, was not even allowed to cross-examine the prosecution's witnesses. Despite her claims that her international liaisons were due to nothing more than her infamous exotic dancing, Mata Hari was found guilty after only 45 minutes of deliberation; she was executed by a French firing squad in October of 1917. Today, for all the infamy that's been associated with her name, sources disagree on whether Mata Hari actually ever spied for anybody, insinuating her actions were merely those of an attention-seeking, romantic gold-digger.

Our last story is more recent and takes us to the United States. Throughout the late 1980s, the Federal Bureau of Investigation began to recognize there was a dangerous double agent in their midst who was posing a threat to national security. The Soviets' own security and intelligence agency, at that time, the KGB, was increasingly in possession of highly-classified information that could only be coming from one or more of the FBI's own agents.

When the Union of Soviet Socialist Republics disbanded in 1991, the KGB's successor became known as the SVR, the new Russian foreign intelligence service. For a while after the transition, it looked as though the leaks had stopped, or at least slowed considerably, but soon the double agent appeared to be at it again. By the late 1990s, forensic profilers within the Bureau were focusing on one particular guy, who both the FBI and Central Intelligence Agency began to investigate. They secretly entered the agent's home, searched his property, bugged his phone, and began round-the-clock surveillance. After two years, however, which eventually included direct questioning of the suspect and his entire family, the FBI and CIA realized they had been watching the wrong man.

The Bureau's next decision was to try and find a Russian, who would be willing to give up the name of the traitor in exchange for money. They settled on a former KGB operative, who said that while he didn't know the double agent's name, he knew where to find it. The man went to Russia's SVR headquarters and actually stole files he knew were related to the American mole. He informed the FBI the materials included correspondence between their double agent and the KGB between the years 1985 and 1991. The Russian agreed to surrender the files in exchange for the FBI's promise of $7 million, along with safe passage to the United States and new identities for his entire family; to this day, that Russian's name remains classified.

In November of 2000, the FBI obtained the package, and while it didn't contain the name of the double agent, it included documents, computer disks, and an audiotape. The recording was of a July 21, 1986 conversation between the FBI mole and a KGB agent. Upon listening to the tape, investigators thought they recognized the voice as agent Robert Hanssen, one of the Bureau's wiretapping, surveillance, and computer experts, who had worked for the FBI for nearly 25 years in their Washington, DC and New York offices. The materials also included a plastic bag, and despite over a decade, the FBI's forensic unit was able to lift two fingerprints off it. They belonged to Hanssen.

Although the FBI was sure they had their man, they had to catch him in the act. Interestingly, to do so, in December of 2000, they promoted him to a new position and gave him an assistant, who was, of course, assigned to watch Hanssen's every move at work. His assistant was able to download Hanssen's palm device, you know, those precursors to smartphones, and he recovered incriminating encrypted data. Agents also searched Hanssen's car while it was parked in the FBI office garage and found white adhesive tape and white chalk in the glove compartment. In the trunk, they discovered seven classified documents Hanssen had downloaded, which they assumed he planned to pass to the Russians.

In Hanssen's suburban DC neighborhood, the FBI even purchased the house across the street, and began 24-hour surveillance. On several occasions, in December of 2000, agents observed Hanssen driving slowly past a local recreational spot called Foxstone Park. By January of 2001, he was driving past the same place in the park almost every night. They realized he was looking for a sign from the Russians that they were ready for his next drop.

On Sunday, February 18, 2001, agents stationed in the park watched Hanssen exit his car and place a piece of white tape on a park sign. Next, he walked off into the woods carrying a package and taped it underneath a wooden footbridge. At that point, 10 FBI agents came at him from all angles and put him under arrest. Hanssen's reply was, "What took you so long?"

Since 1985, using the name Ramon Garcia, and similar drop-style transfers, Hanssen had passed thousands of pages of documents to intelligence officers of the Soviet KGB and later the Russian SVR, apparently only once meeting any of his contacts face to face. He refused Russian offers of overseas travel, instead making them come to the drop sites that he arranged. Hanssen regularly monitored FBI communications to make sure no one suspected him and was apparently aware of the Bureau's suspicions near the end.

During the mid-'80s, Hanssen also had named three Soviets who had been betraying their country as U.S. double agents. Within three years of being exposed, two of those men had been executed and the third wound up in a Soviet prison. In fact, after the FBI learned their Russian operatives had been exposed, Hanssen himself was the man they tasked with finding out who did it, so he was, in essence at that point, looking for himself. Hanssen's expertise in computer forensics gave him not only the know how to get the information, but also enabled him to avoid detection.

His motive was apparently purely financial; over the two decades Hanssen was a double agent, he received nearly $1.5 million in cash and diamonds from the Russians, but was careful to never display any indications he was making more money than a typical agent. On July 6, 2001, 25 years after taking an oath to, "support and defend the Constitution of the United States against all enemies, foreign and domestic," Hanssen pled guilty to espionage and was later sentenced to life in prison, with no possibility of parole.

Motive and Kidnapping

Lecture 22

At the mention of kidnapping, most people think of cases in which someone is held for ransom, but there are many different motives for kidnapping. People have been abducted into slavery, and warring factions have taken prisoners. Human trafficking networks for forced labor or the sex trade are still serious criminal enterprises, and in developed countries, kidnappings of children by noncustodial parents are far more common than any other type of abduction. Motive is often an important consideration in abduction cases because the motive can lead investigators to likely suspects or to locations where victims are being held. In this lecture, we'll look at some historical and modern kidnapping cases.

The Munich Massacre

- Three years after World War II ended, Israel was founded as a Jewish homeland in what was then the territory of Palestine, creating lasting conflict between the Israeli and Palestinian people. Israeli-Palestinian hostility has spawned numerous terrorist organizations, including one known as Black September, formed in 1970. One of its most notorious acts was a mass kidnapping carried out in 1972 during the summer Olympics in Munich, Germany.

- On September 5, 1972, at approximately 4:30 a.m., eight armed members of Black September entered the Olympic Village compound. Within a short time, the terrorists had taken nine Israelis captive and killed two others. They held the hostages in an apartment building of the Olympic Village and dropped a list of demands out the window: They wanted the release of 234 prisoners in Israel's custody and two German founders of a militant neo-Nazi group, the Red Army Faction—all by 9:00 that same morning.

- Among the most striking aspects of this kidnapping was the real-time media coverage it carried, beginning within about a half-hour of the initial attack. This coverage created an additional problem in an already-tense situation: The Palestinian terrorists were also watching everything unfold on TV in the apartment where they were holding the hostages.

- Israel was unwilling to negotiate with the terrorists, fearing further attacks. The Germans offered unlimited money and even a trade of prominent Germans for the captives, but the Palestinians refused.

- One of the keys in hostage situations is to buy time to enable authorities to plan and learn more about the background of the terrorists. Negotiators were about to get four extensions throughout the day, but with all the cameras focused on German police moving into position, by 5:00 p.m., the Palestinians refused any further delays.

- At 6:00, the terrorists demanded to be flown to Cairo, Egypt. German authorities agreed to transport the Palestinians and their hostages to a military airbase outside Munich on two helicopters, followed by a third helicopter of German officials. Authorities told the terrorists they could board a Cairo-bound plane at the airstrip, but in reality, the Germans were planning an ambush. They had snipers positioned when the helicopters landed at the base at approximately 10:30 p.m.

- When two terrorists examined the waiting plane on the airstrip and found no flight crew, they realized that their demands weren't being met. As they ran back to the helicopter, one of the snipers shot and wounded one of the Palestinians. Sharpshooters opened fire and killed two other kidnappers, but the hostages were still bound and unable to get away. The remaining terrorists sought cover and returned fire, and a standoff began.

- By around midnight, the Germans were able to get armed personnel carriers onto the airbase; at that point, the terrorists probably realized they were trapped. At 12:04, a kidnapper opened fire on four hostages in one of the helicopters, then threw in a hand grenade, incinerating the chopper and all those in it.

- Then, the terrorists turned to the second helicopter, where the remaining five hostages were strafed with gunfire. By 1:30 a.m., the battle ended. All of the Israeli captives were dead, along with five members of Black September; three terrorists were captured.

Aftermath of the Massacre

- Three days after the Munich attack, 10 bases of the Palestine Liberation Organization (PLO) were bombed by Israeli airstrikes in Syria and Lebanon, killing an estimated 200 people.
    - A little more than six weeks later, PLO terrorists hijacked a Lufthansa flight and demanded the release of the three surviving Munich kidnappers. German authorities immediately complied, and those terrorists have never been brought to justice.

    - The following April, members of Israel's Mossad intelligence and the Israeli army stormed an apartment building in Beirut, killing three senior PLO members in retaliation for the Olympic executions the year before.

- In 2012, as the 40th anniversary of the Munich Massacre approached, the German magazine *Der Spiegel* published a series of articles, summarizing the results of a comprehensive analysis of formerly classified government documents related to the incident. The analysis alleges that German authorities ignored clear warnings of the Munich attack and points to documentation that German officials intentionally covered up many mistakes they made during the hostage negotiations and the ambush.

- The German government has never publicly acknowledged connections between any German citizen and the Black September group. However, *Der Spiegel* cites a 1973 police report on the Munich Massacre as definitively linking German neo-Nazi criminals to the Munich kidnappers.

- Perhaps most shocking was *Der Spiegel*'s allegation that after the kidnapping, German officials maintained both secret and blatant diplomatic contacts with the masterminds behind the Munich Massacre.

  - In late 1972, a senior German Foreign Ministry official actually sought out the PLO to broker an agreement in which the Palestinians would refrain from attacks on West Germany.

  - Yasser Arafat, then-leader of the PLO, agreed in early 1973 to abide by Germany's request, if the Germans would allow him to install a Palestinian envoy in Bonn. Amazingly, the man Arafat sent—who remained in Germany until 2005—Abdallah Frangi, had been knowingly and directly involved in the Munich Massacre.

- Although German authorities have open warrants for the arrest of the remaining three Munich kidnappers—one of whom has appeared in film documentaries, but two of whom are suspected dead—*Der Spiegel* found no evidence that German officials have ever actually pursued them.

J. Paul Getty Jr.

- In 1957, J. Paul Getty, founder of the Getty Oil Company, was named the wealthiest man in the world, with a net worth estimated between $700 million and $1 billion. One of his five sons was Eugene Paul Getty, who later changed his name to J. Paul Getty Jr.

- After he married and had a son—named J. Paul Getty III—Getty Jr. was sent to Italy to run the Italian branch of Getty Oil. But Getty claimed he wasn't cut out for the oil business, and he and his wife ultimately divorced. He then married a second time, moved to England, and was disinherited by his father.

- After his parents divorced, Getty's son, who was called Paul, stayed in Italy with his mother, but he was a troublemaker and was expelled from school by age 15. On July 10, 1973, at age 16, Paul was kidnapped in Rome.

- His captors called Paul's mother, demanding $17 million, but neither she nor Paul's father could afford the ransom. The estranged couple turned to Paul's grandfather, the original J. Paul Getty, but he refused to pay.

- After months with no resolution, the kidnappers—members of the Calabrian mafia—reduced the ransom to $3 million, but still it was not paid. In November, four months after the kidnapping, a newspaper in Rome received a package containing a decomposing right ear. A note indicated that the ear belonged to Paul, and if the kidnappers didn't receive the money within 10 days—a deadline that had already past—Paul would begin to arrive "in little pieces."

- Authorities wondered if the ear was really Paul's or could have been removed from a corpse. Although 1973 was before the advent of DNA technology for identification, blood typing could have been used to show consistency between Paul's blood proteins and the discovered ear. It's also possible that if an ear print was on record for Paul, it could have been compared to the ear in the package. Rarely, ear prints can be used in forensic science in much the same way as fingerprints.

- After another newspaper received photos of Paul's wounds, his grandfather finally agreed to pay $2.2 million and to loan the balance of the ransom demanded by Paul's captors to Getty Jr. On December 15, more than five months after the kidnapping, Paul was released. The ensuing investigation led to the arrest of nine men, but only two were ever convicted and imprisoned, and most of the ransom money was never found.

Paula Beverly Davis

- When Paula Beverly Davis didn't come home on the night of August 9, 1987, her roommate called Paula's mother at 3:00 a.m. The roommate said that she had last seen Paula earlier that evening at a truck stop along Interstate 70, just outside of Kansas City.

© Photodisc/Thinkstock.

Some hospitals use sensors in wristbands to monitor the movements of infants throughout the facility and prevent kidnappings.

- Mrs. Beverly contacted police and described her daughter, including the fact that she had two tattoos on her chest: a unicorn and a rose. Authorities reminded Mrs. Beverly that Paula was an adult and may have simply disappeared of her own volition, but her family didn't believe Paula would do that.

- The day after Paula disappeared, the recently deceased and nearly nude body of a young, white female, was found along Interstate 70—but it was 600 miles east of Kansas City, in Montgomery County, Ohio. An autopsy was performed, and the cause of death was ruled to be strangulation by ligature. When no identification was made, the woman's body was buried in a cemetery in Dayton.

- For more than two decades, although leads were explored, the identity of the young woman was not known. In the spring of 2009, however, officials from the Montgomery County Coroner's Office entered all of the deceased's pertinent information into the new National Missing and Unidentified Persons System (NamUs).
    - That fall, Paula's sister, Stephanie Beverly Clack, heard about NamUs and visited its website. After just a half-hour on the

system, Stephanie found a match for her sister: a woman found on August 10, 1987, who had two tattoos on her chest, a unicorn and a rose.

- Within two months, DNA testing confirmed the identity of the victim buried in Ohio as Paula Davis. Authorities now believe that her abductor was convicted serial killer Lorenzo Gilyard, although he has yet to be charged with Paula's murder.

Suggested Reading

Concannon, *Kidnapping*.

Douglas and Olshaker, *The Anatomy of Motive.*

Fox, *Uncommon Youth.*

Large, *Munich 1972.*

National Missing and Unidentified Persons System, http://www.namus.gov/.

Primorac and Schanfield, eds., *Forensic DNA Applications*.

Reeve, *One Day in September*.

Questions to Consider

1. What trace evidence might be found at the scene of a kidnapping to lead investigators to the perpetrators and/or victim?

2. Can you name other famous kidnappings and the motives for them?

Motive and Kidnapping

Lecture 22—Transcript

At the mention of kidnapping, most people's first thoughts are of cases where someone is held for ransom money. The West African Republic of Mali, a new stronghold of al-Qaeda, is said to be currently financing terrorism with an estimated $9 million obtained in the past decade from kidnappings for ransom. There are many different motives for kidnapping, though, and the practice is undoubtedly as old as humanity itself.

Abductions have led people into slavery, including countless kidnappings of West Africans during the Atlantic slave trading of the 16th to 19th centuries. Warring factions have taken prisoners, including concubines and children, for political leverage, or simply as a show of power. For instance, since 1986, Joseph Kony, leader of the Lord's Resistance Army in Uganda, has taken over 50,000 children as sex slaves and child soldiers. Today, human trafficking networks for forced labor or the sex trade are still serious criminal enterprises, and especially in developed countries, kidnappings of children by non-custodial parents are far more common than any other type of abduction.

Motive is often an important consideration in forensic cases, because the motive can lead investigators to likely suspects. By following the trail of evidence, whether at an abduction scene or by studying communications made between the hostage takers and those they contact, police may be able to hone in on the criminals involved or the likely location where victims are being held. But sometimes, no matter how much law enforcement studies a situation and tries to negotiate with the perpetrators, kidnappings do not always resolve without bloodshed or death; that goes for victims, kidnappers, or even law enforcement. Let's look at some historical and modern examples.

Three years after World War II ended, a Jewish homeland was founded as the modern state of Israel. The territory selected for Israel's establishment was then Palestine, and controlled by Great Britain by a post–World War I mandate. Problems this decision created have been a source of ongoing conflict between the Israeli and Palestinian people, even to this day. Israeli-Palestinian hostility has spawned numerous terrorist organizations, including one known as Black September, formed in 1970. One of their most notorious acts was a mass kidnapping, carried out in 1972, during the summer Olympics in Munich, Germany.

Now, 1972 was the first time the Olympic Games were held in Germany since 1936, when the Nazis were in power. Hitler wanted to use the Berlin Olympics as an attempted show of Aryan white superiority. By the way, the African-American athlete, Jesse Owens, took four gold medals in Berlin, but I digress. Anyway, the Germans dubbed the 1972 Summer Olympics The Happy Games, and approached the event in a welcoming and relaxed manner, which some, including the head of the Israeli delegation, thought a little too lackadaisical, given the ongoing tensions in the Middle East and the deep wounds the German Holocaust caused the Jewish people, if not the entire world.

On September 5, the 10th day of the games, at approximately 4:30 in the morning, eight Palestinian members of the Black September group, dressed in track suits and carrying duffel bags containing assault rifles, handguns, and grenades, climbed a six-foot fence, unwittingly aided by some Canadian athletes, and they entered the Olympic Village compound. Most of the Israelis were housed on the first floor of a small building. The terrorists entered Apartment 1, where five Israeli coaches and officials were housed. One of those men was shot, taken hostage, and ordered to escort the captors through the building to find more Israelis. This initial hostage bypassed Apartment 2, where sharpshooters, a race walker, and the delegation leader were sleeping. The hostage told the attackers those in Apartment 2 were not Israelis, leading them instead to Apartment 3, where six wrestlers and weightlifters were staying. Forensic analysts suspect the hostage thought those bulky athletes might be better able to overpower the terrorists than the guys in Apartment 2.

The six sleeping men in the third apartment were rounded up and made to return with their captors to the first apartment. Once there, the hostages began to put up a fight, and two were shot and killed by the Palestinians. One body was thrown out the front door of the apartment to show authorities the captors were serious, and the other dead hostage was left on the floor of Apartment 1 with the remaining nine Israeli captives, who had been bound and then tied together. The commotion awoke others nearby, including the men in Apartment 2, and several other Israelis were able to escape the building without capture. Next, the Palestinians dropped a list of demands out the apartment window. They wanted the release of 234 prisoners in Israel's custody, plus the release of two German founders of a militant neo-Nazi group, the Red Army Faction, and all by 9:00 that same morning.

Among the most striking things about this kidnapping was the real-time media coverage it carried all over the world, beginning within about a half-hour of the initial attack. I remember images of German police snipers dressed in tracksuits climbing up the building where the hostages were held and terrorists in ski masks on the balconies. If you're around my age or older, and from the U.S., you probably recall ABC Olympic broadcaster Jim McKay narrating live coverage of the disaster. This technological issue created an additional problem for the tense situation; the Palestinian terrorists were also watching everything unfold on TV in the apartment where they were holding the hostages.

We've all seen television or movie dramas in which hostage negotiators frantically move officers into place, deciding what to tell the captors and what to hold close to the vest, but with that amount of media coverage, the Munich events were completely out in the open. Reporters and their cameras were, in a sense, communicating directly with the hostages and terrorists, which had the potential to circumvent authorities. This was an extremely dangerous situation. Israel was unwilling to negotiate with the terrorists, fearing further attacks if they did, so this put Germany in a really difficult position, because all the hostages were Jewish, and we all know the history there. The Germans offered unlimited money and even to trade prominent Germans for the captives, but the Palestinians refused.

One of the keys in hostage situations is to buy time so that authorities can figure out the best plan and learn more about the background of the terrorists, which can allow more effective dialogue. Negotiators were first able to push the deadline until noon, and then got three more extensions throughout that day, telling the captors they were considering their demands. But with all the cameras focused on German police moving into position and sportscaster

Jim McKay giving blow-by-blow descriptions of the events, by 5:00 pm the day of the kidnapping, the Palestinians refused any further delays.

At 6:00, the terrorists demanded to be flown to Cairo, Egypt. So, German authorities agreed to transport the Palestinians and their hostages to a military airbase outside Munich on two helicopters, followed by a third helicopter of German officials. Authorities told the terrorists they could board a Cairo-bound Boeing 727 at the airstrip, but in reality, the Germans were planning an ambush. They had snipers positioned when the helicopters landed at the base at approximately 10:30 that night.

When two terrorists examined the waiting plane on the airstrip and found no flight crew, they realized things were not going as they had demanded. As they ran back to the helicopter, one of the ambush snipers started shooting, wounding one of the Palestinians. Sharpshooters then opened fire and killed two other kidnappers that were guarding the helicopter pilots, allowing the pilots to run to safety. But the hostages were still bound and unable to get away. The remaining terrorists sought cover and returned fire, shooting out several airport spotlights, and in that partial darkness, a standoff began.

By around midnight, the Germans were able to get armed personnel carriers onto the airbase, and forensic analysts believe the terrorists probably realized they were trapped at that point. So at 12:04 a kidnapper opened fire on the four hostages in one of the helicopters with an assault rifle, and then threw in a hand grenade, incinerating the chopper and everybody in it. Then, when the terrorists turned to the second helicopter where the five remaining hostages were located, they were strafed with gunfire, and killed. By 1:30 a.m., the battle ended, with five members of the Black September terrorist group dead and three captured. One German policeman also lost his life. All of the Israeli captives were dead. I can still vividly remember Jim McKay's broadcast, which was in part, "Our worst fears have been realized tonight. … They're all gone."

In the forensic investigation that followed, tooth marks were found on the ropes that bound the men in the second helicopter, indicating how desperately they were trying to release themselves and each other. Autopsies on some victims in the burned helicopter were problematic, because their bodies were so charred, but at least one of them survived the shooting, because he had soot in his airways, showing he died of smoke inhalation. But forensic analyses don't consist only of physical evidence.

It was later discovered that before the start of the games, three of the Black September members had managed to get jobs in the Olympic compound to gain familiarity with the location. That might also render them more familiar to witnesses, so their presence wouldn't initially raise alarm. Numerous sources have also alleged that neo-Nazis operating in Germany aided the Black September group, and that's why the Palestinians asked for the release of two German neo-Nazi prisoners, among their ransom demands.

Scholars have debated the relationship of the Black September group to the larger Palestinian Liberation Organization, commonly known as the PLO, or it's Fatah branch, controlled by Yasser Arafat back then. But, three days after the Munich attack, 10 PLO bases were bombed by Israeli airstrikes in Syria and Lebanon, killing an estimated 200 people. A little over six weeks later, PLO terrorists hijacked a Lufthansa flight bound for Frankfurt and demanded the release of the three surviving Munich kidnappers. German authorities immediately complied, and those terrorists have never been brought to justice. The following April, members of Israel's Mossad intelligence and the Israeli army stormed an apartment building in Beirut, Lebanon, killing three senior PLO members in retaliation for the Olympic executions the year before.

In 2012, as the 40th anniversary of the Munich Massacre approached, the German magazine *Der Spiegel* published a series of articles summarizing what they've learned from a comprehensive analysis of formerly-classified government documents that are related to the incident. The reporters allege that German authorities ignored clear warnings of the Munich attack, even from one of their own embassy personnel. That person reported about three weeks prior to the kidnapping "An incident would be staged from the Palestinian side during the Olympic games in Munich."

German police were also aware of an article that ran three days before the kidnapping in an Italian magazine, about a "sensational act during the Olympic Games." They knew that was being planned by the Black

September organization. The critical analysis also points to documentation that German officials intentionally covered up many mistakes they made during the hostage negotiations and the ambush, mistakes involving everything from erroneous assumptions about the number of kidnappers to inadequate numbers, preparation, and weapons of their German snipers.

The German government has never publicly acknowledged connections between any German citizen and the Black September group. However, *Der Spiegel* cites a 1973 final police report on the Munich Massacre as definitively linking German neo-Nazi criminals to the Munich kidnappers. This connection was made through a forensic comparison between weapons found on two German neo-Nazis arrested in October of 1972 and those used by the Olympic hostage-takers. Specifically, the grenades the neo-Nazis had were made in Belgium with Swedish explosives and had been manufactured for use in Saudi Arabia. Now, those rare grenades were identical to the ones used to blow up the helicopter in the Munich Massacre. The arrested neo-Nazis were also carrying a threatening letter to the Munich judge involved in the ongoing legal investigation of the Olympic incident.

But most shocking, at least to me, was *Der Spiegel*'s allegation, based on the declassified documents, that after the kidnapping, German officials maintained both secret and blatant diplomatic contacts with the masterminds behind the Munich Massacre. Within six months after the Olympics, the German authorities were attempting to appease the Palestinians in order to prevent any further attacks on German soil. Allegedly, even the BKA, which is the German equivalent of the U.S. Federal Bureau of Investigation, was involved in these secret meetings. In late 1972, a senior German Foreign Ministry official actually sought out the PLO to broker that the Palestinians refrain from attacks on West Germany so that good relationships, including, of course, German oil supplies from the Middle East, be maintained.

Yasser Arafat, then leader of the PLO, agreed in early 1973 to abide by Germany's request, if they would let him install a Palestinian envoy in Bonn. Amazingly, the man Arafat sent, and who remained in Germany until 2005, Abdallah Frangi, had been knowingly and directly involved in the Munich Massacre. The Black September kidnappers had phoned him several times during the day of the Olympic killings. Although German authorities have open warrants out for the arrest of the remaining three Munich kidnappers, one of whom has appeared in film documentaries, but two of whom are thought to be dead, *Der Spiegel* found no evidence that German officials have ever actually pursued them.

Now, when kidnapping is for ransom of a financial, rather than political nature, the wealthy are obvious targets, so let's examine a kidnapping for money. In 1957, Fortune Magazine published its first-ever list of the richest people in the world. Any idea who they named the wealthiest man in the world that year? This guy had a net worth estimated at between $700 million and $1 billion U.S. dollars, and that's 1957 dollars. Would it help if I told you the source of that wealth was oil?

J. Paul Getty, founder of the Getty Oil Company, got his start in his father's oil fields in Oklahoma in the early 1900s. Despite his riches, he was a known tightwad, famously installing a pay phone in his mansion for guests. Among J. Paul Getty's five sons, from as many wives, was Eugene Paul Getty, who later changed his name to John Paul Getty, making him J. Paul Getty Jr. After he married and had a son, who would be John Paul Getty III, Junior was sent to Italy to run the Italian branch of Getty Oil, where he and his wife had three more children. Junior claimed he wasn't cut out for the oil business; he was a drug and alcohol abuser as well as a womanizer. When the couple ultimately divorced, J. Paul Getty Jr. took up with a second wife, moved to England, and was subsequently disinherited by his wealthy father for being a hippie and an embarrassment to the family. And this wasn't the first time Junior had disappointed his daddy. In fact, from a very early age, the Senior Getty would return letters his young son sent him with any misspellings underlined and corrected.

Junior's son, J. Paul Getty III, who went by the name Paul, was a partier like his father. After his parents divorced, Paul stayed in Italy with his mother, but was expelled from school by age 15. The boy had long, red, curly hair, was a known drug abuser, cavorted with older women, and had even posed nude in a magazine. In January of 1973, he was arrested for throwing a Molotov cocktail at an anti-fascist demonstration, but officials dropped the charges. And then, on July 10, 1973, at age 16, J. Paul Getty III was kidnapped in Rome.

His captors called Paul's mother, and they demanded $17 million in ransom. She didn't have that kind of money, and because her ex-husband was disinherited, J. Paul Getty Jr. couldn't afford the ransom, either. The estranged couple turned to Paul's grandfather, the original J. Paul Getty. And his reply was, "I have 14 grandchildren; if I pay one penny now, I'll have 14 kidnapped grandchildren." The older Getty also wondered if the entire thing wasn't a hoax contrived by his grandson to extort money.

The captive Paul sent a pleading letter to his mother, saying he was chained to a stake in mountain cave and was being tortured. Paul believed he would be killed if his mother or his family didn't pay up. After months with no resolution, the kidnappers, members of the Calabrian mafia, reduced the ransom down to $3 million, but still, it was not paid. In November, four months after the kidnapping, a newspaper in Rome received a package containing a decomposing right ear; a postal strike had delayed the delivery by nearly three weeks. A note with the body part indicated the ear belonged to Paul, and if the kidnappers didn't receive the money within 10 days, a deadline that was already past, the other ear would be sent. "In other words, he will arrive in little pieces."

Authorities wondered if the ear was really Paul's, or could have maybe been removed from a corpse. While 1973 was prior to the advent of DNA technology for identification, but blood typing could have been used to show consistency, or lack thereof, between Paul's blood proteins and the discovered ear. The success of such tests would have depended largely on the state of decomposition. It's also possible that if an ear print—yes, you heard right, an ear print—for some reason was on record for J. Paul Getty III, it could have been compared to the ear in the package, again, depending on how intact its features were and its state of preservation.

Rarely, ear prints can be used in forensic science in much the same way as fingerprints, although the first prosecution using ear prints wasn't made until 1998 in England. During a two-year data-collection period, as part of an effort to create an ear-print database, UK authorities detected about 100 ear prints of forensic value at crime scenes they investigated. The study found that even identical twins have distinguishable ear prints. Although not widely used or universally accepted, ear biometrics continue to be studied in forensics. Though in the Getty case, the most compelling evidence that the ear in question belonged to Paul was the presence of a lock of long, curly, red hair that accompanied it.

After another newspaper soon received photos of Paul's wounds, his grandfather finally agreed to pay $2.2 million. True to his frugality, that's the maximum amount his accountants said would be tax-deductible. The balance of the ransom Paul's captors demanded was loaned by the Senior Getty to his son, Junior, at four-percent interest. On December 15, over five months after the kidnapping, Paul was finally released. He was found bruised and bandaged near an abandoned gas station in Calabria, Italy in a rainstorm. The investigation led to the arrest of nine men, but only two of the lower-level gang members were ever convicted and imprisoned, and most of the ransom money was never found. After his recovery, Paul's mother had him call his grandfather to thank him for financing his return, but J. Paul Getty, Sr., wouldn't even come to the phone. Paul went on to a life of drugs and alcohol, and as a result, suffered a massive stroke at the age of only 24; he spent the rest of his life in a wheelchair and died at age 54 in 2011.

There have been many notable kidnappings of children for ransom. The very word kidnapping, coined in the late 17th century, refers to the nabbing of kids. The 1920s and '30s saw a whole rash of child abductions in the United States and elsewhere. The famous snatching of the infant son of American aviator Charles Lindberg prompted the Federal Kidnapping Act of 1932, but that hasn't stopped criminals from grabbing kids from homes, and even newborns in hospitals.

Two cases that made news during the summer of 2012 illustrate how today's technology is solving, and even thwarting, hospital abductions of infants. In July of 2012 a woman was sentenced to 12 years in prison for stealing a baby from a New York hospital in 1987 and raising that baby girl as her own. The case came to light because in 2010, the 23-year-old kidnapping victim needed to get a birth certificate, ironically, because she was pregnant herself and was unable to find a birth certificate in her name. She got suspicious and reached out to the National Center for Missing and Exploited Children, a group I've interacted with as part of my work with the National Missing and Unidentified Persons System. DNA testing proved the young woman was actually the baby girl who had been abducted from the hospital 23 years prior. The kidnapper claimed two stillbirths in the late 1980s prompted her to steal the baby.

The second case happened in August of 2012, when a woman disguised herself as a nurse and attempted to steal a baby from a California hospital in a tote bag, allegedly to try to win her estranged husband back by claiming she had borne him a child. High-tech security methods, now in place in most hospitals, at least across the U.S., thwarted the crime. The hospitals place a sensor on the baby's wrist or ankle that monitors the infant's movements throughout the hospital. If the baby is moved to a part of the hospital where he or she doesn't belong, or if the sensor is tampered with, an alarm goes off and automatically locks down exit doors until authorities can investigate.

But once a child goes home, the hospital security ends. In November of 2012, a Chicago woman kidnapped a one-month-old infant from a neighbor's home. This time the unbelievable motive was to go to court and purport the baby as the child of her boyfriend, who had been arrested on drug charges. The kidnapper figured that the judge would be more lenient in sentencing her man, after hearing the convict was a father. This time, simple but effective good-old police legwork solved the case. While they were canvassing the neighborhood the morning after the abduction, officers heard muffled cries, and they found that baby girl in the trash, where the kidnapper had put her once the story of the missing child had gone public.

In addition to infant abductions, we've all heard of horrific cases in which a mother is actually cut open in an attempt to steal her unborn child. In October of 2012, a woman visited a Brazilian clinic because she thought she was pregnant. After finding out her belly bulge was actually a tumor, and not a baby, she was afraid her husband was going to leave her. So the woman returned to the clinic later and lured a pregnant woman back to her home under the ruse of giving the visitor some baby clothes. Once at her home, the kidnapper knocked the pregnant woman unconscious and cut the 37-week fetus out of her uterus. She left the victim for dead and proceeded to parade the baby in the street, telling neighbors it was hers. They didn't buy her story, and stormed the woman's house and found the real mother in a pool of blood on the bedroom floor, and they called police. Later, scene investigators found a wooden cutting board which had been used to beat and disable the mother, and also found the razor blade that was used to cut her belly open. If the situation hadn't been so obvious, DNA testing would have certainly resolved the relationships. Luckily, the case did have a happy ending for both the mother and child, as they both survived the attack.

Our last case in this lecture is the August 9, 1987 abduction of 21-year-old Paula Beverly Davis from Kansas City, Missouri. It involves an entirely different motive. Paula had a one-year-old son and was employed as a clerk at a local store, but she was also alleged to be involved in prostitution and had a drug problem. When she didn't come home that night, her roommate called Paula's mother at 3:00 am saying she had last seen Paula earlier that evening at a truck stop along Interstate 70, just outside of Kansas City. Mrs. Beverly contacted police and gave them all the details she could think of about her daughter, including that she stood 5′ 5″, weighed 125 pounds, was Caucasian, had brown hair and brown eyes. Mrs. Beverly also described the two tattoos on Paula's chest: a unicorn and a rose. Authorities reminded Mrs. Beverly that Paula was an adult and may have simply disappeared of her own volition, but her family just didn't believe Paula would do that.

The day after Paula disappeared, the recently-deceased and nearly-nude body of a young, white female with brown hair and eyes was found along Interstate 70 but 600 miles, or a nine-hour drive east of Kansas City in Montgomery County, Ohio, about an hour north of where I live. Now that was the year after I began my forensic consulting with the Montgomery County Coroner's Office, but this wasn't an anthropology case; the body wasn't skeletonized, there was no bony trauma, so I wasn't directly involved. The pathologists determined—I believe by analyzing the vitreous fluid from the woman's eyes—that she had been dead about 14 hours. They did a thorough autopsy, and the cause of death was ruled a strangulation by ligature, but when no identification was made, the woman's body was soon buried in a cemetery in Dayton, Ohio.

For over two decades, although leads were explored, the identity of the young woman was not known. In the spring of 2009, though, officials from the Montgomery County Coroner's Office entered all of the deceased's pertinent information into the brand-new National Missing and Unidentified Persons

System, otherwise known as NamUs. That fall, when Paula Beverly Davis' sister, Stephanie Beverly Clack, whom I've met, heard about a public service announcement regarding this new technology to resolve cold ID cases, she went to the NamUs website. All Stephanie entered into the database's search function was that the missing person was a white female, aged 21, and then 10 suggested possibilities popped up. The first 9 were clearly not her sister, but after just a half an hour on the system, Stephanie found the 10th possible match listed was that of a woman found on August 10 of 1987, the day after her sister's disappearance, with brown hair, brown eyes, and two tattoos on her chest—a unicorn and a rose.

Stephanie knew she'd found her sister. You might wonder, how on Earth those cases weren't connected sooner, but in those days we just didn't have the same kind of computerized databases we have today. But within two months, DNA testing from her family was compared to DNA from tissues retained by the Montgomery County Coroner's Office, and the victim's identity was confirmed. As for who abducted Paula Davis, authorities now believe convicted serial killer Lorenzo Gilyard, also known as the Kansas City Strangler, is responsible for her kidnapping and murder, although at this point, he has yet to be charged with Paula's untimely death.

Identification Matters

Lecture 23

In this lecture, we will look at six cold cases from Hamilton County, Ohio, spanning the years 1986 to the present. These cases have all been investigated by Project Identify, a cooperative effort among the Hamilton County coroner's and sheriff's offices, local media, and a team of students from Mount St. Joseph University. Project ID has reopened these cold cases using the National Missing and Unidentified Persons System (NamUs), a database that employs sophisticated computer algorithms to try to match missing and unknown persons. In this lecture, we'll follow the steps of Project ID team members in gathering information on these unknown victims.

The Settler

- On July 8, 1986, the Cincinnati homicide squad found the badly decomposed, partially skeletonized remains of what looked to be a middle-aged male in a vacant building in a rundown area of the city. There appeared to be no signs of foul play. Because the man had barricaded himself inside the building, he is referred to as The Settler.

- The body was transported to the Hamilton County coroner's office for an autopsy, but because of the degree of decomposition, the cause and manner of death were ruled undetermined, and fingerprints couldn't be obtained.
    - His remains weighed only 94 pounds, but that

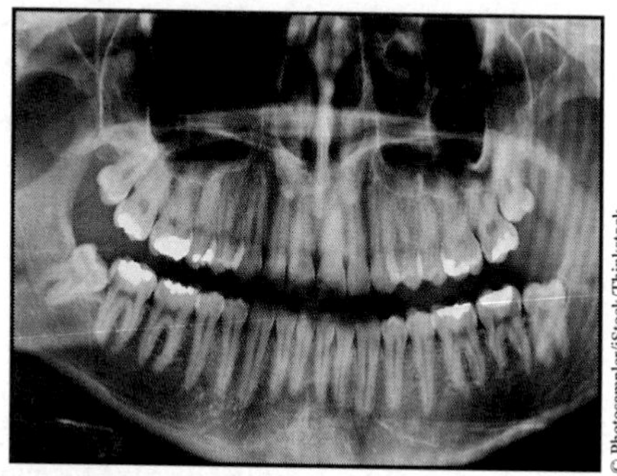

Some methods related to human identification haven't changed much over the last 30 years; visual observation of scars and tattoos, fingerprints, medical details from autopsies, and X-rays of skeletons are still used.

can't be used to estimate living weight. Both visual cues and mathematical formulas suggested that the victim was of African ancestry, and using the length of his femur, the man's height was estimated to be between 5′ 6″ and 5′ 10″. The Settler was probably more than 50 years old. The poor state of his dental health was thought to reflect a low socioeconomic status.

- Local missing persons files were consulted in 1986, and five possibilities were checked out, but no identification was made. The final note in the record indicates that full-body X-rays were taken and clothing was fully photographed before disposal, but these records were later destroyed in routine purges.

- The Settler's remains were buried in Wesleyan Cemetery in Cincinnati, a facility with a history of poor recordkeeping.
  - In the summer of 2013, students from Project ID teamed up with civil engineering interns from Cincinnati State to try to map the portion of the cemetery where records indicated The Settler was buried. Unfortunately, there are few headstones in the area, and his grave probably never had one.

  - Further, even if the grave could be located and a DNA sample could be obtained from the skeleton, it seems unlikely that a relative would come forward and contribute a sample for comparison.

  - Given that The Settler was probably homeless and may have barricaded himself in a building because he didn't want to be known by others, he will probably remain unidentified.

Broadway Doe

- On October 27, 1988, the crumpled body of a young man was discovered at the bottom of a stairwell leading to the basement of an apartment building on Broadway in downtown Cincinnati. The man was taken to University Hospital and pronounced dead on arrival.

- Although the exact time of death wasn't known, Broadway is a busy downtown street, and the man probably hadn't been dead for more than a few hours. He was fully dressed but had no identification.

- An autopsy revealed the cause of death as blunt-force trauma to the head of an undetermined manner, meaning that his death could not be ruled a homicide, suicide, or accident. There was no way to know whether he had fallen down the stairwell or had been pushed, but toxicology revealed that his blood-alcohol level was .08 percent, which is the current legal limit for intoxication.

- This 20- to 30-year-old man, referred to as Broadway Doe, had a completely recognizable face. He had been healthy, weighed 137 pounds, and was 5′ 8″ tall. A full set of fingerprints was taken, his blood type was noted as O-positive, and he was fully photographed. A police artist's sketch of the man's face was released to the local media, but no one came forth with information. His body was kept in storage at the Hamilton County Coroner's Office for five months before being buried in Wesleyan Cemetery in March 1989.

- In 2009, Broadway Doe's autopsy report, dental chart, and photographs were still available in the coroner's office. Amazingly, in 2013, a DNA analyst at the coroner's crime lab found an envelope containing a blood card from Broadway Doe—a paper record impregnated with drops of blood that can be used for testing—and the NamUs DNA lab was able to generate his genetic profile.

- The profile has been uploaded into the Combined DNA Index System (CODIS), a national DNA registry. CODIS is a self-searching database, with algorithms that automatically search all of its records against each other. Unfortunately, there has not yet been a hit for Broadway Doe.

Tracks

- On May 27, 1992, at about 10:35 a.m., train yard workers phoned the Cincinnati Fire Department to report a man lying along the tracks. An emergency squad took the man—who reportedly smelled of alcohol and was unresponsive—to University Hospital, where he was admitted in a coma. Physicians determined that he had suffered a stroke; the man never regained consciousness and was removed from life support.

- An autopsy showed that the man, who came to be called Tracks, suffered from numerous illnesses. He was a black man, estimated to be in his 50s or 60s; he was 6′ 1″ and weighed 206 pounds. Other than the stroke, there was no evidence of injuries; thus, the man's death was ruled to be from natural causes.

- In 1992, missing persons records from both Hamilton County and the City of Cincinnati were checked, but no one came forth with information. When Project ID team members reopened the case, an autopsy report was available, but there were no dental records, fingerprints, DNA samples, or even photographs. A record from the emergency responders seems to indicate that the man's first name was Bill, but his last name is unknown.

- In March of 2014, the grave of Tracks was opened, and his DNA profile is now being generated. In addition, a forensic artist will reconstruct his facial appearance.

Riverman

- On May 17, 2003, a tugboat operator reported a body floating in the Ohio River near the city's public landing. The deceased was an unknown black male, who was badly decomposed.

- The body was brought to the Hamilton County morgue for an autopsy, but there were no obvious injuries, and neither the cause nor manner of death could be determined. Riverman was estimated to be between 20 and 40 years old; he was 6′ tall and weighed 220 pounds. The toxicology report showed no evidence of drugs or alcohol, but these results can be affected by decomposition.

- A forensic dentist conducted a complete dental examination; a full DNA profile was uploaded to CODIS; and one viable fingerprint was uploaded to the FBI's Integrated Automated Fingerprint Identification System (IAFIS).

- One entry in the death investigator's portion of the record indicates that five days before Riverman's body was discovered, a pair of concerned citizens had reported seeing a black male jump from one of the bridges that span the Ohio River in Cincinnati. Although that man was probably this victim, his identity is still not known.

Pearl

- On November 29, 2006, another body was found in the Ohio River, this time, by workers at a grain shipping facility. The body was that of a woman, who was estimated to have been dead a day or two at most.

- The woman, known as Pearl, was white and 50 to 70 years old; she was 5′ 2″ and weighed 134 pounds. Her death was due to drowning, but no one knows whether she fell in the water or was pushed. Investigators believe that she may have been a passenger on one of the local gambling riverboats or sightseeing cruises.

- Why no one reported Pearl missing remains a mystery. Perhaps the person who might normally be looking for her—a husband or boyfriend—is the one responsible for her death.

The Traveler

- On January 5, 2009, completely skeletonized remains were found under some brush in a wooded area near a major highway in Hamilton County. Some badly tattered clothing and two rings were associated with the remains.

- The remains were identified as belonging to a Caucasian female, who stood between 5.1′ 5″ and 5′ 6″ tall. Using methods developed to assess aspects of the ribcage and pelvis—two areas that demonstrate specific age-related changes—she was estimated to be between 40 and 65 years old when she died.

- Other noteworthy features included a poor state of dental care and rugged landmarks on the muscle-attachment sites of her lower limbs, which suggested that she spent a great deal of time walking; possibly, she was homeless. The remains had probably been by the highway for many months, if not years. DNA samples were submitted to the NamUs DNA lab for upload to CODIS.

- The Traveler's remains are still available at the Hamilton County morgue. In the spring of 2013, members of Project ID and other students visited the site of her discovery. Surprisingly, a few more small bones were found, as well as an upper denture that the forensic dentist was able to fit to the woman's skull.

- The findings of an anthropology exam, along with the skull and denture, were provided to a forensic artist in Florida who does reconstructions for unidentified dead in the NamUs system. By the summer of 2013, an image was available for a new press release that was heavily covered by the media. Sadly, no new information has been forthcoming.

An Identification Success Story

- On June 24, 1975, the decomposing body of an unknown drowning victim was discovered in the Scioto River in Ross County, Ohio. He was a white male; 5′ 11″ tall; weighing around 150 pounds; and with an age estimated between 30 and 65 years. His body was autopsied the next day at the Hamilton County coroner's office, and when no identity could be confirmed after four months, the remains were released for burial as a John Doe.

- In late 2009, a NamUs record was started for this man, and in November 2011, his body was exhumed for DNA testing and a thorough dental exam. His skull was sent to a forensic artist, and the image of the clay reconstruction she created was released to the media.

- A woman saw the image in the newspaper and contacted authorities. In September of 2012, based on DNA comparison with a family member, the man was identified as 40-year-old Arthur Raymond Flowers, a veteran who had last been seen at a VA hospital, where he was being treated for depression. Mr. Flowers's niece was finally able to bring Arthur home to his family, 37 years after he disappeared.

Suggested Reading

Bass and Jefferson, *Death's Acre*.

Mallet, Blythe, and Berry, eds., *Advances in Forensic Human Identification*.

Murray, *Forensic Identification*.

National Missing and Unidentified Persons System, http://www.namus.gov/.

Primorac and Schanfield, eds., *Forensic DNA Applications*.

Rathbun and Buikstra, *Human Identification*.

Steadman, *Hard Evidence*.

Questions to Consider

1. Had you heard of the National Missing and Unidentified Persons System prior to this course?

2. What are some of the ways forensic scientists reopen and reinvestigate cold cases involving unknown persons?

3. What types of specialists/specialties do you think are involved in working together to find missing persons and identify unknown persons?

Identification Matters

Lecture 23—Transcript

In this lecture, I'd like to introduce you to six people I've become fairly close to over the past few years, even though I don't know their names. All of these unidentified victims were discovered in my home county of Hamilton in Ohio, and all of their cases have gone cold. Their stories help highlight some of the most remarkable developments I've witnessed in forensic science over my nearly 30 years of practice, from my graduate student internship with my mentor, forensic anthropologist, Dr. Anthony Perzigian from the University of Cincinnati, up until today.

Unknown persons are commonly known as John or Jane Doe in the U.S., but called Joe Bloggs in the U.K. and Ashok Kumar in India. Unfortunately, some can be called Baby Doe, since the unidentified dead also include abandoned newborns and young children. During my career, some methods related to human ID haven't really changed much; we still use visual observations, like scars, marks, and tattoos; we use fingerprints, medical details from autopsy; and for skeletons, the same specific measurements and X-rays as when I started.

But there have been changes. For instance, computer technology has definitely changed how we analyze measurements of the skeleton. When I first started in anthropology, we used some fairly cumbersome mathematical formulas to help assess which sex, ancestry, and height was a best fit to an unknown person. Today, much of that is done by computer software, particularly a package known as Fordisc 3.0.

The following six cases date from 1986 to the present, the exact span of years I've been involved in forensic anthropology. My focus on them relates to a partnership I organized in February of 2012, called Project Identify. It's a cooperative effort between the Hamilton County Coroner's and Sheriff's Offices, a local news reporter, and a small team of my students from Mount St. Joseph University. We have reopened these cold cases using the National Missing and Unidentified Persons system, NamUs, which is a database that uses sophisticated computer algorithms to try to match missing and unknown persons. We've given each victim a codename to help remind us of the person's humanity, and since those are much easier than case numbers in discussions among Project ID members. Together, our group hopes to find answers to some long-standing mysteries.

In the heat of the summer, July 8 of 1986, a woman reported a foul odor coming from a vacant building on Republic Street in a rundown area of Cincinnati. The Cincinnati homicide squad found the badly decomposed, partially skeletonized remains of what looked to be a middle-aged male, who had perhaps been there a week or so. His body was clothed, and there appeared to be no signs of foul play. The man had barricaded himself inside the building, so we refer to him as The Settler.

The body was transported to the Hamilton County coroner's office for an autopsy, but due to the degree of decomposition, the cause and manner of death were ruled undetermined, and fingerprints couldn't be obtained. His remains weighed 94 pounds, but that really can't be used to estimate any type of living weight. Both visual cues and mathematical formulas suggested the victim was a man of African ancestry, and using the length of his femur or thighbone, the man's height was estimated in the range of about 5′6″ to 5′10″. His age was likely over 50 years, based on degenerative changes in the skeleton, including some arthritis in his spine. A very poor state of dental health was thought maybe to reflect a low socioeconomic status. Some badly healed fractures suggested the man had, at one time, suffered a broken nose and rib.

Records indicate that local missing persons files were consulted in 1986, and five specific possibilities were checked out, but nothing. The death investigator's notes say that several months before the body was discovered, a neighbor had chased a "vagrant" from the building. Although that could have been the victim, the neighbor knew nothing about the man's identity. The final note in the record indicates that full-body X-rays were taken to aid any future inquiries, and clothing was fully photographed before being disposed.

This was 1986, before computers were at all commonplace, so the autopsy records I obtained to develop this man's NamUs record were from microfilm. Some of you younger folks may not even know what microfilm is, but it was a long-standing archival process in which documents were photographed as tiny film records for easier storage. In order to view the records, you need to pull the needed film out of a file and magnify it on a special machine.

You can print from the viewer, but the quality is often poor. In fact, when The Settler's records were microfilmed, the camera must have been slightly out of focus, because the documents are terribly difficult to read, and there's nothing that can be done about that. Once microfilmed, the original paper records were destroyed to save space, as was common practice. Unfortunately, his X-rays and images of his clothing were also destroyed during routine record purging over the years.

How about DNA? Well, the first forensic use of DNA technology occurred in 1986, the same year The Settler's remains were discovered, but DNA sampling didn't become routine until over 10 years later. So what options are left now? The man's remains are buried in Wesleyan Cemetery in Cincinnati, so perhaps could be exhumed for DNA, but that cemetery has a notorious past of some managers that were less than reputable; it had poor record keeping and even had an office fire at one point.

So in the summer of 2013, students from Project ID and I teamed up with some civil engineering interns from Cincinnati State, a local technical college, and we tried to map the portion of the cemetery where records indicated The Settler was buried. But there are very few headstones in the area, and his grave probably never had one. Besides, it's sad to say, what are the odds that someone is still looking for that poor man, who was likely homeless?

Remember, even if we could locate his grave and get a DNA sample from his skeleton, we would need a relative to come forward and likewise contribute a sample for comparison. At this point, I believe since The Settler barricaded himself in a building, maybe indicating he didn't want to be discovered or known by others, that's probably the way he will stay, unidentified, and with only limited information in his NamUs record.

On October 27 of 1988, around 8:30 in the evening, a woman who lived on Broadway, in downtown Cincinnati, was leaving her home and saw the crumpled body of a young man at the bottom of the stairwell leading to her apartment building's basement. Emergency responders drove the man to University Hospital where he was pronounced dead on arrival. Cincinnati police officers brought his body to the Hamilton County Coroner's office, which is literally across the street from the hospital.

Although the exact time of death wasn't known, Broadway is a busy downtown street, and the man likely hadn't been dead for more than a few hours. He was fully dressed, wearing a red hooded sweatshirt, blue jeans with a brown belt, white socks, and gray and white gym shoes, as well as a Timex wristwatch. The man had no identification and only 70 cents in his pocket. An autopsy revealed his cause of death was blunt force trauma to the head of an undetermined manner, meaning his death couldn't be ruled as homicide, suicide, or an accident. There was no way to know whether his injuries were from falling down the stairwell or having been pushed, but toxicology revealed the man's blood alcohol level was 0.08 percent, which is the current legal limit for intoxication. There were no other drugs in his system.

This 20- to 30-year-old man, who Project ID refers to as Broadway Doe, had a completely recognizable face. He had wavy brown hair about four inches long, blue eyes, and was healthy, well-nourished, and well-groomed, including a neatly trimmed mustache and clean-cut fingernails. He weighed 137 pounds and was 5′ 8″ tall. A full set of fingerprints was taken, his blood type noted as O+, and he was fully photographed, including the scars on his forehead, shoulder, hands, and leg, two of which showed signs of having been stitched.

All of these details about this young man, including a police artist's sketch of his face, were released to the local media, but no one came forth with information, or if they did, nothing is recorded in the coroner's office paperwork. Our forensic odontologist, who, like me, started as a consultant to the office in 1986, generated a complete dental chart for Broadway Doe. His body was kept in the cold locker at the Hamilton County Coroner's Office for five months, before being buried in Wesleyan Cemetery at the end of March in 1989.

At the time of Broadway Doe's discovery, I had no involvement in his case. Anthropology services weren't needed; he wasn't decomposed. But when I went to the coroner's office to get unidentified persons records for input into NamUs in 2009, I was able to get his autopsy report, his dental charting, and numerous photographs. Images of his clothing, though, had been purged; just about everything else was in his record, though, including a notation that his fingerprints were on file with the Cincinnati Police Department.

Unfortunately, no one has been able to locate that fingerprint record; it appears it was inadvertently discarded over the years. In 1988, DNA sampling was still very rare, and I figured that would be a dead end. But amazingly, in March of 2013, the DNA analyst at the coroner's crime lab found an envelope with Broadway Doe's case number on it in their frozen archives. It contained what's called a blood card; that's a small paper record that's impregnated with a few drops of blood that can be used for testing. And as unbelievably good luck would have it, 25 years after his death, the NamUs DNA lab was able to generate a genetic profile for Broadway Doe.

The DNA doesn't stop there, though, because the lab can upload encoded DNA information from unidentified and missing persons into a national DNA registry, known as CODIS. That stands for the Combined DNA Index System, which is a self-searching database with algorithms that routinely and automatically search all of its records against each other. Unfortunately, so far, we haven't gotten a hit for Broadway Doe, but perhaps someone will see this lecture, or the media releases we continue to push, or maybe look on NamUs and say to themselves, I wonder if that man could be my brother, father, or son that I haven't seen since 1988.

It's just so hard to imagine that no one, back in 1988 or to this day, seems to be looking for this young man. And because that is so hard for me to imagine, I'm going to believe that someone is looking for him, and will someday find him. At least that way his friends and family will know what happened. It's so sad to think a grieving mother, wife, brother, or some other relative may be out there right now thinking that young man simply stopped caring and disappeared.

Of the six unidentified persons found in Hamilton County since 1986, surprisingly, the one we knew the least about was alive when he was found. On May 27, 1992, about 10:35 in the morning, train yard workers phoned the Cincinnati Fire Department to report a man lying along the tracks. An emergency squad took the man, who reportedly smelled of alcohol and was unresponsive, to University Hospital, where he was admitted in a coma. Physicians determined he had suffered a stroke in a vital area of his brainstem. The man never regained consciousness, was declared brain dead two days later, and they removed life support.

The next day, the Hamilton County Coroner's Office conducted an autopsy, which showed the man suffered from numerous illnesses. In addition to the stroke, he had heart disease, hardening of the arteries, fluid and clots in his lungs, kidney problems, and cancer in both his prostate and parathyroid glands. He was a black man, estimated to be in his 50s or 60s, who stood 6′ 1″ tall, and he weighed 206 pounds. He had brown eyes, curly black and gray hair, and a full beard and mustache. The man had no upper teeth and few lower teeth, and his nails were noted to be dirty. Other than the stroke, there was no evidence of injuries, so the man's death was ruled as from natural causes.

In 1992, missing persons records from both Hamilton County and the City of Cincinnati were checked, but nobody came forth with information. Maybe the man had come from somewhere else, perhaps by jumping on the train. The forensic dentist was apparently not consulted to generate a chart, so all we had to go on were the notes about his dental pattern from the autopsy report. Records indicate fingerprints were taken and provided to the local police for comparisons with both missing persons and criminal fingerprint records, but nothing panned out.

As time went on and Tracks, as we've come to call him, was not readily identified, his fingerprint records were apparently discarded. My theory is that since his death was ruled from natural causes, and it wasn't the responsibility of the police to identify him, the records were purged. I assume the coroner thought the police had the prints on file, and law-enforcement probably assumed the coroner's office had kept a copy, but that left us with next to nothing to help identify this man, no dental records, no fingerprints, and no DNA sample was taken. There was only the most basic description of his clothing, and unbelievably, not even any photographs. This man was alive when discovered, and we don't even have a picture of his face. Now, that was over 20 years ago, and I'm confident that wouldn't happen today; things were just so different back then. That's one of the lessons I've learned from this project.

In 2013, I was able to find one piece of new information. The Cincinnati Fire Department Emergency Medical Squad still had the microfilm version of the EMS run sheet from the day they transported this man. It's very difficult to read, but my students and I believe the name Bill is listed in the box marked "first name" on the paperwork. There's a question mark in the box for a last name. Is it possible the man was still conscious when EMS arrived and provided his first name? It's not much help, but at least it was a new direction to be pursued, so that was added to his NamUs record. With so little to go on, though, the Coroner decided Tracks would be exhumed. In March of 2014, his grave was opened, and as I give this lecture, this man's DNA profile is being generated; I'm excited to say we're arranging for a forensic artist to reconstruct his facial appearance.

For over 10 years, Hamilton County did not have another chronically-unidentified victim, and personally, I don't think that's an accident. Tracks was found in 1992, and two years later, Hamilton County got its first new coroner in over 30 years. Things quickly started to change and modernize. The office got a new centralized computer system, and records since then are practically impeccable. Photos were either taken with a digital camera or promptly digitized for archiving. DNA has become routine in unidentified cases, though, as we'll see, that doesn't guarantee a positive ID.

On May 17 of 2003, a tugboat operator contacted the Cincinnati Police Department to report a body floating in the river near the city's Public Landing. The deceased was an unknown black male who was decomposed to the point that his facial features were too distorted to be of any help. The body was brought to the Hamilton County morgue for an autopsy, but there were no obvious injuries, and neither the cause nor manner of death could be determined. The man's age was estimated as between 20 and 40 years; he was 6′ tall, and weighed 220 pounds. No distinctive bodily features could be seen, like scars or tattoos, and although he had brown hair, his eyes were already too cloudy to be certain of their natural color. The man was wearing blue jeans, and we know the specific brand, size, and length. The toxicology report showed no evidence of drugs or alcohol, but results can be affected by decomposition.

The office's forensic dentist conducted a complete dental examination, with X-rays, on Riverman, as we like to call him. A full DNA profile was developed and uploaded into the national CODIS database. The one viable fingerprint that could be rolled was submitted to the FBI and uploaded into their Integrated Automated Fingerprint Identification System, IAFIS. This computerized fingerprint database, launched in 1999, allows comparisons between unknown prints and any records in its files. Because the coroner's office was fully computerized by this time, all information about this case is easily accessible and permanently archived.

One entry in the death investigator's portion of the record, indicates that five days before Riverman's body was discovered, a pair of concerned citizens had contacted Cincinnati Police to report that a black male, dressed in a gray shirt and blue jeans, with his shoes in his hand, jumped from one of the bridges that span the Ohio River in Cincinnati. Although that man was probably our victim and lost his shirt in the fall in the water, we still do not know his identity. But we do know exactly where he's buried, in the event that some new technology is developed that could further aid his identification. Since we can't predict what the future holds, and because some religions object to cremation, unidentified remains should never be incinerated.

Three years later, on November 29 of 2006, another body was found in the Ohio River by workers at a grain shipping facility, but this time, west of the city, very near the university where I teach. The body was that of a woman who was estimated to have been dead a day or two at most, since she was still in rigor mortis. She was white and estimated as 50 to 70 years old, with hazel eyes and light brown hair with dyed-blonde highlights. She was 5′ 2″ tall and weighed 134 pounds. Her death was due to drowning, but no one knows whether she fell into the water or was pushed, perhaps off one of the gambling boats that traveled that stretch of the river.

We call this victim Pearl, because of the pair of pearl-style necklaces she was wearing, in addition to a black skirt, black blouse, pantyhose, bra, and panties. Her white tennis shoes suggested she wanted to walk comfortably, but she was otherwise dressed up, complete with nail polish and lipstick. That's another reason investigators think she may have been on one of the local gambling boats or one of the sightseeing cruises. Although those boats require tickets, they don't take names, nor do they necessarily count heads when guests are departing the boat and returning to shore. Pearl didn't have the type of massive internal injuries that would signify she went off a bridge, nor did her body show the typical signs of damage seen when a body travels far in the river, colliding with floating debris.

The woman showed no evidence of prior surgeries or any significant medical conditions. She had a touch of arthritis in spine and her hands, consistent with a woman of that age, as well as stretch marks on her abdomen, suggesting that, at least at one time, she had given birth. I imagine that somewhere, her children, maybe even grandchildren, are wondering where she is, and why she hasn't reached out to them in years now. Perhaps the person who might be looking for her normally, like a husband or boyfriend, is the one responsible for her death; otherwise, why didn't anyone report her disappearance?

On the fifth of January in 2009, two people picking up aluminum cans along the road discovered the last of our six unidentified Hamilton County cases. The remains were completely skeletonized, so normally I would have gone to the scene, but I was out of town teaching at a death-investigator's seminar. The bones were scattered under some brush in a wooded area near a major highway. Some badly tattered clothing was associated with the remains, with other items, including two rings, nearby. There was evidence of carnivore scavenging on the bones, and the skeletal elements were scattered around the area. Several agencies assisted in the recovery, and cadaver dogs were used in the search.

The following week, I examined the bones at the Hamilton County morgue. Based on visual observations and measurements that I input to Fordisc software, I told them the person was a Caucasian female who stood between 5′ 1.5″ and 5′ 6″ tall. Now that's a wide range, but anthropologists have learned much more about normal human variation over the years, and that has to be taken into consideration. It's too limiting to report a narrow window, since people searching for a missing person could easily dismiss a victim if we didn't include the full statistical range. Using methods developed to assess aspects of the ribcage and pelvis, two areas that demonstrate specific and well-studied age-related changes, I estimated the woman was between around 40 and 65 years of age when she died.

Other noteworthy features included a poor state of dental care and very rugged landmarks on the muscle-attachment sites of her lower limbs, which suggested to me she spent a lot of time walking. Taken together with the mix of remaining clothing, which was the waistband of men's cotton briefs that had disintegrated under polyester sweat pants, plus a bra and remnants of a sequined top, I thought she was probably a homeless person, who maybe created a makeshift encampment near the highway, perhaps to rest, or maybe she wasn't feeling well. The condition of her clothing and moss growing on her inexpensive gym shoes, suggested to me the remains had been there for many months, if not years. The forensic dentist was consulted, and DNA samples submitted to the NamUs DNA lab for upload into CODIS.

The remains of this woman, whom we've dubbed The Traveler, due to apparently being homeless and on the move a lot, are still available at the Hamilton County morgue. So in spring of 2013, I requested we submit her skull to a forensic artist for a facial reconstruction. I also asked if my Project ID students, and some others from my forensic science class, could visit the site to learn more about the location. To our surprise, over four years after the remains were discovered, we not only found a few more small bones that were likely scattered by the carnivore activity, but we also found an upper denture that the forensic dentist was able to fit to this woman's skull.

The findings of my anthropology exam, along with the skull and a denture, were provided to a forensic artist in Florida, who was doing reconstructions for unidentified dead in the NamUs system. By summer of 2013, we had an image to use in a new press release that was heavily covered in the local media. We knew it was a long shot, but had hopes that as the years passed, people might now start to be wondering why they hadn't seen this woman in so long, maybe a relative of hers, or someone from a local social services agency. Sadly so far, nothing has panned out.

The Traveler's facial reconstruction was so good, though, we asked the artist to do additional work for Project ID. We provided her the morgue photographs of both Broadway Doe from 1988 and Pearl from 2006. The artist was able to use computer software, along with the known eye colors of both victims, to generate a life-like appearance for presentation to the public, both in NamUs and in posters that Project ID students created to be distributed throughout the area. The problem with local distribution, though, is obvious. If any of these people were from the greater Cincinnati area, they'd probably have been identified by now.

Given that DNA profiles of four of the six—Broadway Doe, Riverman, Pearl, and The Traveler—are already uploaded to the CODIS national database, as Tracks soon will be, and, since fingerprints of Riverman and Pearl are entered into the IAFIS national database, well, we thought maybe no one was actively looking for any of these people. Since the NamUs database is publicly-searchable, it also seemed unlikely that relatives were using NamUs to find any of these six individuals.

But amazingly, as I stand here delivering this lecture, as of just last week, it appears one of our six unknown persons will very soon be identified. A DNA profile is currently being generated for someone who believes she's a family member of one of our Project ID cases. Due to the ongoing investigation surrounding the victim's death, I can't reveal which case, but I am fairly certain, and very hopeful, that by the time you hear this, one of our six persons will once again have a name.

Since I can't talk about that case, let me tell you about a different identification success story. On June 24 of 1975, the decomposing body of an unknown drowning victim was discovered in the Scioto River in Ross County, Ohio. He was a white male, 5′ 11″ tall, weighing around 150 pounds, with an age estimated between 30 and 65 years. The man had brown eyes, was clean-shaven, and had brown hair, about six inches in length. His body was autopsied the next day at the Hamilton County Coroner's Office, and when no identity could be confirmed after four months, despite dissemination of the information in the media, the remains were released for burial as a John Doe.

In late 2009, when I was employed part time as a consultant with NamUs, I learned of this case and obtained the man's information to start a NamUs record. The current death investigator in Ross County got very interested in the case, and he worked really hard to figure out new angles, despite the almost 35-year time lag. We were able to use federal resources to exhume this man's body for DNA testing and get a thorough dental exam. His skull was sent to a forensic artist, the same one who helped in the Project ID cases, and the image of the clay reconstruction she created was released to the media.

A woman saw the image in the newspaper and contacted authorities. In September of 2012, based on a DNA comparison with a family member, the man was identified as 40-year-old Arthur Raymond Flowers, a veteran who had been missing since June of 1975, having been last seen at a VA hospital, where he was being treated for depression. Mr. Flowers' niece was finally able to bring Arthur home to his family 37 years after he disappeared.

The Past, Present, and Future of Forensics

Lecture 24

Neither criminal motives nor the tendency of some members of society to engage in aberrant behavior has changed much over human history; it is the materials and technology used—both to commit crimes and to solve them—that transform over time. Because humans are by nature scientific beings, we continue to explore new technologies to enhance our understanding of what goes on around and inside us. Forensics is no exception and has prompted and capitalized on many scientific advances. In this lecture, we'll look at three essential tools used heavily today in forensics and see how they've changed over time and continue to move science forward: fingerprinting, DNA analysis, and computer technology.

Fingerprinting

- Fingerprints have the longest history as a physical means of identifying both offenders and victims in forensics. In the 1880s, two British scientists who had been studying prints, Henry Faulds and William Herschel, published papers on their individuality. This led the British anthropologist Francis Galton to start promoting the acceptance of fingerprints as forensic evidence in court.
    - Around this same time, an Argentine police officer named Juan Vucetich used fingerprint evidence to solve a murder in which a woman killed her two sons and was identified because her prints were left behind using their blood.
    - A few years later, London Police Commissioner Edward Henry developed a fingerprint classification system that was adopted by Scotland Yard and is, in essence, still used today.
- The oldest method of revealing latent fingerprints (those that aren't obvious to the eye) goes back to 1863, when French professor Paul-Jean Coulier published a paper on the use of iodine fuming. Iodine fuming is still used today, especially on valuable paper items that would be damaged by other methods. Much later, in 1977, a

Japanese police scientist named Fuseo Matsumur discovered that the cyanoacrylate in superglue developed latent fingerprints.

- In 2008, British scientist John Bond discovered that acids in a person's sweat can etch a fingerprint into metal, especially brass, by corrosion. The etching causes a permanent change in the surface of the metal, meaning that fingerprints may still be present on, for example, bullet shell casings, even if the shells were wiped clean before being loaded into a gun. The heat from firing a bullet or from the explosion of a bomb actually enhances the corrosion. One negative issue here is that the process used to develop metal corrosion fingerprints creates problems for collecting DNA present on the metal.

- In 2013, the University of Leicester announced the testing of a new latent print development method for metals that should allow prints to be recovered from weapons months or perhaps years after a crime, even if the surface has been wiped clean.
    - At present, only about 10 percent of recovered prints are of sufficient value to be used in court, but if this new method works, researchers believe that it will allow twice as many fingerprints to be recovered, while not damaging the potential for DNA testing.

    - The technique involves putting the object into a liquid containing a fluorescent dye that, when an electric current is applied, will stick to metal that wasn't touched, rather than the print itself. Essentially, the residue from the fingerprint insulates the metal beneath it, and investigators are left with a negative of the latent print that can be reversed digitally.

- Nanotechnology is projected to be the new frontier of fingerprint development and analysis. Chemical methods will be downsized to the nano level to improve the sensitivity of existing methods. Even more intriguing uses of nanotechnology are also proposed. For instance, antibody-like nanoparticles could be engineered to link to and detect nicotine, cocaine, marijuana, or heroin in a user's sweat.

That would give investigators a window into a person's life that could help narrow the search for either a criminal or a victim.

Developments in Serology

- Today's DNA analysis has its historical roots in serology, which is the general study of body fluids. Microscopy is one important aspect of forensic serology, and microscopes—in one form or another—have been around since the 1600s. The first forensic application came in the late 1830s, when French forensic pathologists perfected the use of the microscope to reliably detect sperm.

- In 1853, a Polish anatomist working in Germany developed the Teichmann test, which is named after him. This method crystallizes certain components of the hemoglobin in blood, and these crystals can be viewed under a microscope. The Teichmann test is confirmatory, meaning that if crystals form, the substance is definitely blood. In the 1860s, scientists from the Netherlands and Germany generated presumptive tests for blood; these are quicker tests that might be done in the field but require follow-up testing in a lab.

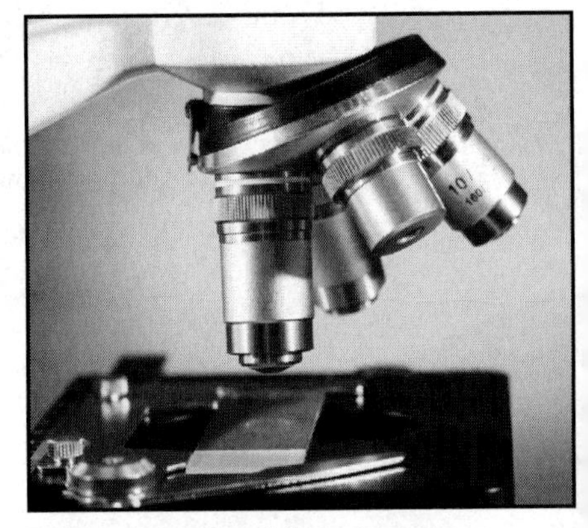

Both presumptive field tests and confirmatory lab tests are important in detecting the presence of blood.

- Another major milestone on the way to DNA technology occurred in 1901, when Karl Landsteiner, an immunologist from Austria, identified the ABO system of human blood types; his colleague, Max Richter, determined how to detect blood type on stains for use in forensics.

- Soon after, a Japanese researcher named Masao Takayama created another reliable test for the presence of hemoglobin in bloodstains. And by 1925, a second Japanese scientist first recognized that some people display their blood type in other body fluids, such as semen or saliva. These individuals became known in forensics as *secretors*.

- The chemical luminol, used at crime scenes to detect blood that is not visible to the naked eye, was developed by a German forensic scientist in 1937. Throughout the 1960s, research that spanned the globe led to further refinements in serology.

DNA Analysis

- The 1953 discovery of the chemical structure of DNA by James Watson and Francis Crick was essential to all forensic DNA analytical methods to come. It wasn't until 1980, however, that genetic researchers discovered what are known as the *hypervariable regions* in human DNA that are different enough among people to be used in forensics.

- Soon after that came nuclear DNA fingerprinting by Alec Jeffreys of the United Kingdom. In 1986, American biochemist Kary Mullis invented the polymerase chain reaction method that's still used today in a modified form to make mass copies of desired segments of DNA, allowing even small samples to be compared.

- DNA testing methods changed again in 1992 when a biochemistry professor in Texas determined that regions of DNA known as *short tandem repeats* (STRs) could be used in forensics. The resulting STR type of DNA fingerprinting means that very small and degraded samples can still net a genetic profile. By 1996, the first criminal case to use mitochondrial DNA analysis was tried in the United States.

- The most recent advance in DNA testing is *touch DNA*. With this method, only 5 to 10 skin cells are needed to generate a useful DNA profile. This method was used to exonerate the parents and brother of JonBenet Ramsey in her murder in 1996 and played a role in the investigation of Caylee Anthony's death in Florida in 2008.

Computer Technology

- As in countless other fields, computers have changed much of the work in forensic science. Today's forensic labs are filled with computerized equipment for toxicological studies of body fluids and tissues and analyses of suspected drug samples, accelerants from arson cases, automotive paints from hit-and-run accidents, and more.

- In such fields as forensic anthropology, accounting, and engineering, computer software has greatly enhanced and sped up assessments. Digital methods have also revolutionized forensic art, not only in terms of the speed with which a facial reconstruction or police sketch can be made but also in the ease of adjusting the results to show variations in appearance.

- Computers have been used in forensics to create *demonstrative evidence*, such as an accident reconstruction, the projected path of a bullet, or a model of a fall from a tall building. Forensic engineers use computer modeling to help assess why a building or bridge collapsed or where a fire started.

- With global positioning systems, computers allow the precise mapping of evidence, such as the locations of body parts in a mass disaster. Three-dimensional digitization and printing allow replicas to be made of skulls and other evidence. Digital X-rays and scanning enable scientists to rapidly transfer detailed images. Computer power can also enhance recognition of criminal patterns, such as might be used by a forensic profiler or behavioral analyst.

- Research conducted by experts in digital forensics has led to advancements in crime scene photography and video documentation of scenes and suspects. Digital experts can enhance poor-quality images and sounds and determine whether stored digital information has been altered. They can develop latent digital evidence, such as data on a computer hard drive that may have been erased.

- One significant application of computers in forensics has been in the realm of comparisons, such as of fingerprints and other image-based evidence. Facial recognition software can pick a person out of a crowd photograph, and digitization allows comparison of voice and other sound patterns. Large databases containing comparative evidence are now routinely used in all types of forensic investigations.

Computers and Crime

- Of course, computers have also changed and enhanced crime. *Cybercrime* includes bank thefts and corporate embezzlement, identity theft, and hacking and virus attacks on governmental, financial, and media entities. Just as the Internet has provided worldwide assistance to solve crimes, it has also allowed individuals to essentially cross borders and other jurisdictional lines to commit crimes without ever leaving home.

- In one cybercrime that took place between July and October of 1994, more than $10 million was stolen from a U.S. bank. This robbery was perpetrated by a worldwide group of hackers who tapped into the bank's cash management system. Ultimately, the leader of the operation, a Russian named Vladimir Levin, was extradited to the United States and pled guilty at trial. This case is thought to be the first-ever online bank robbery.

- In mid-May of 2014, law enforcement officials in 19 countries announced a global takedown of cybercriminals. This sting was related to malware known as the Blackshades Remote Access Tool (RAT), which had been used by thousands of hackers to steal passwords and access codes from unsuspecting computer users.

- Finally, a case that illustrates how innovative and diverse computer crimes can be was the coordinated attempt to hack into a pay-per-view TV system over the course of several years in the 1990s.

Allegedly, the motive behind the attempts was to enable the hackers to watch *Star Trek* reruns for free. Undoubtedly, computers will continue to provide new opportunities both for those who commit crimes and for those who solve them well into the future.

Suggested Reading

Mallet, Blythe, and Berry, eds., *Advances in Forensic Human Identification.*

Moore, *Cybercrime, Second Edition.*

Primorac and Schanfield, eds., *Forensic DNA Applications.*

Thies, *Cybercrime: Digital Forensic Investigations.*

U.S. Department of Justice, National Institute of Justice, *Latent Print Examination and Human Factors.*

Questions to Consider

1. What are some of the other major innovations in forensic science other than the three referred to in this lecture?

2. Do you have predictions for the future of forensic science?

The Past, Present, and Future of Forensics
Lecture 24—Transcript

In my opinion, neither the thinking of criminal perpetrators, nor their motives, has really changed much over human history. It's the materials and technology used, both to commit crimes and solve them, that has transformed over time. Humans are, by our very nature, curious and scientific. So we continue to explore new technologies to aid our lives and help us better understand ourselves and our world. As part of this, forensics has prompted and capitalized on many scientific advances.

In the first half of the 20th century, when forensic science was really coming into its own, France's forensic pioneer, Dr. Edmond Locard, set forth what's become known as Locard's exchange principle. It essentially states that each and every contact between two persons or things leaves a trace on both. We've come a long way since Locard's time and can now detect and analyze exchanges that are infinitely small, tinier than our earliest forensic forefathers could possibly have imagined. Likewise, the everyday technologies of our modern lives, like digital communications, leave detectable signatures that I'm sure would have been far beyond the comprehension of forensic science's pioneers. Let's look at three heavily-used essential tools in forensic science to see how they've continued to move us forward. One of these is very old, but still has much to offer; another has completely revolutionized our forensic capabilities; and the third has modified just about everything we do in terms of analysis and comparison.

My choice for the oldest technology that's still critical in forensics goes to fingerprints. Those traces of contact definitely have the longest history as a means of both identifying offenders and victims. Fingerprints were used to establish identity even before the Common Era. The Chinese have employed them since at least 200 B.C.E., including in criminal investigations, like burglary. By the 7th century, prints were being used in the Middle East to secure loans and prove debts. The ridge patterns in fingerprints have been studied since the late 1600s in Europe, and by the early 1800s, nine different major pattern types had been described.

In the 1880s, just a handful of years before Jack the Ripper took his first victim, two different British scientists who had been studying prints, Henry Faulds and William Herschel, published on their individuality. This led the British anthropologist, Francis Galton, to start promoting the acceptance of fingerprints as forensic evidence in court. Researchers also began studying fingerprint minutiae; those are the fine details that go beyond simply the basic arch, loop, and whorl patterns and really hone in on the uniqueness of prints.

Around this same time, a Croatian-born Argentine anthropologist turned police officer, named Juan Vucetich, used fingerprint technology to solve a crime in which a woman murdered her own two sons and even cut her own throat to make it look like an intruder was the killer. She was identified because her prints were left behind in her sons' blood. A few years later, London Police Commissioner, Edward Henry, enhanced on Galton's work and developed a classification system, adopted by Scotland Yard, which in essence, is still used today.

What about work with latent fingerprints? You know, prints that aren't obvious to the eye, the way those made in blood or ink are? The oldest method of revealing invisible fingerprints goes back to 1863 when France's Professor Paul-Jean Coulier published on the use of iodine fuming. He was actually working on techniques to reveal forgeries in questioned documents, but noticed that his iodine method was developing his own fingerprints on the paper. Coulier even figured out a chemical means of preserving fumed prints, since the iodine image is transient. Iodine fuming is still used today, especially on valuable paper items that would be damaged by other methods. In fact, I use it in my forensic science class, but we digitally photograph the resulting print in order to preserve it.

Much later, and also quite by accident, in 1977, a Japanese police scientist named Fuseo Matsumur, was examining hairs from a crime scene by mounting them on glass microscope slides, using Super Glue. He discovered that the cyanoacrylate in the glue was also developing his latent fingerprints on the slides. As a result, Super Glue fuming is very commonly used today to detect fingerprints. The science of revealing prints didn't stop there, though.

In 2008, it was discovered that simply by touching something metal, especially brass, the acids in a person's sweat could actually etch a fingerprint into the metal by corrosion, like a bullet shell casing, or even the metal of a pipe bomb. This was discovered by U.K. scientist, Bond, that's John, not James Bond! It works even if the person wiped the bullet clean before loading the gun, and even when the metal surface that's been touched is painted over. The etching causes a permanent change in the surface of the metal. And the heat from the firing a bullet, or from the explosion of the bomb, that actually enhances the corrosion.

Here's an example. On a Texas evening, December 9, 2007, 68-year-old Marianne Wilkinson got up from watching TV with her husband to answer the front door. She was fatally shot four times. Police think it was a case of mistaken identity and that the intended victim was another woman who lived down the block at a similar address and was involved in a nasty divorce dispute. Three months later, investigators found a discarded gun, which had been wiped clean of prints. They were able to match test-fired ammunition to the bullets recovered from Mrs. Wilkinson's body. The registered owner of the gun was deceased, and the weapon had apparently been passed around some.

With little to go on, in late 2008, Texas authorities contacted Dr. John Bond and requested he use his new corrosive technique on a spent bullet casing from the Wilkinson crime scene. Even though the casing had also been wiped clean of prints, Bond was able to develop a latent partial print. But, the person who loaded the gun, so assumed to be the shooter, still hasn't been identified, since the print matches no known prints on record. If that person is caught someday, though, his or her prints will be tied back to this crime, since they're now on record, and then justice will hopefully be served.

One downside to the metal corrosion development method is it requires a certain powder to be applied. The powder will stick to the latent print once a 2,500-volt electrical current is passed through the metal object. That's problematic for collecting any DNA present on the metal. So in 2013 in the University of Leicester in the U.K., that's the same lab that pioneered forensic DNA analysis and developed the corrosive technique, well, they announced the testing of a new latent print development method for metals.

It should allow prints to be recovered from weapons, like knives or guns, even months or years after a crime, even if the surface was wiped clean.

At present, only about 10 percent of recovered prints are of sufficient value to make it to court. But if this new method works, researchers believe it will allow twice as many fingerprints to be recovered, and importantly, while not ruining the potential for DNA testing. The technique involves putting the object into a liquid containing a fluorescent dye that, when an electric current is applied, will stick to the metal that wasn't touched, rather than the print itself. Essentially, the residue from the fingerprint insulates the metal beneath it, so investigators wind up with a negative of the latent print that can be reversed digitally.

So, what does the future hold for fingerprints? Well you've probably heard of nanotechnology; that's the science of how molecules and atoms can be manipulated and utilized in all types of fields. Nanotechnology is projected to be the new frontier of fingerprint development and analysis. Chemical methods will be downsized to the nanotech level to improve the sensitivity of already-existing methods. But even more intriguing uses are proposed.

For instance, antibody-like nanoparticles could be engineered to link to and detect nicotine, cocaine, marijuana, or heroin, and I'm not talking about drugs the person may have handled before touching a surface, but the actual metabolites of those drugs in a user's sweat. That would give investigators much more than a print; it would be a window into someone's life that could be used to help narrow the search for either a criminal or a victim. It's amazing to me that someday a single touch might reveal aspects of a person's habits or lifestyle.

Now, the technique that gets my vote for most revolutionizing the capabilities of forensic science is DNA technology, and I'm sure you'd agree. Like fingerprints, DNA is unique to each of us, except for identical twins, but while prints are only on our hands and feet, every millimeter of our bodies contains countless cells housing our genetic material. Today's DNA analysis has its historical roots in serology, which is the general study of body fluids. Microscopy is one important aspect of forensic serology, and microscopes, in one form or another, have been around since before the 1600s. But the first real forensic application wasn't until the 1830s, when French forensic pathologists perfected the use of the microscope to reliably detect the presence of sperm. A spinoff of that work was microanalysis of fabric types, since those two types of evidence can often be linked in a single case, like a rape case.

In 1853, a Polish anatomist working in Germany developed the Teichmann test, named after him. This method crystallizes components of the hemoglobin in blood, so the crystals can be viewed under a microscope. It's what's called a confirmatory test, meaning if the crystals form, the substance is definitely blood. But hemoglobin is pretty ubiquitous in vertebrate animals, and even in some invertebrates, so the Teichmann test couldn't discriminate whether blood discovered at a crime scene even belonged to a human. Still, the Teichmann test was extremely important for a long time and was used on clothing and other types of evidence in forensic cases.

Scientists from the Netherlands and Germany next generated a couple of what are called presumptive tests for blood. These are quicker tests that might be done in the field to either eliminate a stain as being blood or to show that it probably is. But follow-up confirmatory testing is always required in the lab. Although more advanced chemical tests are used today, both presumptive field tests and confirmatory lab tests are vitally important in detecting the presence of blood.

Another major milestone on the way to DNA technology occurred in 1901, when Karl Landsteiner, an immunologist from Austria, identified the ABO system of human blood types. His colleague, Max Richter, figured out how to detect type from bloodstains for use in forensics. Around the same time, three German scientists independently developed immunological tests that could differentiate among different species of animals. Italian forensic professor Leon Lattes used immunological testing for ABO blood types and then adapted it for paternity testing. He published an extensive reference on the use of blood groups in both forensic and clinical medicine.

Soon after that, a Japanese researcher named Masao Takayama created another reliable test that could detect hemoglobin in bloodstains, and this was also used for many years. By 1925, a second Japanese scientist first recognized that some people display their blood type in their other body fluids. These individuals became known in forensics as secretors. This discovery meant that a person's blood type could be found in semen from a rape or saliva from a bite, even if no blood was shed. Keep in mind, though, that blood type could only be used to eliminate suspects whose type didn't match the evidence. Too many of us share the four major blood groups, A, B, AB, and O, for them to be used as identification.

Forensic TV shows often illustrate the use of luminol at crime scenes to detect blood, even sometimes after it's been cleaned up and is no longer visible to the naked eye. You may not realize, though, that luminol dates all the way back to 1937, when a German forensic scientist came up with a chemical formula that causes blood stains to luminesce, or glow, when it's applied.

Throughout the 1960s, research that spanned the globe led to further refinements to serology, especially regarding more individuating features of blood testing, like enzyme tests. By that time, though, the groundwork for DNA analysis had already been laid down. The 1953 discovery of the chemical structure of DNA by American James Watson and U.K.'s Francis Crick, well that ultimately turned out to be essential to all forensic DNA analytical methods to come.

It wasn't until 1980, though, that genetic researchers from the US discovered what are known as the *hypervariable regions* in human DNA. These are areas different enough among people to be used in forensics. Soon after that came nuclear DNA fingerprinting by Alec Jeffreys of the U.K. That was quickly followed by the invention of the polymerase chain reaction by American biochemist Kary Mullis. This so called PCR is still used today in a modified form to make mass copies of desired segments of DNA, so even small samples can be compared. And by this time, DNA analysis wasn't only being used to convict; it was also already being used to exonerate, including in the U.K. by 1986, and the US by 1989.

DNA testing methods changed again in 1992 when a biochemistry professor in Texas figured out that regions of DNA, known as *short tandem repeats*, could be used in forensics. The resulting so called STR type of DNA fingerprinting is what's in use today and means that very small and degraded samples can still net a genetic profile. And by 1996, the first criminal case to use mitochondrial DNA analysis was tried in the United States.

The most recent advance in DNA testing is what's known as *touch DNA*. Back in the 1980s, DNA analysts needed a sample of blood or sperm about the size of a U.S. quarter. By the 1990s, if the sample could be seen, it was big enough to analyze. Now, with the touch DNA method, fewer than 10 skin cells are needed to generate a useful DNA profile. This means that if a criminal merely touches a surface, like a doorknob or a steering wheel, forensic scientists may be able to swab the surface and get a genetic fingerprint. The method also works on clothing, so an area where sweat or cells would accumulate can be cut out and sampled.

Touch DNA has been used in quite a few high-profile cases in the US. In 1996, the body of six-year-old JonBenet Ramsey, who has been referred to by some as the current generation's Lindbergh baby, she was found in the basement of the family's Boulder, Colorado home. There was evidence that the little girl's body had been wiped down, but later, touch DNA was recovered from her clothing. The tests ended up exonerating her two parents and her older brother, who were the initial suspects. Many still believe, though, that the family was involved, especially given that the elaborately-written ransom note and other items related to the murder all came from the home. The touch DNA testing also eliminated another man, John Karr, an American teacher, who was ultimately found in Thailand; he had falsely confessed to killing JonBenet, likely for attention. The DNA profile recovered in three different locations and from two different pieces of the murdered girl's clothing is that of a man, but to this day, his identity remains unknown, and the case is not yet solved.

The killing of two-year-old Caylee Anthony in Orlando Florida, in 2008, is another notorious case where touch DNA was employed. The toddler's maternal grandmother was the one that first brought her own daughter, Casey, the child's mother, to the attention of authorities. She reported that Casey came home after a 31-day absence with no legitimate explanation for her two-year-old daughter's disappearance. That's got to be an unthinkable task for a mother and grandmother, not to mention the fact that a few days before Casey returned to her parent's home, where she and little Caylee had lived, her father was called to an impound lot to pick up Casey's car, and it smelled of death.

When the girl's skeletal remains were found months later in the woods near the family's home, duct tape was associated with the skull. Investigators were surprised not to find Caylee's skin cells on the adhesive side of the tape, as would be expected if it was used to cover the little girl's mouth. Nor did they find DNA from Casey, her mother. Allegations were, though, that Casey had put a chloroform-soaked cloth between the tape and the girl's mouth, which could have either degraded the DNA or simply blocked the tape from sticking to the child's face. The Dutch couple who pioneered touch DNA, forensic scientist Richard Eikelenboom and his coroner wife, Selma, they were called to testify in Casey Anthony's trial regarding their touch DNA methodology. Apparently, though, it was determined that the only DNA found on the duct tape was contamination from an FBI agent.

Okay, early in this lecture I mentioned that one technology has modified everything we do, particularly in terms of comparisons. I should add that the same technological advance has also likely changed much of what you do in your own daily live and work. Of course, I'm talking about computers. Today's modern forensic labs are filled with computerized equipment. Applications include toxicological studies on body fluids and tissues, the analysis of suspected drug samples, possible accelerants from arson cases, automotive paints from hit-and-run accidents, inks from questioned documents, gunshot residues, and that's just naming a few examples. DNA testing and comparison would also not be possible without computer technology.

In forensic fields like anthropology, accounting, and engineering, computer software has greatly enhanced and sped up our assessments, partly because computers can do more mathematical calculations in a fraction of a second than could ever be humanly possible. Digital methods have also revolutionized forensic art, not only in terms of the speed by which a facial reconstruction or police sketch can be performed, but also the ease of quickly adjusting the results, if needed. Such as to illustrate different "looks" the depicted person may have had, like hairstyles and eyeglasses.

Computers have been used in forensics to create what's known as *demonstrative evidence* as well. That refers to things like graphics and digital reconstructions that are often used in the courtroom to help the judge and/or jury better visualize evidence, or maybe comprehend an expert's opinion. You've likely seen things like this on TV, like a vehicular accident reconstruction, or the path of a bullet, or a computerized representation of someone falling from a tall building. Forensic engineers can use computer modeling to help assess why a building or bridge collapsed, and arson investigators can use computerized representations to diagram where a fire likely started.

Applications, like global positioning systems, or GPS, have allowed the precise mapping of evidence, like locations of body parts and personal effects in a mass disaster, or to document blood spatter patterns at crime scenes. Three-dimensional digitization and printing have allowed replicas to be made of skulls and other evidence to preserve the originals while still allowing experimentation or manipulation, such as for an artist's facial reconstruction.

Digital X-rays can allow a pathologist or forensic dentist to rapidly send detailed images across huge distances. Digital scanning can permit fingerprints or microscopic images of hairs and fibers to be transferred among experts. Computer power can also enhance recognition of criminal patterns or statistical trends, like might be used by a forensic profiler or behavioral analyst.

In fact, the American Academy of Forensic Sciences, which is a national organization of thousands of forensic scientists to which I've belonged since 1990, well, the most recent group they've accepted into the fold is what's called the Digital and Multimedia Sciences section. Experts in digital forensics specialize in a huge array of computer and digital applications. Research conducted by such scientists has led to advancements in crime scene photography and video documentation of scenes and suspects. Digital experts can enhance images and sounds of poor quality and determine whether stored digital information has been altered. They can develop latent digital evidence, like data on a computer or hard drive that a person thought they erased. Computer forensic experts can figure out what devices may have been connected to a computer and when, or whether someone hacked into a computer from a remote location.

One huge application of computers in forensics has been in the realm of comparisons. Point-by-point, and pixel-by-pixel, digitization allows the comparison of fingerprints and other image-based evidence, including footwear, tire tracks, handwriting, bullets, and shell casings. Facial recognition software can pick a person out of a crowd photograph, and digitization allows voice and other sound patterns to be compared using computers. Large databases containing comparative evidence are now routinely used in all types of forensic investigations. Some are very specialized, but others are used on a daily basis by analysts all over the world.

The three particular forensic databases are used most often in the U.S. One is the National Crime Information Center, or NCIC, that was launched in 1967. A second is the Combined DNA Index System, or CODIS, which was established by Congress in 1994. And the third is the Integrated Automated Fingerprint Identification System, IAFIS, which came online in 1999. All of those are under the auspices of the FBI and help to integrate jurisdictions into one large crime-fighting group.

Of course, computers have also changed and enhanced crime. The term *cybercrime* has been given to these kinds of high-tech offenses, which include bank thefts, and corporate embezzlement, identity theft, and a variety of frauds and scams. Governmental, financial, and media entities have all been the targets of hacking and viruses. People, including kids, have been subjected to cyberbullying via text and social media, and computers have been used in child pornography and other sexual exploitation crimes. Even drug traffickers have used computers to more effectively peddle their wares. Just as the Internet has provided worldwide assistance to solve crimes, it's also allowed the means for individuals to essentially cross borders and other jurisdictional lines to commit crimes without ever leaving home.

In 2013, the *Wall Street Journal* estimated that in the U.S. alone, computer crime costs the nation over $100 billion a year. We've come a long way, baby, although computers were first pioneered in the 1930s and '40s, the first wide-scale users were the U.S. Government and its military. It wasn't until the 1970s that the first personal computers came about, and as many of you may know, they were a far cry from today's sophisticated laptops and hand-held devices. Let's take a look at some cybercrime cases.

Between July and October of 1994, over $10 million was stolen from a U.S. bank. This robbery, though, didn't involve a threatening note, a weapon, or even any hold-up men. It was all perpetrated by a worldwide group of hackers who tapped into the bank's cash management system, which was completely computerized and allowed interbank transfers. The FBI figured out that the thieves had been able to gather user IDs and passwords from the bank's telecommunications network. The Bureau narrowed in on two of the ringleaders, a Russian couple who had lived for a time in the United States.

After learning some of the money was being funneled into a San Francisco bank, the FBI was able to apprehend the wife of the Russian pair when she came to San Francisco and tried to take money out of the account. With cooperation from authorities in Russia, they were able to nab the kingpin of the operation, Vladimir Levin, who was captured in London through a sting operation. He had been orchestrating the whole heist from his own laptop. Levin was extradited to the U.S., where he pled guilty in 1998. The case is thought to be the first-ever online bank robbery. In fact, the FBI didn't even have a cybercrime unit at the time; their so called white-collar crime division was the one that unraveled the case. This heist was one of the first wakeup calls to the banking industry that computerization was paving the way for new types of robberies.

Twenty years later, in mid-May of 2014, law enforcement officials in 19 countries announced a global takedown of cyber criminals. It was related to a software marketed as the Blackshades RAT, or Remote Access Tool, which cost only around $40. This malware had been sold through PayPal and even discussed in various hacker chat rooms and other forums, so the crooks should have seen the bust coming. Using a method known as spear phishing, cybercriminals sent out what the FBI terms socially engineered emails to unsuspecting computer users, maybe offering some product or something else the person might find attractive. If the victim clicked on the attachment, it could launch the Blackshades software platform to their PC, and the criminal could take over the person's computer without them even knowing it.

The FBI said the malware was purchased by thousands of cyberhackers in over 100 countries since about 2010. The RAT platform can remotely hijack computers, digging into hard drives, stealing passwords and other access codes, hacking into social media accounts, capturing photos, and even turning on webcams. At the time I wrote this lecture, over 700,000 potential victims had been identified worldwide.

The FBI first discovered the Blackshades software while investigating rings that were using and selling stolen credit card numbers, as well as other personal data. Once they identified the ringleaders who had created and sold the Blackshades malware, they didn't stop there; they went after the purchasers and the users as well. The FBI, along with other global police agencies, decided to apprehend the attackers pretty much simultaneously in mid-May of 2014, because they knew if they didn't, the cybercriminals would be able to instantly alert each other that the bust was on. The U.S. Federal Bureau of Investigation said the worldwide cooperation in this takedown endeavor was literally unprecedented.

Now, I'll leave you with an example that will hopefully end this lecture on a lighter note and serve to illustrate how innovative and diverse computer crimes can be. In the late 1990s, a group of coordinated hackers spanning Europe made numerous attempts over the course of several years to break in to the encrypted News Corp. satellite, Sky-TV, pay-per-view system. Why? Allegedly so the attackers could commit the diabolical offense of watching *Star Trek* re-runs for free. Now, there's an example of the present state of technology colliding headlong with the once-imagined future.

Bibliography

Note: Some of the following references contain graphic images.

Albergotti, Reed, and Vanesse O'Connell. *Wheelmen: Lance Armstrong, the Tour de France, and the Greatest Sports Conspiracy Ever*. New York: Gotham, 2013. Written by two *Wall Street Journal* reporters that followed Armstrong's career, this book discusses evidence for blood-doping scandals of professional cycling in America.

Bartz, Scott. *TYMURS: The 1982 Tylenol Murders* (book 1); *Tylenol Man: A 30-Year Quest to Close the Tylenol Murders Case* (book 2). Scotts Valley, CA: CreateSpace Independent Publishing Platform, 2012. This self-published series examines the 1982 Tylenol poisonings but has been criticized for its bias against Johnson & Johnson and those who investigated the case (as well as for being a not-very-interesting read, particularly book 1). The second part focuses on the investigation of suspect James Lewis and the 2009 reopening of the forensic case.

Bass, Bill, and Jon Jefferson. *Death's Acre: Inside the Legendary Forensic Lab the Body Farm Where Dead Men Do Tell Tales.* New York: Berkley Books, 2003. A history of the decomposition facility at the University of Tennessee, using a case-based approach.

Begg, Paul. *Jack the Ripper: The Definitive History.* Oxford: Taylor and Francis, 2005. Discusses the so-called "canonical five" Ripper victims, as well as others often associated with the series of killings. Also covered are the general climate of the time in London's East End, forensic evidence, media connections, and possible suspects.

Blum, Deborah. *The Poisoner's Handbook: Murder and the Birth of Forensic Medicine in Jazz Age New York.* New York: Penguin Books, 2011. The history of toxicology in the United States.

Bondeson, Jan. *Blood on the Snow: The Killing of Olof Palme*. Ithaca, NY: Cornell University Press, 2005. Covers the 1986 assassination of the Swedish prime minister and the wounding of his wife and attempts to debunk the many conspiracy theories surrounding the murder. The author examines missteps in the forensic investigation and presents his own theory regarding the assassination.

Borchart, Edwin M. *Convicting the Innocent: Sixty-Five Actual Errors of Criminal Justice*. Scotts Valley, CA: CreateSpace Independent Publishing Platform, 2010 (republished from a 1932 edition by Garden City Publishing/ Yale University Press). A perhaps-surprisingly old treatment of many cases of wrongful conviction; Borchart studied the causes of such injustices and pushed for compensation for the falsely imprisoned.

Britton, Nan, with a preface by Sam Sloan. *The President's Daughter.* San Rafael, CA: Ishi Press, 2008 (originally published in 1927). This is a reprint of a book written by President Warren G. Harding's mistress, who allegedly bore him a child. (The original copyright by Elizabeth Ann Guild, Inc., was not renewed; thus, Sloan has rereleased the book under his name and through his Ishi Press.)

Brown, Arnold. *Lizzie Borden: The Legend, the Truth, the Final Chapter.* New York: Dell, 1992. A somewhat controversial work that presents an alternative theory regarding the killing of the Bordens; written by a man from Fall River.

Butler, William S., and L. Douglas Keeney. *Secret Messages: Concealment Codes and Other Types of Ingenious Communication.* New York: Simon & Schuster, 2001. This book is more about the types of codes spies have used throughout history, but it contains a few pages that cover the Velvalee Blucher Dickinson story.

Canseco, José. *Juiced: Wild Times, Rampant 'Roids, Smash Hits, and How Baseball Got Big.* New York: William Morrow, 2005. This book about performance-enhancing drugs in U.S. Major League Baseball is mentioned in Lecture 9 (mostly for its title) but is notoriously poorly written.

Concannon, Diana M. *Kidnapping: An Investigator's Guide.* 2nd ed. London; Waltham, MA: Elsevier, 2013. This professional reference is a detailed guide to the typology (including motive) and resolution of kidnapping; based on 100 U.S. cases, but none of those specifically presented in the lecture on kidnapping.

Davis, Don. *Bad Blood: The Shocking True Story behind the Menendez Killings.* New York: St. Martin's Paperbacks, 1994. This reference covers the Menendez family background, the crime, and the forensic evidence but goes only as far as the first trials.

Dillow, Gordon, and William J. Rehder. *Where the Money Is: True Tales from the Bank Robbery Capital of the World.* New York: W. W. Norton & Company, 2004. Written by an FBI agent who spent his 30-year career investigating Los Angeles bank heists, this book covers the modus operandi and minds of several of LA's most notorious bank robbers in a very entertaining way.

DiMaio, Dominick, and Vincent J. M. DiMaio. *Forensic Pathology.* 2nd ed. Boca Raton, FL: CRC Press, 2001. An advanced reference for those with an understanding of the human body. Covers most causes of death, although the source has been criticized because it is lacking in some critical areas, such as gunshot wounds.

DiMaio, Vincent J. M., and Susanna E. Dana. *Handbook of Forensic Pathology.* 2nd ed. Boca Raton, FL: CRC Press, 2001. A condensed reference about forensic pathology, including information related to natural, homicidal, suicidal, and accidental deaths.

Di Mambro, Dina. *True Hollywood Noir: Filmland Mysteries and Murders.* Classichollywoodbios.com Publications (self-published), 2013. Includes discussions of the deaths of George Reeves and Bob Crane, among many others, as well as longstanding Hollywood scandals and other controversies spanning the years 1922 to 2001.

Dinsio, Amil. *Inside the Vault: The True Story of a Master Bank Burglar.* I Love You, Brother, LLC (self-published), 2013. This account is written by the perpetrator of the 1972 United California Bank heist, the largest bank vault robbery in history.

Douglas, John, and Mark Olshaker. *The Anatomy of Motive.* New York: Simon & Schuster, 1999. Written by a member of the FBI's behavioral analysis unit, this is an examination of the motives that help agents track violent sociopaths.

Ellroy, James. *The Black Dahlia.* New York: Mysterious Press, 1987. A novel that centers on the actual facts of the 1947 Black Dahlia case, as told through the fictional eyes of one of the detectives on the case and his partner. Gives the reader the "feel" of the times.

Emsley, John. *Molecules of Murder: Criminal Molecules and Classic Cases.* Cambridge: Royal Society of Chemistry, 2008. Covers 10 major chemicals historically used in poisonings; discusses the Tylenol murders and the assassination of Georgi Markov. Also reviews polonium, which is germane to the alleged poisoning of Yasser Arafat. Although interesting, this book is more about the chemistry and toxicology than the forensic cases themselves.

Farquhar, Michael. *A Treasury of Deception: Liars, Misleaders, Hoodwinkers, and the Extraordinary True Stories of History's Greatest Hoaxes, Fakes and Frauds.* New York: Penguin Books, 2005. A fun romp through short pieces about a great number of historical frauds and hoaxes, including Piltdown man, Clifford Irving's fake biography of Howard Hughes, political sex scandals, and Anastasia Romanov imposters.

———. *A Treasury of Great American Scandals: Tantalizing True Tales of Historic Misbehavior by the Founding Fathers and Others Who Let Freedom Swing.* New York: Penguin Books, 2003. There are plenty of historic political sex scandals in this fun collection of short works, but the book also includes many more types of political scandals, including those of President Warren G. Harding.

Federal Bureau of Investigation. *Serial Murder: Multi-Disciplinary Perspectives for Investigators (True Crime—Serial Killers).* Scotts Valley, CA: CreateSpace Independent Publishing Platform, 2014. This short reference came out of a 2005 FBI symposium directed toward better understanding of serial murder and murderers (motivation, causality, and so on).

Finley, Peter, Laura L. Finley, and Jeffrey J. Fountain. *Sports Scandals.* Portsmouth, NH: Greenwood, 2008. Discusses the evidence behind a host of sports scandals, with sections subdivided by the type of scandal (drug-related, gambling-related, and so forth). The book does not cover the Lance Armstrong blood-doping case, but it includes the Black Sox scandal, Tonya Harding, Pete Rose, and other topics mentioned in Lecture 9.

Firstman, Richard, and Jay Salpeter. *A Criminal Injustice: A True Crime, a False Confession, and the Fight to Free Marty Tankleff.* New York: Ballantine Books, 2008. This source offers start-to-exoneration coverage of the Marty Tankleff case covered in Lecture 14 from retired NYPD detective Jay Salpeter, whose forensic investigation ultimately freed Tankleff from his wrongful conviction.

Fisher, Jim. *The Mammoth Book of Murder: True Stories of Violent Death.* Scotts Valley, CA: CreateSpace Independent Publishing Platform, 2014. Includes more than 200 stories about murders, both historical and more recent, organized by overarching topics and written by a former FBI agent. The stories cover the cases, relevant forensic science, and courtroom trials.

Fox, Charles. *Uncommon Youth: The Gilded Life and Tragic Times of J. Paul Getty III.* New York: St. Martin's Press, 2013. Covers the life of Paul Getty, but the writing has received poor reviews and the work has been called overly sensational. Not necessarily recommended, but it does offer an overview of Paul's tragic life.

Gantt, Paul H. *The Case of Alfred Packer, the Man-Eater: An Unsolved Mystery of the West.* Whitefish, MT: Literary Licensing, LLC, 2013 (originally published by University of Denver Press, 1952). An early historic account of the case of the Colorado Cannibal.

Gilmore, John. *Severed: The True Story of the Black Dahlia Murder.* Los Angeles: Amok Books, 2006. This volume focuses on the life of the victim,

Elizabeth Short, as well as the crime, evidence, forensic investigation, and suspects. Be aware, however, that aspects of this book have been subject to some serious criticism.

Graysmith, Robert. *The Murder of Bob Crane: Who Killed the Star of Hogan's Heroes?* New York: Crown Publishing, 1993. Interesting coverage of Crane's life, including his sexual addiction and conquests. However, the book was published before the trial of prime suspect John Carpenter; thus, it lacks any information about the case beyond 1992.

Guinn, Jeff. *The Last Gunfight: The Real Story of the Shootout at the O.K. Corral and How It Changed the American West.* Fort Worth, TX: 24Words, LLC, 2011 (reprinted by Simon & Schuster, 2012). Puts the gunfight into a historical context, covering all the players and the time in which they lived and providing rich detail about the shootout itself.

Henderson, Jan Alan. *Speeding Bullet: The Life and Bizarre Death of George Reeves.* 2nd ed. Grand Rapids, MI: M. Bifulco, 2007. The republication of a 1999 book about George Reeves, including coverage of his controversial death. This revision contains more information about the forensic case than the first edition.

Hodgson, Ken. *Lone Survivor.* Hanover, MA: Pinnacle, 2001. Historical fiction based on the story of Alferd Packer; told from Packer's point of view.

Hynd, Alan. *Passport to Treason: The Inside Story of Spies in America.* Whitefish, MO: Kessinger Publishing, LLC, 2005 (originally published by Robert M. McBride and Company, 1943). A photocopied version of the 1943 original, this book includes the story of William Sebold, who was a double agent for the United States against Germany. It covers the details of Sebold's training by the Hamburg spy school and information about his infiltration of the Duquesne spy ring.

Innes, Brian. *Fakes and Forgeries: The True Crime Stories of History's Greatest Deceptions: The Criminals, the Scams, and the Victims.* Pleasantville, NY: Readers Digest, 2005. A basic look at the forensic analysis of frauds and forgeries, including Piltdown man, artwork, and documents.

Innocence Project. http://www.innocenceproject.org/. One of the oldest of the major exoneration agencies in the United States, the Innocence Project was founded in 1992 at the Benjamin N. Cardozo School of Law at Yeshiva University in New York. (It has since been reorganized as a nonprofit.) The group's focus is on using forensic DNA testing of evidence to exonerate those who have been falsely convicted and on studying the causes of wrongful convictions to help develop new public policies, reform the criminal justice system in the United States, and avoid future miscarriages of justice. The website includes the stories of those the project has helped to exonerate.

Irving, Clifford. *Fake: The Story of Elmyr de Hory, the Greatest Art Forger of Our Time.* New York: Dell, 1971. This book, mentioned in Lecture 8, covers the life and art forgery of Elmyr de Hory. The author himself subsequently perpetrated a major fraud.

———. *The Hoax.* London: Franklin Watts, 1981 (republished in paperback by Miramax, 2007). This memoir is the confession of Clifford Irving following his attempt to write a fake biography of Howard Hughes.

James, Stuart H., Jon J. Nordby, and Suzanne Bell. *Forensic Science: An Introduction to Scientific and Investigative Techniques.* 4th ed. Boca Raton, FL: CRC Press, 2014. A comprehensive textbook covering all aspects of forensic science.

Kiernan, Ben. *Blood and Soil: A World History of Genocide and Extermination from Sparta to Darfur.* New Haven, CT: Yale University Press, 2009. Covers genocides from ancient to modern times, including Guatemala and the Holocaust, among many others. The author describes the historical and political climates that led to the various atrocities.

Large, David Clay. *Munich 1972: Tragedy, Terror, and Triumph at the Olympic Games.* Lanham, MD: Rowman & Littlefield Publishers, 2012. A history of the events leading to, surrounding, and following the kidnapping and killing of Israeli Olympians in 1972, written in time for release on the 40th anniversary of the attack.

Lee, Linda. *Life and Tragic Death of Bruce Lee.* Singapore: Star Publishing, 1975. Written by Lee's wife shortly after his death. Does not cover any reanalysis of the forensic death investigation.

Loftus, Elizabeth. *Eyewitness Testimony.* Cambridge: Harvard University Press, 1996. An update of the 1979 book that was the seminal work about the problems with eyewitness testimony; although much has been written since, Loftus is still considered one of the world's prominent experts on the topic.

Loftus, Elizabeth, and Katherine Ketcham. *Witness for the Defense: The Accused, the Eyewitness and the Expert Who Puts Memory on Trial.* New York: St. Martin's Press, 1991. Continues the presentation of Loftus's studies on eyewitness testimony, with special attention to criminal court cases.

Macur, Juliet. *Cycle of Lies: The Fall of Lance Armstrong.* New York: Harper, 2014. Written by a sports reporter, this book analyzes the blood-doping cyclist's scandal-ridden career and is based on hundreds of interviews with Armstrong and those involved with him.

Mallet, Xanthé, Teri Blythe, and Rachel Berry, eds. *Advances in Forensic Human Identification.* Boca Raton, FL: CRC Press, 2014. This comprehensive work (from a U.K. perspective) covers the range of methods used in human identification—from coroner's cases to mass disasters and genocides; includes recent advances in fingerprinting, anthropology, analysis of genetic material, forensic art examination, and more.

Maples, William R., and Michael Browning. *Dead Men Do Tell Tales.* New York: Doubleday, 1994. This casebook of forensic anthropology has a chapter devoted to the skeletal analysis of the Romanov remains and their comparison to photographs and other records.

Massie, Robert K. *The Romanovs: The Final Chapter.* New York: Random House Publishing, 1995. This reference covers the life and times of the Romanov family, as well as the scientific investigation into their deaths and the subsequent imposters, particularly Anna Anderson. It reports on the discovery and forensic testing (including DNA) of the remains discovered in 1991.

McCrery, Nigel. *Silent Witnesses: The Often Gruesome but Always Fascinating History of Forensic Science.* Chicago: Chicago Review Press, 2014. A global history of forensic science over the past two centuries and its major players, written by a police officer who became a BBC correspondent.

Mee, Charles L., Jr. *The Ohio Gang: The World of Warren G. Harding.* New York: M. Evans & Company, 2014. Covers Harding's political/sex scandals and his cronies and relatives, including Albert Fall, Henry Daugherty, Jesse Smith, Nan Britton, and Harding's wife, Flossie, among others.

Michigan Innocence Clinic. https://www.law.umich.edu/clinical/innocenceclinic/Pages/default.aspx. The website for the Michigan Innocence Clinic of the University of Michigan Law School. Provides information about people the group has exonerated, as well as the causes of wrongful convictions (police/legal issues and misconduct, faulty forensic science, false convictions, mistaken eyewitness testimony, jailhouse convictions, poor lawyering, and so on). The clinic specializes in cases for which no DNA evidence exists for testing, which are some of the most difficult to exonerate.

Montejo, Victor, and Victor Perera. *Testimony: Death of a Guatemalan Village.* Willimantic, CT: Curbstone Books, 1995. A schoolteacher's firsthand account of the decimation of his Mayan people in the 1980s.

Moore, Robert. *Cybercrime: Investigating High-Technology Computer Crime.* 2nd ed. Newark, NJ: Anderson, 2006. A good introductory text; covers the history of cybercrime but a bit dated.

Münsterberg, Hugo, and Mark Hatala, with a foreword by Elizabeth Loftus. *On the Witness Stand: Essays on Psychology and Crime.* Greentop, MI: Greentop Academic Press, 2009. This volume was originally published in 1908 as a groundbreaking look at problems that result in wrongful convictions, such as faulty memory and false confessions.

Murray, Elizabeth A. *Death: Corpses, Cadavers, and Other Grave Matters.* Minneapolis, MN: Lerner Publishing Group/Twenty-First Century Books, 2010. Examines the science of death, including decomposition, autopsy, cause and manner of death, and forensic death investigation. It is written for a young adult audience (grades 7–12), but adults have also appreciated the book for its ability to bring the topic to an accessible level.

———. *Forensic Identification: Putting a Name and Face on Death.* Minneapolis, MN: Lerner Publishing Group/Twenty-First Century Books, 2012. Covers the science of human identification, including aspects of skin and other soft tissues (such as fingerprints, tattoos, scars, and hair and eye color), skeletal and dental analysis (including forensic anthropology and odontology, medical implants, and forensic art), and the cellular level of identification (such as blood type and body fluid enzymes, isotope testing, and DNA analysis). The book is geared toward grades 6–12 but is a basic primer on the scope and science of human identification for all and uses a case-based approach.

———. *Overturning Wrongful Convictions: Science Serving Justice.* Minneapolis, MN: Lerner Publishing Group/Twenty-First Century Books, 2015. Written for grades 7–12 but suitable for adults, this book introduces the reader to the legal system, the nature of scientific evidence, the exoneration process, and the causes of false convictions. The cases of those exonerated help illustrate major points.

National Missing and Unidentified Persons System. http://www.namus.gov/. Publically accessible website where the U.S. missing and unidentified persons databases can be found and explored. The site is maintained through a program of the Office of Justice, is administered by the U.S. Department of Justice's National Institute of Justice, and is devoted to matching the records of missing and unidentified persons to resolve cold cases.

National Registry of Exonerations. http://www.law.umich.edu/special/exoneration/Pages/about.aspx. A cooperative effort between the University of Michigan Law School and the Northwestern University School of Law. The site lists all known cases of cleared wrongful convictions dating back to 1989 in the United States and includes graphs and statistical data (such as age at conviction and exoneration, years spent in prison, gender, race, and so on) that can be sorted and reviewed. The site allows users to browse and examine case profiles of exonerees.

Nelson, Anne. *Murder under Two Flags: The U.S., Puerto Rico, and the Cerro Maravilla Cover-Up.* Boston: Ticknor & Fields, 1986. This book covers the 1978 killing of two young dissidents by police on a Puerto Rican mountain and the allegations of a subsequent U.S. cover-up. It was written by one of the major journalists who investigated the case.

Nilsen, Anna. *Art Fraud Detective: Spot the Difference, Solve the Crime!* New York: Kingfisher, 2000. Although this comic-strip-style book is geared toward grades 3–5, it's a fun way to learn about art forgery.

Northwestern University School of Law Center on Wrongful Convictions. http://www.law.northwestern.edu/legalclinic/wrongfulconvictions/. The center aids exonerations and studies the causes of false convictions. Its website profiles the center's exonerees and includes resources to better understand problems with eyewitness identification, coerced confessions, perjury, police and other official misconduct, improper analysis or presentation of forensic evidence, and poor legal representation. Its mission—"representation, research, and reform"—led to a moratorium on executions in Illinois (2000), followed by the abolition of the death penalty in Illinois (2011), and has prompted many reforms in police and legal processes in the state.

Parish, James. *The Hollywood Book of Death: The Bizarre, Often Sordid, Passings of More Than 125 American Movie and TV Idols.* New York: McGraw-Hill. 2001. Covers the murder of Bob Crane and many other Hollywood figures; arranged by manner of death.

Petro, Jim, and Nancy Petro. *False Justice: Eight Myths That Convict the Innocent.* New York: Kaplan Publishing, 2011. Written by the former attorney general of Ohio, this reference looks at the many challenges to, and failures of, the U.S. justice system that contribute to wrongful convictions.

Pietras, Davis. *Unanswered Evidence.* Scotts Valley, CA: CreateSpace Independent Publishing Platform, 2014. In this work, a prolific true-crime writer (who often gets mixed reviews) covers five high-profile unsolved cases, including those of Lizzie Borden, Bob Crane, and the Black Dahlia.

Porter, Edwin H. *The Fall River Tragedy: A History of the Borden Murders*. Edited by Michael W. Paulson. Scotts Valley, CA: CreateSpace Independent Publishing Platform, 2011 (Fall River, MA: J. D. Munroe, 1893). This work covers the killings of Andrew and Abby Borden, including the forensic evidence, case investigation, and trial of Lizzie Borden. Porter was a reporter who lived in Fall River at the time and wrote this account just after the courtroom trial.

Primorac, Dragan, and Moses Schanfield, eds. *Forensic DNA Applications: An Interdisciplinary Perspective*. Boca Raton, FL: CRC Press, 2014. An advanced, technical, and highly comprehensive reference on the uses of DNA and other genetic material in the many subfields of forensic science (and medicine), including a brief history of the use of DNA in forensics. Techniques for recovering genetic material and using it in all manner of forensic cases are covered from a U.S. and a European perspective.

Prouse, Lynda, Kyle Torke, and M. Stefan Strozier. *The Tonya Tapes*. New York: World Audience, 2008. This interview-style work covers Harding's life and the assault on Nancy Kerrigan but is transcribed from audiotape and not necessarily told as a coherent story.

Przyrembel, Alexandra. "Transfixed by an Image: Ilse Koch, the 'Kommandeuse of Buchenwald.'" *German History* 19, no. 3 (2001): 369–399. This article covers the social psychology that may have led to the unprecedented level of vilification of Ilse Koch in the public eye.

Ramsland, Katherine. *Beating the Devil's Game: A History of Forensic Science and Criminal Investigation*. New York: Berkley, 2014. Covers the history of numerous subfields in forensic science, dating back to the 13th century, and introduces many pioneers of the science.

Rathbun, Ted A., and Jane E. Buikstra. *Human Identification: Case Studies in Forensic Anthropology*. Springfield, IL: Charles C. Thomas Publishing Ltd., 1984. A classic reference in the field of forensics; takes a case-based approach to the science of identification from the skeleton.

Reeve, Simon. *One Day in September: The Full Story of the 1972 Munich Olympics Massacre and the Israeli Revenge Operation "Wrath of God."* New York: Arcade Publishing, 2000. Covers the 1972 kidnapping at the Olympics in Munich, as well as the historical context in which the incident occurred. The story goes beyond the actual events of that September to the incidents of retaliation and the cover-up that followed.

Robenault, James David. *The Harding Affair: Love and Espionage during the Great War.* Basingstoke, UK: Palgrave Macmillan, 2009. This reference (mentioned in Lecture 10) discusses the affair between President Warren G. Harding and Carrie Phillips, as documented by their history of love letters. The book also examines the question of whether Phillips was a German spy.

Rose, Pete, and Rick Hill. *My Prison without Bars.* Emmaus, PA: Rodale Books, 2004. This book (mentioned in Lecture 9) tells Pete's story but as directed toward his quest for the Baseball Hall of Fame; it is more about his problems and downfall than his baseball career.

Rumbelow, Donald. *The Complete Jack the Ripper.* London: Virgin Books, 2013. An updated version of a classic work in "Ripperology," released for the 125th anniversary of the 1888 Whitechapel killings.

Saferstein, Richard. *Forensic Science: From the Crime Scene to the Crime Lab.* 2nd ed. Prentice Hall, 2012. A good beginning textbook on forensic science, including evidence collection and processing.

Scheck, Barry, Peter Neufeld, and Jim Dwyer. *Actual Innocence: When Justice Goes Wrong and How to Make It Right.* New York: New American Library Trade, 2003. Scheck and Neufeld are the cofounders of the Innocence Project in New York, and Dwyer is a reporter for *The New York Times*. This book focuses on the problems that cause wrongful convictions and makes recommendations for change.

Schroeder, Barbara, and Clark Fogg. *Beverly Hills Confidential: A Century of Stars, Scandals and Murders.* Santa Monica, CA: Angel City Press, 2012. Brief coverage of many Hollywood crimes, sex scandals, and other cases from 1910 to 2010; includes the Menendez brothers.

Shelton, Donald. *Forensic Science in Court: Challenges in the Twenty-First Century (Issues in Crime and Justice).* Lanham, MD: Rowman & Littlefield Publishers, 2010. Written by a prominent judge, this book (often used as a textbook) examines many current forensic science methods and discusses how they can be used and misused in U.S. courtrooms.

Shipman, Pat. *Femme Fatale: Love, Lies, and the Unknown Life of Mata Hari.* New York: Harper Perennial, 2007. This book, written by an anthropologist, presents the life of Mata Hari, as well as evidence regarding whether or not she was really a spy.

Siegel, Jay A., and Kathy Mirakovitz. *Forensic Science: The Basics.* 2nd ed. Boca Raton, FL: CRC Press, 2010. This is the entry-level text I use in my own forensic science survey course; however, I am disappointed with the numerous typographical and printing errors in the book (which I hope will be remedied in subsequent editions). It does a nice job of briefly covering most forensic science methods but lacks any coverage of the behavioral sciences.

Sifakis, Carl. *Encyclopedia of Assassinations.* New York: Skyhorse Publishing, 2013. A collection of more than 400 short works about politically motivated acts of violence worldwide, including many assassinations. The majority of these cases are likely fairly obscure to all but the most serious history buffs, but pertinent to this course is the coverage of Yasser Arafat, Jesse James, Georgi Markov, Olaf Palme, and Nicholas II and the Romanov family.

Slater, Wendy. *The Many Deaths of Tsar Nicholas II: Relics, Remains and the Romanovs.* New York: Routledge, 2007. Examines the myths surrounding the death of the czar and his family; the group of researchers who discovered the remains in the 1970s, only to rebury them; and the many Alexei imposters, not often discussed in other accounts.

Soble, Ron, and John H. Johnson. *Blood Brothers: The Inside Story of the Menendez Murders.* New York: New American Library–Penguin Books/ Onyx True Crime, 1995. Because of its publication date, this book does not include full coverage of the second trial or conviction but presents the brothers' defense in the first trial and evidence against them.

Spencer, Frank. *Piltdown: A Scientific Forgery.* New York: Oxford University Press, 1990. An interesting account that examines the hoax, as well as possible perpetrators.

Spiering, Frank. *Lizzie: The Story of Lizzie Borden.* New York: Dorset Press, 1991. This book has received mixed reviews, mostly because of the suspect it picks, but is generally well received owing to the background it provides on the family and its coverage of the courtroom trial.

Spitz, Werner U., and Daniel J. Spitz. *Spitz and Fisher's Medicolegal Investigation of Death: Guidelines for the Application of Pathology to Crime Investigation.* 4th ed. Springfield, IL: Charles C. Thomas Publishing, 2005. One of the most comprehensive references on death investigation available; recommended primarily for those with background knowledge in anatomy and physiology (or for those who want to learn).

Steadman, Dawnie W. *Hard Evidence: Case Studies in Forensic Anthropology.* 2nd ed. Upper Saddle River, NJ: Pearson, 2009. A case-based approach to the field of forensic anthropology, as used in my own casework, the Romanov case, mass graves from genocide, and human identification.

Stiles, T. J. *Jesse James: Last Rebel of the Civil War.* London: Vintage/ Penguin Random House, 2003. The author seeks to debunk the "cult hero" status of James through this biography, which places James in the context and times in which he was raised, equating him to a "homegrown terrorist" of his day.

Sugden, Philip. *The Complete History of Jack the Ripper*. London: Robinson Publishing, 1994. Covers the investigation of the evidence in the Whitechapel murders in good detail. The author also examines and compares some of the theories about the killer's identity and debunks some of the myths about this series of cases. More recent republications of this work are available.

Sylado, Remy. *My Name Is Mata Hari.* Translated by Dewi Anggraeni. Dalang Publishing, LLC (an Indonesian publisher with books made available in the United States by Ingram Book Company), 2012. Follows the evolution of the young Dutch woman into the infamous exotic dancer.

Tefertiller, Casey. *Wyatt Earp: The Life behind the Legend.* Hoboken, NJ: Wiley, 1999. Primarily focuses on the years Earp and his brothers spent in Tombstone, Arizona, and provides much information about the events leading up to the gunfight at the O.K. Corral.

Thies, Charles. *Cybercrime: Digital Forensic Investigations.* Dulles, VA: Mercury Learning & Information, 2014. A technical reference/textbook for those with a background in computers; it includes case studies.

Thompson-Cannino, Jennifer, and Ronald Cotton, with Erin Torneo. *Picking Cotton: Our Memoir of Injustice and Redemption.* New York: St. Martin's Griffin, 2009. The compelling cowritten story of a rape victim and the man she mistakenly put behind bars as her rapist as a result of faulty eyewitness testimony.

Totten, Samuel, and William S. Parsons, eds. *Centuries of Genocide: Essays and Eyewitness Accounts.* New York: Routledge, 2012. Systematically covers worldwide genocides, using a political and sociological approach. Includes full chapters on the Holocaust and the Guatemalan genocide.

Triplett, Frank. *Jesse James: The Life, Times, and Treacherous Death of the Most Infamous Outlaw of All Time.* New York: Skyhorse Publishing, 2013 (originally released in 1882). Written just after James's death by an unknown author (at the time); covers the life and death of James and the James Gang.

U.S. Department of Justice, National Institute of Justice. *Latent Print Examination and Human Factors: Improving the Practice through a Systems Approach: The Report of the Expert Working Group on Human Factors in Latent Print Analysis.* Scotts Valley, CA: CreateSpace Independent Publishing Platform, 2014. This reference came from a scientific working group charged with investigating the problems with fingerprint methodology and interpretation.

U.S. National Library of Medicine. *Handbook: Help ME Understand Genetics 2014.* Scotts Valley, CA: CreateSpace Independent Publishing Platform, 2014. A primer to understand DNA, though the applications are largely focused on genetic testing in health and medicine.

Warden, Rob, and Steven A. Drizin, eds. *True Stories of False Confessions.* Chicago: Northwestern University Press, 2009. This reference covers more than 40 cases of false confessions, some of which involve brainwashing the person being interrogated, police coercion, and other means by which people are manipulated into confessing to crimes they did not commit.

Weiner, J. S. *The Piltdown Forgery.* Oxford: Oxford University Press, 2004 (50th anniversary edition; originally published in 1955). An early account of the Piltdown fossil hoax, rereleased with an introduction and afterword by physical anthropologist Chris Stringer.

Whitlock, Flint. *The Beasts of Buchenwald: Karl and Ilse Koch, Human-Skin Lampshades, and the War-Crimes Trial of the Century.* Brule, WI: Cable Publishing, 2011. The focus in this volume is specifically the Buchenwald work camp and how it was run. The book covers the American liberation, the postwar trial of Ilse Koch, the forensic investigation regarding her alleged involvement in production of artifacts made of human tissues, and her suicide in prison.

Wilson, Colin. *A Criminal History of Mankind.* 2nd ed. San Francisco: Mercury Books, 2005. This reference covers the history of violence from early human existence to the present, putting the events in temporal context, along with changes in sociology and technology.

———. *The Mammoth Book of True Crime: A New Edition.* New York: Carroll & Graf Publishers, 1998. This topic-based book focuses more on European crimes that are typically lesser known in the United States, but many have complained that this work is dated and suffers from poor writing and editing.

Wise, David. *Spy: The Inside Story of How the FBI's Robert Hanssen Betrayed America.* New York: Random House, 2002. The detailed account of the life of FBI-agent-turned-spy Robert Hanssen.

Notes